P9-CQS-849

Formal Methods for Interactive Systems

Computers and People Series

Edited by
B. R. GAINES and A. MONK

Monographs

Communicating with Microcomputers: An introduction to the technology of man–computer communication, *Ian H. Witten* 1980

The Computer in Experimental Psychology, *R. Bird* 1981

Principles of Computer Speech, *I. H. Witten* 1982

Cognitive Psychology of Planning, *J-M. Hoc* 1988

Formal Methods for Interactive Systems, *A. J. Dix* 1991

Edited Works

Computing Skills and the User Interface, *M. J. Coombs and J. L. Alty (eds)* 1981

Fuzzy Reasoning and Its Applications, *E. H. Mamdani and B. R. Gaines (eds)* 1981

Intelligent Tutoring Systems, *D. Sleeman and J. S. Brown (eds)* 1982 (1986 paperback)

Designing for Human–Computer Communication, *M. E. Sime and M. J. Coombs (eds)* 1983

The Psychology of Computer Use, *T. R. G. Green, S. J. Payne and G. C. van der Veer (eds)* 1983

Fundamentals of Human–Computer Interaction, *Andrew Monk (ed)* 1984, 1985

Working with Computers: Theory versus Outcome, *G. C. van der Veer, T. R. G. Green, J-M. Hoc and D. Murray (eds)* 1988

Cognitive Engineering in Complex Dynamic Worlds, *E. Hollnagel, G. Mancini and D. D. Woods (eds)* 1988

Computers and Conversation, *P. Luff, N. Gilbert and D. Frohlich (eds)* 1990

Adaptive User Interfaces, *D. Browne, P. Totterdell and M. Norman (eds)* 1990

Human–Computer Interaction and Complex Systems, *G. R. S. Weir and J. L. Alty (eds)* 1991

Practical Texts

Effective Color Displays: Theory and Practice, *D. Travis* 1991

EACE Publications

(Consulting Editors: *Y. WAERN and J-M. HOC*)

Cognitive Ergonomics, *P. Falzon (ed)* 1990

Psychology of Programming, *J-M. Hoc, T. R. G. Green, R. Samurçay and D. Gilmore (eds)* 1990

Formal Methods for Interactive Systems

ALAN JOHN DIX

Department of Computer Science
University of York

ACADEMIC PRESS

Harcourt Brace Jovanovich, Publishers
London San Diego New York
Boston Sydney Tokyo Toronto

ACADEMIC PRESS LTD
24/28 Oval Road,
London NW1 7DX

United States Edition published by
ACADEMIC PRESS INC.
San Diego, California 92101-4311

This book is printed on acid-free paper

A catalogue record for this book is available from the British Library

ISBN 0-12-218315-0

Printed and bound in Great Britain by the
University Press, Cambridge

CONTENTS

Preface

This book is the product of over six years of research in the Human–Computer Interaction Group at the University of York. This group includes members of the Computer Science and Psychology Departments and it has been a privilege to work in such a stimulating environment.

HCI is itself a cross-disciplinary field and this book brings together aspects of HCI with the sort of formal methods found in software engineering – at first sight, an unlikely marriage. Computers, in general, and formal methods, in particular, define strict laws of operation, yet people cannot usually be fitted into such straight-jackets. On the other hand, so many systems I have seen and used, in banks, shops, and including all that I have used to produce this book, have user-interface faults that could have been prevented by relatively simple formal analyses. So, aware of both these facts, this book contains both eulogies of the usefulness of formal techniques and warnings of their misuse.

The recurrent theme of the book is to take a class of interactive systems and to define a formal model which captures critical aspects of them. These models are then used to frame formal statements of properties relating to the systems' usability. The emphasis is on describing the systems and defining the properties. The intention is not to produce a formal notation or architecture for constructing interactive systems, but rather to develop frameworks which, among other things, guide the use of existing notations and architectures. The formal models can be used in a strictly rigorous way, as part of a formal design process, however, they also, and more importantly, act as conceptual tools to aid our understanding of the systems.

There will be few readers in this strange border country between HCI and formal methods, so I assume that most will be in one camp or the other. I hope that it will prove useful to both camps: to those working in HCI who feel the need for a more rigorous framework for aspects of their work; and also to those software engineers working within increasingly formal methodologies who recognise a need to address the people who use their systems. There may be a temptation for the latter to notice only the eulogies and the former the warnings, however, I hope that you will heed both.

Those who work in HCI, whether drawn from psychology, sociology or computing tend to share an interest in people. Is this because they are self-selecting, or because their jobs emphasise an empathy with their users? I'm not sure, but it places them in a position of responsibility within their workplace and within the information technology community, to reflect humanity amid mechanism. Their job actually asks them to balance the law of the machine with love, mercy and care for its users.

The balance between law and mercy is never easy, and naturally involves a whole wealth of ethical, philosophical and religious issues. In the work described here, there is a clear stance, which I never deliberately took, but gradually noticed in my work. I have a disinclination to build models of the user and distrust of excessively task-oriented analyses. Instead, the models are formal models *of* the system *from* the user's viewpoint. The principles primarily address issues of observability and control. That is, they emphasise the user as being in charge of the interaction, and don't presume to determine precisely what the user will do.

Several parts of this book have appeared in modified form in previous published papers. I would like to thank the following for permission to reuse material: Cambridge University Press for "The myth of the infinitely fast machine" which appeared in *People and Computers III* (1987), "Abstract, generic models of interactive systems" in *People and Computers IV* (1988) and "Nondeterminism as a paradigm for understanding problems in the user interface" in *Formal Methods in Human-Computer Interaction* (1989); and Butterworths for "Interactive systems design and formal refinement are incompatible?" which appeared in *The Theory and Practice of Refinement*.

Finally I would like to thank my family, Fiona, Esther and Ruth, for their love which counterbalanced the law of the ever present PC (even on holiday!); and to thank God, our Father, who brought us to York and has sustained us since, and in whom law and love meet in costly but complete harmony.

CHAPTER 1

Introduction

1.1 Formal methods and interactive systems

This books looks at issues in the interplay between two growth areas of computing research: formal methods and human–computer interaction. The practitioners and styles of the two camps are very different and it can be an uneasy path to tread.

1.1.1 Formal methods

As the memory of computers has increased, so also has the size and complexity of the software designed for them. Maintaining and understanding these systems has become a major task. Further, the range of tasks under direct control of computers has increased and the effects of failure, in say a space station or a nuclear power plant, have likewise increased. There may be little or no opportunity for direct control if the software malfunctions, for instance, in a fly-by-wire aircraft. Further, the costs of producing software have increased dramatically, and the possibility of maintaining code has diminished. This cumulation of factors has been termed the *software crisis* (Pressman 1982) and has led to a call for software design to become more of an engineering discipline. In particular, there is the desire for a more rigorous approach to software design, possibly including elements of mathematical formalism and having the possibility of being proved or at least partially checked.

Several varieties of formal methods have been developed for different purposes:

- *Graphical methods* – Such methods include Jackson Structured Programming (JSP) (Jackson 1983) for the design of data processing (DP) systems and various dataflow methods (Yourdon and Constantine 1978). Typically, only a portion of the required information is held in the graphs.

The rest may be informally annotated or there may be additional non-graphical notations, as is the case with JSP. Again the notation may be self standing, or be part of a larger methodology.

- *Program proof rules and semantics* – Another strand of formality concentrates on proving properties of programs or program fragments (Hoare 1969, Dijkstra 1976), implicitly defining a meaning for the language. Others search for more explicit expression of program language semantics (Stoy 1977).
- *Specification notations* – These are languages and notations specifically designed for the formal specification of software. Examples include Vienna Definition Method (VDM) (Jones 1980), Clear (Burstall and Goguen 1980) and Oxford's Z notation (Sufrin *et al.* 1985, Morgan 1985). Often these notations have explicit rules for the correct transformation of a specification towards an implementable form (Jones 1980).
- *Methodologies* – Instead, or as well as defining precisely how a design is to be specified or proved, there are many methodologies aimed at defining what should happen in the design *process*. For example, JSP, mentioned above, is part of a complete design process. These formal methodologies may incorporate parts of the design process that are beyond what could be expected of an entirely rigorous approach. These approaches may involve graphical and textual notations and may be amenable to computer verification of consistency (Stephens and Whitehead 1986).

There is an underlying assumption to much of this book that software is being developed using some formal notation of the third category. Various sections of the book propose parts of a semi-formal design methodology.

1.1.2 Interactive systems

In the early days of computing the modes of interaction with the user were severely limited by the hardware available: initially cards and switches, later teletypes. The more limited the capabilities the greater the need for effective interface design, but early users were usually experts and there was little spare processing power for frills. Even as processing power increased and the interface hardware improved there was still a strong pull from the experts, who were the major users, for power and complexity. Two major influences have pushed the computer industry towards improved user interface design:

- *Technology push* – The realisation that there was the possibility of a computer society prompted research using state-of-the-art technology into futuristic scenarios. The Xerox work on Smalltalk (Goldberg 1984) and the Star interface (Smith *et al.* 1983) are examples of this.

- *Large user base* – The plummetting cost of hardware has led to a huge growth in computers in the hands of non-computer-professionals. The personal computer boom has taken the computer out of the hands of the DP department, and the new users are not prepared for inconsistent and obscure software.

The two strands are not independent. The Xerox Star has led to the very popular Macintosh, and the WIMP (windows, mice and pop-up menus) interface has become standard in the market-place. If one were to balance the two, it is perhaps the latter strand which is really of most significance in the current high prominence of issues of human–computer interaction (HCI).

The perceived importance of HCI is evidenced by the large number of conferences dedicated to it: the CHI conferences in USA, HCI in Britain, INTERACT in Europe, and HCI International. The "man–machine" interface was also a major strand of the Alvey initiative (Alvey 1984) and (under a different name) of its successor, the IED program. HCI also has a prominent role in the European Community's Esprit program.

1.1.3 The meeting

There is a certain amount of culture shock when first bringing together the concepts of formal methods and interactive systems design. The former are largely perceived as dry and uninspiring, in line with the popular image of mathematics. Interface design is, on the other hand, a more colourful and exciting affair. Smalltalk, for instance, is not so much a programming environment as a popular culture. Also it is hard to reconcile the multi-facetedness of the user with the rigours of formal notations. Some of these problems may be to do with misunderstandings about the nature of formalism (although even I, a mathematician, find a lot of computer science formalism very dry). However, this is not a problem just between formalisms and users: mathematical and formal reasoning typically is performed by people and does therefore have a more human side. The shock really occurred when living users met dry unemotional computers and must therefore be dealt with in any branch of HCI. However, the gut reaction still exists and is a reminder of the delicate balance between the two.

No matter how strong the reaction against it, there is clearly a necessity for a blending of formal specification and human factors of interactive systems. If systems are increasingly designed using formal methods, this will inevitably affect the human interface, and if the issue isn't addressed explicitly the methods used will not be to the advantage of the interface designer. If we look again at some of the reasons for needing formal methods, large critical systems where the crisis is most in evidence clearly need an effective interface to their complexity. The penalty for not including this interface in the formal standards will be more accidents due to human error such as at Chernobyl and Three Mile Island, and the more powerful the magnifying effect of the control system the more damaging the possible effects.

The need for more formal design is seen also in more mundane software. Many of the problems in interactive systems are with awkward boundary cases and inconsistent behaviour. These are obvious targets for a formal approach.

1.1.4 Formal approaches in HCI

There are several approaches taken to the formal development of interactive systems:

- *Psychological and soft computer science notations* – These include the layered approach of Foley and van Dam (1982), or the more cognitive and goal-oriented methods such as TAGPayne 1984. The uses of these vary, for instance improving design, predicting user response times and predicting user errors. They are not intended for combination with the formal notations of software engineering.
- *Specifying interactive systems in existing notations* – Several authors use notations intended for general software design to specify interactive systems. Examples of this include Sufrin's elegant specification of a text editor using Z (Sufrin 1982), a similar one by Ehrig and Mahr (1985) in the ACT ONE language, and no less than four specifications in a paper by Chi (1985) in which he compares different formal notations for interface specification. Sometimes it is some component of the interface that is specified rather than an entire interactive system, as is the case with the Presenter, an autonomous display manager described by Took (1986a, 1986b, 1990). Pure functional languages have also been used to specify (and implement) interactive systems. Cook (1986) describes how generic interface components can be specified by using a pure functional language and Runciman (1989) has developed the PIE model, described later in this book, in a functional framework.
- *Notations for specification* – A general-purpose notation is not necessarily best suited to specifying the user interface, and various special purpose notations have been developed for interface, and especially dialogue, design. Hekmatpour and Ince, for instance, have a separate user interface design component in their specification language EPROL (Hekmatpour and Ince 1987). Marshall (1986) has merged a graphical interface specification technique with VDM in order to obtain the best of both worlds. Alexander (1987a, 1987b) has designed an executable specification/prototyping language around CSP and functional programming.

- *Modelling of users* – Another strand of work concerns the formalisation of the user. This may take the form of complex cognitive models using techniques of artificial intelligence, such as the expert system for interface design described by Wilson *et al.* (1986). Another proposal is *programmable user models,* an architecture for which programs can be writen that simulate the use of an interface. The approach is advocated by Young *et al.* (1989) with the intention of studying user cognitive processes. It has also been advocated by Runciman and Hammond (1986) and Kiss and Pinder (1986) with the aim of using the complexity of the user programs to assess the complexity of the interface.
- *Architectural models* – Any specification or piece of software has some architectural design, and specific user interface architectures have been designed with the aim of rationalising the construction of interactive systems and improving component reuse. These may be structuring techniques for existing languages such as PAC (Coutaz 1987) (Presentation–Abstraction–Control) an hierarchical agent-oriented description technique, or the MVC (Model–View–Control) paradigm used in many Smalltalk interfaces; or may be part of an overall system as is the case with UIMS (Pfaff 1985) (User Interface Management Systems). Architectural techniques are often combined with notations for dialogue design and (more rarely) interface semantics. Production rules, for example, are frequently used as the dialogue formalism in UIMS. On the other hand, interface design notations may implicitly or explicitly encourage particular architectural styles.

More extensive reviews of these different areas can be found in Alexander's thesis (Alexander 1987c) and in a report on formal interface notations and methods produced collaboratively between York and PRG Oxford. (Abowd *et al.* 1989) A recent collection of essays on the subject of formal methods in HCI edited by Harrison and Thimbleby (1989) contains papers in most of the above categories.

An additional category has become characteristic of the "York approach" to HCI, which is the main subject of this book:

- *Formal, abstract models of interaction* – These are formal descriptions of the *external* behaviour of systems. They are not models of specific systems, but each covers a *class* of interactive systems, enabling us to reason about and discuss interactive systems in the abstract. As well as a large body of work originating in York, the approach has been taken up by Anderson (1985, 1986), who uses a blend of formal language and denotational semantics to describe interactive systems, and by Sufrin and He (1989), who cast in Z, a model similar to the *PIE* model presented in Chapter 2.

We can lay out the formal approaches to interactive systems design in a matrix classified by concreteness and by generality (*fig.* 1.1). The concreteness axis distinguishes between the internal workings of the systems and the specification of their external behaviour. The former are more useful for producing systems, the latter for reasoning about them. The generality of a method may lie between those which can be realised only in the context of a specific system and those that have some existence over a class. Laid out like this, it is obvious that abstract models fill a crucial gap.

	generality	
concreteness	specific	generic
specification	notations for specification *task and goal descriptions*	abstract models
implementation	prototypes of the actual system *programmable user models*	architectural models *cognitive architectures*

figure 1.1 *formal methods matrix*

In drawing up the matrix (and making my point!) I have rather overplayed the gap filled by abstract models. Specifications of particular systems may be deliberately vague in places, and thus begin to encroach on the generality barrier. Similarly, architectural models, although aimed at implementation, may be given a suitable form for us to use for specification and reasoning, and hence begin to move up towards the domain of formal models. Cockton's work (Cockton 1986) is a good example of this. He uses a description technique drawing on an analysis of UIMS. The notation is used to express properties of interface separability and comes close in spirit to the idea of an abstract interface model.

From the other side, the abstract models in this book are supplemented by examples of specifications of parts of actual systems, hence bridging the generality barrier from their side. Also, especially in Chapters 8 and 9, there is a movement towards more architectural descriptions, that is, a movement towards concreteness. Abowd (Abowd 1990) has produced a notation which attempts to sit in this middle ground between the formal models of this book and architectural models. It is, of course, no good describing useful properties of systems in a highly abstract manner, if these cannot be related to more concrete and specific situations, and thus these areas where the various techniques overlap are most important.

The more psychologically based formalisms sit rather uneasily in the matrix, but I have included them as, to the extent that they do fit the classifications, a similar gap is seen on their side. Now the abstract models we will deal with are primarily descriptions of the system *from* the user's point of view. (But definitely *not* in the language a typical user would use!) They do have then an implicit abstract, generic model of the user, purely because of the perspective from which they are drawn. It is though a rather simple model, and a more explicit model might be useful. On the other hand, I find myself feeling rather uneasy about the idea of producing generic models of users: individuality is far too precious.

1.2 Abstract models

We have seen that abstract models fill a niche in the range of available HCI formalisms, but we also need to be sure that it is a gap worth filling. We shall take a quick look at why we need abstract models, and at the philosophy behind them.

1.2.1 Principled design

There are many principles for the design of interactive systems. Some are very specific (e.g. "error messages should be in red") and others cover more general properties (e.g. "what you see is what you get"). Hansen (1984) talks about using user engineering principles in the design of a syntax-directed editor Emily. Bornat and Thimbleby (1986) describe the use of principles in the design of the display editor ded.

Thimbleby (1984) introduced the concept of *generative user engineering principles* (GUEPS). These are principles having several properties:

- They apply to a large class of different systems: that is, they are *generic.*
- They can be given both an informal *colloquial* statement and also a *formal* statement.
- They can be used to constrain the design of a given system: that is, they *generate* the design.

The last requirement can be met at an informal level using the colloquial statement – as was the case with the development of ded. While not superceding this, it seems that at least some of the generative effect should be obtained using the formal statements. The authors cited above who have specified particular interactive systems have proved certain properties of their systems, by stating the properties they require in terms of the particular specification and then proving these as theorems. The same approach could be taken for GUEPS; however, there are some problems.

- *Commitment* – The statement of the principles cannot be made until sufficient of the design has been completed to give an infrastructure over which they can be framed. This means that the principles cannot be used to drive this early part of the design process, which lays out the fundamental architecture, and hence may be crucial for the usability of the system.
- *Consistency* – Because the principles are framed in the context of a particular design, there is no guarantee that a given informal principle has been given equivalent formal statements in the different domains.
- *Conflict* – Expanding on the last point, there is a conflict between the desire for *generic* principles and the requirement for *formal generative* principles.

Formal abstract models can be used to resolve this conflict. The principles required can be given a formal statement using an abstract model. Because they do not represent a particular interactive system, but instead model a whole class of systems, we can reason about the effects and interactions of these principles as they apply to the whole class. Then, when a particular system is being designed, the abstract model can be mapped at a very early stage onto the design – or even refined into the first design. The principles defined over the abstract model can then be applied to the particular system.

We see that this tackles the problem of commitment because the principles are stated before the particular system is even conceived, and the problem of consistency because the principles are only stated once and then the same statement is applied to many systems.

1.2.2 Meta-models

Why is it that existing notations cannot be turned to this purpose? If we use a standard specification notation, the principles are stated within the context of a particular design – with the drawbacks described above. A principle stated using an abstract model is a requirement of what sorts of specifications are allowable within the notation. That is, it is really a statement about the *meta-model*.

Of course, any notation has, either explicitly or implicitly, a meta-model within which it works. What is wrong with these? Couldn't they be used instead of designing specific abstract models for the principles? The problem is that the notations are designed for the complete specification of the interface and therefore do not satisfy the principle of *structural correlation* necessary for effective requirements capture. This is not to say that an abstract model cannot be expressed within a particular notation. Many general-purpose notations are quite capable of expressing their own meta-model, and therefore an abstract model is no problem. An example is Sufrin's use of Z for models similar to those within this book. However, when this is done the abstract model is stated in the notation: we are not using the meta-model of the notation itself. The abstract models of this book are stated within the notation of general mathematical set theory, but can be translated into whatever notation is being used for a particular design.

1.2.3 Abstract models and requirements capture

Designing a system involves a translation from someone's (the client's) informal requirements to an implemented system. Once we are within the formal domain we can *in principle* verify the correctness of the system. For example, the compiler can be proved a correct transformer of source code to object and we can prove the correctness of the program with respect to the specification. Of course, what cannot be proved correct is the relation between the informal requirements and the requirements as captured in the specification. This gulf between the informal requirements and their first formal statement is the *formality gap* (*fig.* 1.2).

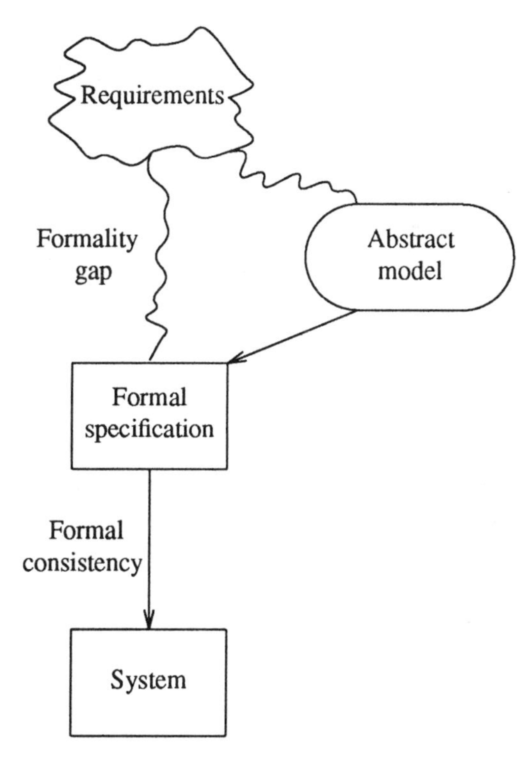

figure 1.2 *formality gap*

For a DP application like a payroll, this may not be too much of a problem. The requirements are already semi-formal (e.g. pay scales, tax laws) and they are inherently formalisable. The capture of HCI requirements is far more complex. Not only are they less formally understood to start with, but it is likely that they are fundamentally unformalisable. We are thus aiming to formalise only some aspect of a particular requirement and it may be difficult to know if we have what we really want.

If the abstract model is designed well for a particular class of principles, it can form a bridge between the informal requirements and the formal specification. To achieve this, there must be a close correspondence of structure between the abstract model and the informal concepts. This principle of *structural correlation* is an important theme.

1.2.4 Surface philosophy

The abstract models must reflect the structures of the requirements in order to bridge the formality gap. What does this mean for interactive systems and user-centred design? For particular classes of properties there will be particular structures of importance that must be reflected in the abstract model. For example, when considering windowed systems in Chapter 4, two informal characteristics are focused on, *contents* and *identity*. These are incorporated into the formal model. I will later argue that the identification of such informal characteristics is not only a precursor to formalism, but is part of the formal method itself.

Not all the features of the abstract model are dependent on the particular principles required. If we are searching for a user-centred design, then we should not be concerned with parts of the system that are not apparent to the user. We should adopt a *surface philosophy* when designing abstract models. We are not interested in the internal details of systems, such as hardware characteristics, languages used, or even specification notations! The models will be, as far as possible, *black-box models* concerned only with the user's input, the system's output, and the relation between them defined as abstractly as possible.

This approach, prompted by our interest in the human-computer interface, corresponds exactly to the desire in formal specification not to overspecify or, as it is often put, to specify "what", not "how". Using a formal specification notation does not prevent overspecification. Jones (1980) warns that a specification may contain *implementation bias,* that is, it may accidentally specify some aspects of a system that should be left undetermined. He gives formal checks for this. Even if a specification satisfies these conditions, it may still have aspects which although not constraining the design totally, do tend to push it in fixed directions. This is another example of *structural correlation*: the structure of the abstract model/specification strongly affects the eventual system.

Sometimes it may be necessary to break the black box slightly. Often the user may perceive some of the internal structure Where this is so, the model should reflect it. This may arise because the system is badly designed and you can "see the bits between the pixels". For instance, the user may be aware of implementation details, such as recirculating buffers or event polling, by the way the system behaves, or even worse be presented with error messages such as "`stack overflow`"! It may also occur in a more acceptable fashion. For instance, it may not be unreasonable for the user to be aware of the filing system as a separate layer of abstraction within an operating system.

Opening up the box does have dangers, however. Users' models may differ, and in particular they will differ from that of the designer. For example, the distinction between the operating system and a programming environment may not be evident to the novice. Even with the example cited above, of the filing system, there may be confusion when using an editor, between the editor's internal buffer and the filing system state. This may lead to errors such as removing a disk without writing the file. Because of these dangers, we will try to retain the black box view as far as possible, and only break it where it is absolutely necessary and when the user model is very clear.

This argument also leads us away from methods based around a user's goals, or cognitive models of the user. This is not because they are not useful. Applied as an evaluation or prediction tool on existing systems, they too complement the approach taken here. Just as the desire for a user-centred approach leads to a surface philosophy with respect to the computer, the need for a *robust* analysis leads to a similar view of the user.

When stating principles over the abstract models, care is again taken to ensure there is no overcommitment to a particular implementation technique or view of the system. So, for example, when we consider sharing properties of windows we reject *explicit* definitions of sharing using the data base or file system, and instead look for an *implicit* definition based on the input/output behaviour. Similarly, in Chapter 5, we define the buffering of user events not in terms of implied internal structures, such as blocks of memory, but in *behavioural* terms at the interface.

1.3 Formalities

So we are going to use formal methods to help us design interfaces. What do we mean then by formal methods, and what sort of results are we likely to get?

1.3.1 What are formal methods?

First and foremost, formalism is not about a particular notation. That much is obvious. Further, it is not the use of *a* notation or even general mathematical notation. When I think formally, εs, δs and Σs rarely cross my mind. These may be used for the communication or recording of the thought, but they do not constitute the formalism in themselves. Not only is this important in that we shouldn't demand such notations before we call something formal, but also just because we see mathematical notation, it is no guarantee of the soundness of the arguments. Even mathematically logical arguments can be spurious when related to the real-world entities they describe.

It is particularly common in computing circles to associate formal methods with a very strict and rigorous form of mathematical logic. The majority of mathematics operates at nothing like this level, always being prepared to leave gaps where the correctness is obvious. In a similar vein, Millo *et al.* (1979) argue that proof is essentially about raising confidence. This is in line with the argument presented earlier to justify abstract models as a bridge between requirements and formal specification. Also, I will later argue that structural correlation between abstract model and specification and between specification and implementation is crucial to raise confidence in the correctness of the final system – whether the relation is strictly proved or not.

So if proof is just a way of raising confidence, what of the *process* of formalisation? In essence, to formalise is to *abstract,* to examine features or parts of a whole and hence understand the whole better. This means that when in Chapter 4 I say that the critical aspects of a windowed system are content and identity, this is not an informal statement about to be formalised, but a formal statement about to be captured in precise notation.

Some of this abstraction can be captured in ordinary English, and indeed the language is rich in terms to describe abstractions of objects. So, for instance, I am a man, a human, a mammal, an animal and a living thing. I am also a biped and a mathematician. Common English is less good for discussing relations between objects in the abstract, and other higher-order concepts. In particular, it is very difficult to describe some of the real-time phenomena in Chapter 5.

In short, I use mathematical notation because it is useful, but the formalism is not restricted to the mathematics and the mathematics does not guarantee formalism. Despite all this, and for want of a better term, I will usually talk of the precise mathematics as being formal, and common English statements as being informal, although the boundary may not be clear.

1.3.2 Applicability of formal methods to design

Thinking of formalism in terms of precisely written rules and formulae, what is its range of applicability for interface design? If we asked instead more definitely – "Will we be able to generate a precise definition of usability?" – we could answer "NO" with little fear of contradiction. However, just where formal methods lie, between being useless and essential, is not obvious.

The use that is most obvious and most "formal" for the abstract models and principles defined in this book is as a *safety net*. We can see this if we look at the role of formalism in the design of a building. We have two putative designs, one a thatched cottage, and one a concrete block of flats.

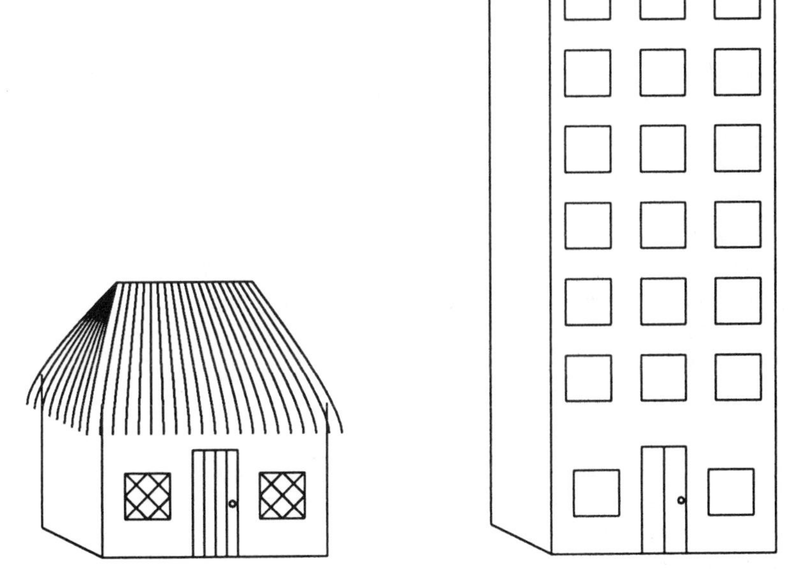

Formal properties of building materials can tell us things like the thermal conductivity of thatch, whether a concrete lintel can bear the required weight, whether the structures are water-tight. These may be codified into formal requirements, either explicit like building regulations on insulation and drainage, or implicit – it mustn't fall down! The formal analysis does *not* tell us which is the best design. Even if we formalise more of the requirements for a particular context – How many people will live there? Are they scared of heights? – we still have an incomplete picture. A formal method cannot tell us that uPVC windows on the cottage would be barbaric, or that leaded panes on the flats would be plain silly. So the formalism of building design can tell us whether the building will fall down or leak, but not whether it will be beautiful or pleasant to live in.

Going back to interface design, by defining suitable principles we will be able to stop interfaces being fundamentally unusable, but not ensure that they *are* usable. So we can't make a bad interface good, but we can stop it being abysmal. The challenge, of course, is to do this and still allow the good interface designer to be creative. That goes beyond this book, but what I have found is that the uses of abstract models for principled design go way beyond being just a safety net. However, even in that capacity only, the experience of other disciplines shows that they would still be worthwhile.

1.4 Editors

1.4.1 Universality of editing

Many of the motivating examples used in this book will be about text editing. Does this limit the applicability of the results?

I am not alone in this emphasis. Moran in a keynote address at Interact'84 referred to text editors as the "white rat" of interface design. Rasmussen at Interact'87 (Rasmussen 1987) similarly termed the word processor the "Skinner box", although he thought that HCI ought to move on from there. Despite this latter view, there are strong arguments for testing a new technique on an old field: if the technique proves useful despite much existing analysis, then this speaks highly in its favour.

I should like to extend that somewhat by saying that editing is universal. Almost all applications can be viewed as editing. For instance, an operating system can be seen as a file-system editor – compilers and other tools are just sophisticated commands. The drive towards direct manipulation emphasises this point still further. The exceptions to this view are non-closed domains like mail systems, but even they have a very large editing and browsing component.

Although editors provide motivating examples, the interaction models are intended to be quite general. This assertion is validated by the way the windowing model of Chapter 4 and the temporal model of Chapter 5 relate easily back to the general PIE model which was designed very much with editors in mind. Some of the facets will require a more general non-deterministic version of the PIE model, which would be suitable for a mail interface.

1.4.2 Some example editors

Occasionally I shall use some actual editor as an example. I describe briefly the nature of those used.

Ed (Thompson and Ritchie 1978) is the standard Unix editor. It is a line editor, like all line editors just more so!

Vi (Joy 1980) has become a standard display editor on Unix systems. It does not demand special keys of the terminal, but uses the normal typing keys in two main modes. Normally they mean some editing command, so for instance "k" means move up one line. If you ask to insert in a line, or add to the end of a line then there is a mode change and the keys generate characters, so "k" would insert as a character in the text. Even experts forget which mode they are in, and it is not uncommon for users to type editing commands into their text, or (more disastrously) type text which is treated as commands. This latter at best generates beeps and errors, and at worst destroys your text – hence the pun in Chapter 6.

Wordstar (Wordstar 1981) is the *de facto* standard word processor for personal computers. It reserves ordinary keys for their plain text meaning, and uses control key combinations for commands such as cursor movement. It has many inspired features, but seems to be looked on as rather a *bête noire* by the human interface community.

Ded (Bornat and Thimbleby 1986) is a display editor working under Unix. It inspired Sufrin's specification. Bornat and Thimbleby concentrated on the principles that underly its design, and on not simply adding features. It has a characteristic command line at the bottom of the screen, which is used for editing commands that cannot be mapped to single keystrokes.

Spy (Collis *et al.* 1984) is a multi-file editor for bit-mapped workstations. Its commands are all operated by menus or mouse clicks. It is based on the paradigm of a selected region of text, like most such editors.

1.5 A taste of abstract modelling

To give a flavour of what is to come we will take a quick look at a simple abstract model, designed to express the property of "what you see is what you get". This model, slightly reformulated, was my first use of an abstract model and it is offered as a first taste to the reader also. Although it is very simple it exposes a serious and pervasive problem, *aliasing*, and thus demonstrates the power of the technique.

1.5.1 A simple abstract model to express WYSIWYG

"What you see is what you get" (WYSIWYG) is one of the most well known principles of interface design. A large number of word processors, CAD systems, etc. claim to be WYSIWYG. Thimbleby (1983) points out that this is usually referring to superficial aspects such as having the correct fonts, whereas there is a much deeper principle expressing the way the system can be manipulated and viewed via its display. This he terms "what you see is what you *have* got". The principles I will try to formalise are based upon this notion rather than the presentation one.

An early suggestion to give his principle a precise formal statement in the context of simple text editors was to say an editor is "WYSIWYG" if the text it represents can be recreated as the concatenation of the various displays generated. This was only intended as a starter to more general definitions, being rather domain specific and restricting the possible displays to be simple windowing. Another problem is that it relates display to internal state, rather than describing the interface properties purely in terms of the external interface. This last point is easy to correct: the displays should (in the case of a text editor) be related to the final printed form. Depending on whether there is any formatting of output the map from internal state to printed form may be trivial or complex, so it is important that we retain the external view as much as possible. In fact, we could go further and express the relation between different facets of the interface in terms of the user's perception of them and the resulting internal models to which they give rise. From this we get a simple picture of the interface (*fig*. 1.3).

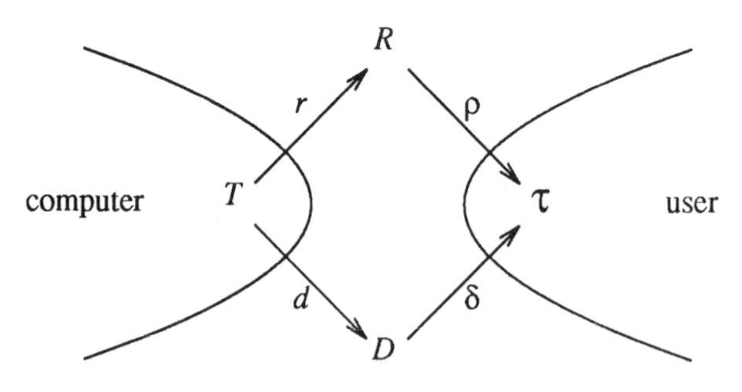

figure 1.3 *WYSIWYG model*

The internal representation of the object being edited (from T) can be printed (using the map r) to yield the result (from R) which is then viewed by the user (with a possibly noisy perception function ρ) to yield an internal idea of the object (from τ). Similarly, the object can be displayed (using d to generate a display in D) and is also viewed (using δ). The viewing of the result is potential (you don't print the entire text after each command!), whereas the viewing of the display is continual. Further, although there is only one result for a particular object, there are many possible displays, so, strictly, the display should be parametrised over some set Λ, giving the set of displays $\{d_\lambda\}$.

This cannot be read as a simple commuting diagram, as a single display will not in general yield all the information about the object, just part. The map δ is not therefore a simple function but instead provides *partial information.* This partial information could be represented by the set of objects in τ consistent with a particular display, or more generally by a function to a lattice based on τ. We can then give simple definitions of WYSIWYHG in terms of the information ordering $\leq$:

consistency:

$$\forall\ \lambda \in \Lambda : \forall\ t \in T : \quad \delta \circ d_\lambda(t) \quad \leq \quad \rho \circ r(t)$$

That is, the information provided about the object from the display is consistent with the text that would be obtained by printing.

It is no good the display being consistent if it doesn't show you the bits you want to see (a blank screen is consistent with anything). We really want the stronger principle "what you can see is *all* you have got": that is, there should be nothing in the result that is not viewable via the display:

sufficiency:

$$\forall\ t \in T : \quad \bigvee_{\lambda \in \Lambda} \delta \circ d_\lambda(t) \quad = \quad \rho \circ r(t)$$

Clearly, since the individual $\delta \circ d_\lambda(t)$ are less defined than $\rho \circ r(t)$, then the least upper bound is also less well defined. Unfortunately, in even apparently innocuous cases equality is not obtained and unexpectedly stringent requirements are made on the system if sufficiency is required to hold.

1.5.2 Aliasing

Consider the following example. Let T be sequences of binary digits and the display (D) is either a pair of binary digits or a "before the start" symbol (») or an "after the end" symbol («). The individual displays are obtained as adjacent pairs of digits from the object and all such displays are possible. The result map and perception functions are all identity. Thus t = "01" has the following possible displays:

"»0", "01", "1«"

In this case there is only one possible consistent object, that is, t itself. However, if t="0100" then the possible displays are:

"»0", "01", "10", "00", "0«"

In this case not only is τ="0100" possible, but so is τ="0010" (and this is assuming the length is known!). This is the problem of *aliasing* and it will arise repeatedly in different guises: essentially, the *content* of a display does not unambiguously define the *context*. We can, of course, overcome this by explicitly coding the context in the display. For instance, in the previous example, we could augment the display by adding a position so that the displays would be:

"[0]»0", "[1]01", "[2]10", "[3]00", "[4]0«"

This time there is only one candidate, τ="0100", as required.

This is not an artificial problem. For instance, large files of data gathered for statistical analysis may contain tracts of identical (or very similar) information. Again when editing programs, several functions may have very similar form and the user may mistake which function is being edited. This second example is more problematical, as in the former case the user is likely to be aware of the possibility of confusion, whereas in the latter, guards will be down.

Some editors do in fact display the line number (or byte number!) of the current cursor position, or alternatively a scroll bar with the current display highlighted presents the same information. However, it is not intuitively reasonable to make all systems display such contextual information, and later definitions of WYSIWYG properties will be given that are more liberal. Because of aliasing these definitions will be considerably more complex and lack the conceptual clarity of a purely information theoretic approach.

1.5.3 Implications for future models

The class of problems that considerations of WYSIWYG lead to I will call *observability* and will be a major interest when considering various models.

Another interesting point that is exemplified in this simple context is the conflict between pedagogic statements and more complex problems of perception. Consider the scroll bar. We might argue that the maximum document size that avoids aliasing is when each pixel of the scroll bar corresponds to a screen of text. More circumspectly, we realise that the user's perception (represented by δ) is unlikely to be perfect, and although it may be reasonable to regard it as identity when reading text, it is unlikely to be up to resolving differences of one pixel between successive displays. After further reflection, one realises that it is likely that users may ignore the positional information entirely unless they are consciously aware that aliasing may arise. So, it may well be that an editor with explicit contextual information loses this in perception, whereas another editor with no explicit information, but which helps the user develop an internal sense of context, perhaps by allowing only small moves and emphasising direction by smooth scrolling, may be better.

If we continue to include the user's perception in our models, then our formal statements can include such considerations: however, this is unlikely to be totally correct even when considering a single user and certainly not for an entire user community. For this reason most formal statements will be made at the interface level (e.g. about d and r here), effectively assuming users are perfect perceivers and reasoners. This is in line with the "surface philosophy" applied to users and the limitations of formal methods described above. At various stages the system should be checked against more informal statements of principles (and/or formal models of perception). The two processes of formal and informal reasoning complement one another and yield a more robust total method.

1.6 About this book

The flow of chapters in this book reflects a partly historical, partly logical progress of different models and methods. The material ranges from highly formal to anecdotal, and may move rapidly between the two. For those without a rock-hard constitution a simple cover-to-cover reading is unlikely to be successful. I will give a brief overview and suggest ways of reading.

1.6.1 The landscape

The book has two main phases of model building. The first five chapters (2–6) deal with some fully abstract black-box style models; the primary purpose is to express usability principles for different contexts. Chapters 7–9 "open up" the black box slightly, reflecting the way users know more about a system than its physical surface. The last three chapters round off the book in various respects.

Principles – abstract models

We begin with what is in some ways the simplest and most general model, the PIE model. It is simple in that its mathematical formulation is sparse, and in that it assumes very little in the way of detail about the systems it describes. However, the analysis in Chapter 2 is probably the most mathematical and detailed in the entire book. By being very rigorous on the simplest model, we are freed to take a slightly looser approach in other areas knowing that we may perform similar activities if it were required.

A good example of this is the use of state-based models. The PIE model takes a very external view of the system. Not only does it not describe the state of the system, it shies away from even mentioning the state! One important result that comes out from the analysis is that it is always possible to give a state-based description of a system that is no less abstract than such a behavioural one. Further, for any view of the system, we can obtain an effective state (called the *monotone closure*) which captures all that's needed to predict the future behaviour of that view. In later chapters, we use state-based formulations where appropriate in the knowledge that we can assume this state to be the minimal "behavioural" one.

The next chapter is about red-PIEs. These augment the very sparse PIE model by distinguishing the display (what you see) from the result (what you will get). The properties of interest are primarily issues of observability: what can we tell about the result from the display. The key phrase is the "gone away for a cup of tea" problem. I use my computer, come back, and wonder where I've got to. The chapter investigates what I can infer about the system upon my return.

Multiple windowed systems are increasingly common. Chapter 4 suggests a model that is capable of describing the special behaviour of such systems. It is particularly concerned to address issues of interference between windows. This is important because of the possibility that different windows are concerned with different tasks and therefore interference between windows means interference between tasks.

In common with many interface descriptions, the basic red-PIE model describes the interleaving of input and output, but not detailed temporal behaviour. The model defined in Chapter 5 can be used to describe behaviour such as *buffering* and *intermittent* and *partial update*. In order to counter some of the problems raised by real-time behaviour it proposes that certain information is available to tell the user when the system is in *steady state* and when commands will be ignored. The model is so constructed that steady-state behaviour can be specified independently of the real-time behaviour.

Chapter 6 tries to bring some of the models together, but finds that a *non-deterministic* model of interaction is required. This leads on to a discussion of the meaning and role of non-determinism in the interface. It is found to be a useful descriptive technique and unifying paradigm. An example of its constructive use is the proposal of a new efficient display update strategy.

Where appropriate these chapters discuss the implications of the principles discussed. For instance, Chapter 5 discusses the various forms that the information on real-time behaviour might take depending on where the user's focus of attention is. It also discusses the requirements that its proposals put on support software such as operating systems and window managers.

Design – opening up the box

Chapter 7 begins this phase by focusing on the design of layered systems: where the inner layers represent the functional core or task activities of the system, and the outer layers the physical level such as keyboards and screens. It investigates various constructive ways of building up the entire system given the inner functional core. Two major conclusions arise. The first is that the relationship between the functional core and the entire system is one of abstraction rather than of being a component. The second is that any understanding of manipulative systems must include a very close connection between the interpretation of users' actions and the presentation of the system's responses. The user's input is mediated by the display context. As a sort of last fling at a constructive model Chapter 7 looks at the use of *oracles* in designing and prototyping systems.

Chapters 8 and 9 take two possible routes to display-mediated interaction. Chapter 8 looks at *dynamic pointers*. Position is regarded as the crucial feature in the interpretation of user commands in a display-oriented system. This leads on to a detailed investigation of the properties of positional information in interactive systems. Dynamic pointers are thus aimed at translating the user's display-level commands into operations on the underlying objects. Chapter 9 takes a complementary approach. It is assumed that users' primary perception is manipulating the view they have of the system. The important issue then becomes how to change the underlying state to keep it consistent with the user's view. The rather odd conclusion of this chapter is that when updating using views, it is what you do *not* see which is of central importance.

Rounding off

Chapter 10 focuses on events (things that happen) and status (things that are). This is partly as an admission of how poorly the majority of the models presented here deal with this distinction, and partly as an attempt to rectify the situation. It also forms a link to the study of multi-person interaction, in contrast to the one-user-one machine assumption of most of the book.

The models and methods in this book aren't much good if they cannot be applied to real situations. The most important application is the way that such models open up new ways to think about interactive systems. I hope that readers will find this for themselves as they move through the book. Chapter 11 gives more explicit examples of how the models can be applied. These range from "informal" applications of the formal models in a real situation, to the way they integrate into a formal development paradigm of software engineering.

Finally, Chapter 12 reflects back on the models presented and some of the threads that emerge using abstraction as a unifying theme.

Appendices – notation

Appendix I describes the notation used in this book and also how to read function diagrams such as the one we have already seen in §1.5 and found in various places, especially in Chapter 3. Appendix II contains a specification of a simple text editor which is analysed in Chapter 11.

A little history

The PIE model is the starting point both historically and logically of many of the other models in this book. It has also become the key term associated by many with the whole spectrum of models and techniques generated at York. I considered writing the relevant chapter for this book using a slightly different formulation, but decided that any gain in comprehensibility of such a new perspective would be outweighed by the confusion for those familiar with the original formulation. In addition, I would have been in grievous danger of losing the acronym entirely!

1.6.2 The route

As I have said, it is probably a bad idea to start at the beginning and work through. The chapter on PIEs is the logical starting point but is one of the most complex mathematically. In order to ease this somewhat I have tried to summarise the primary points raised by the chapter in its introduction. This can be read in order to give the reader enough background to move on to other chapters. In general, I would encourage the reader to skip parts which they find too formal (but **not** *vice versa*). Dependencies are typically loose enough that later parts need only a general understanding of previous material, remembering always that it is the concepts that matter, not the particular formulations.

Quite a lot of the chapters can be read on their own. In particular, Chapter 5 on temporal behaviour has appeared on its own elsewhere (Dix 1987a) and is quite readable on its own. Also, a slightly extended version of Chapter 6 appeared in (Harrison and Thimbleby 1989) and, if you are prepared to believe that non-deterministic models occur naturally when considering formal models, the latter part of it can also be read on its own. The second phase of the book, "opening up the box", requires familiarity only with the PIE model (and red-PIE), and Chapters 8 and 9, although naturally leading on from this, are each self-contained. Chapters 10 and 11 both refer back to previous chapters, but both are largely readable with perhaps a passing knowledge of earlier material.

For the reader who doesn't want to get bogged down in heavy formalisms too early I would suggest starting with Chapter 5, and (with the caveat above) Chapter 6, most of Chapter 10, and the first two sections of Chapter 11. For the more confident reader ... well I'll let you decide.

CHAPTER 2

PIEs – the simplest black-box model

2.1 Introduction

We are interested in formalisation of various properties of interactive systems. We want to state these properties over abstract models of interactive systems, in order to make them generic and to make the correspondence between the informal and formal statements as easy as possible.

This chapter presents a very simple model called the PIE. This is probably one of the simplest models that can be defined, and yet many useful properties can be given some expression in terms of the model. Further, the more specific models given in the succeeding chapters will have very much the same flavour as PIEs and studying them will make these later models easier to understand.

2.2 Informal ideas behind PIEs

Although the PIE model is the simplest in terms of the number of elements in it, the treatment is probably the most complicated in this book. Many things are dealt with in detail here which have parallels in the more specific models developed later. When these issues arise later they can be skimmed over because of the experience gained in the PIE model. This first section gives an overview of the rest of the chapter so that the reader can pick up a good enough idea of the contents to read further chapters. The remaining, more formal, sections can thus be skipped on the first reading and returned to later.

2.2.1 PIEs

As we are interested in the human interface to computer systems, we want our models to refer only to what is visible (or audible, etc.) to the user. That is, we want a *black-box* model. The PIE model describes a system in terms of the possible inputs and the effect these have. Typically the input will consist of a sequence of commands (we will call the set of such allowable commands C, and the set of all such sequences the programs P). The relation between these inputs and the output is described using an *interpretation* function, what in control theory would be called a transfer function. The output we will call the *effect*:

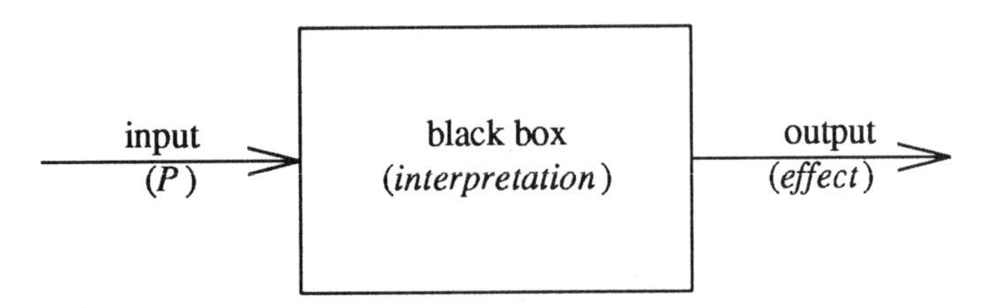

The three together – inputs (P), interpretation (I) and effect (E) – give the model its acronym.

The effect is the most complex part of the model. We may wish to focus on the screen display, or some part of it. Perhaps the effect is some external activity, or it may be interpreted internally, perhaps the database or spreadsheet being manipulated. We deliberately leave this vague so that we may apply our analysis retrospectively to many different situations. However, when we define properties we will usually have some idea of what we mean by the effect in that circumstance, and we would not expect the same principles of usability to apply to all levels of analysis.

The input may similarly be applied at various levels. For instance, at one level we may wish to regard it as the physical keystrokes of the user, but at another we may want to think about the abstract application-level commands.

2.2.2 Properties of interest

Even this sparse model will be sufficient to give formal expressions of certain classes of principles. These same principles will arise in different forms throughout succeeding chapters:

- *Predictability* – Can we work out from the current effect what the effect of future commands will be? This can be thought of as the "gone for a cup of tea" problem. When you return and have forgotten exactly what you typed to get where you are, can you work out what to do next?
- *Observability* – Although the system is seen as a black box, the user can infer certain attributes of its internal state by external observation. How

much can the user infer? How should the user go about examining the system? We will define the idea of a *strategy*: a plan that the user follows in order to uncover information about the system. Predictability can be viewed as a special case of observability when we are interested in observing the entire state as it affects subsequent interaction.

- *Reachability* – This is concerned with the basic functionality of the system. Is it possible to get to all configurations? Are there any blind alleys, where once you've entered some state there are other states permanently inaccessible? Related to the general notion of reachability, will be the specific case of *undo* and the problems associated with this.

2.2.3 Internal state and external behaviour

Although the primary definition of PIEs will make no reference to internal state, we will find it useful to distinguish a (not necessarily real) minimal state that can be inferred from external observation. It will be minimal in that it distinguishes fewer states than any other state representation that describes the same functionality. It is a very important concept, as it is often easier and clearer to give certain properties using the idea of state, and yet this would otherwise risk producing a definition that "knows too much" about the inside of the black box. We can always assume that such properties refer to this minimal state. Similarly, in later chapters when models are defined using state, these will be assumed to be minimal in the same way.

This operation of obtaining just enough information on the state we will call *monotone closure*, and given any view of a system V we will refer to the monotone closure of this as $V^\dagger$. This can be thought of as the part of the state which is just sufficient for predicting the future behaviour of V. The user cannot necessarily observe $V^\dagger$ directly, but if two systems differ in this respect there is some command sequence that the user can type which would expose the difference. Thus the user needs to know $V^\dagger$ in order to predict the behaviour of the system (and hence control it). Unfortunately, in general, the user would have to be able to try, and undo all possible command sequences to know for certain the state of $V^\dagger$. An example of this would be a pocket calculator. After having entered "3+4" the display would read "4". On its own the display is insufficient to predict what the effect of pressing the "=" key would be. The monotone closure of this view would include the pending operations and data.

The user can and does find out about the internal state of the system by experimenting with it. The simplest example of this is scrolling up and down in a word processor to see all of the document. By following such a strategy the user obtains a partial idea of the internal state. We call this the *observable effect*. Obviously, different strategies are useful in different circumstances and yield different observable effects. The observability principles of the next chapter are framed largely in terms of the way the observable effect derived from the display tells us about the monotone closure of the result.

2.2.4 Exceptions and language

We will also consider some issues arising when the system as implemented differs from some "ideal" system, and what principles apply at these boundaries (the exceptions). We will consider two system models, one of the "ideal" system and one of the system as it actually is. The places where mismatch occurs are the exceptions. We will look at principles which demand that the user is made aware of exceptions (e.g. by a bell or message) and that the user is able to predict when exceptions will occur. Depending on the type of exception and the recovery principle used (see below), the designer may choose to apply the former or the latter (stronger) principle. These observability and predictability principles are supplemented by a discussion of recovery principles determining what sort of response the system should make to exceptions. The strongest is "no guesses", which asserts that when an exception occurs the system state is left unchanged. A weaker variant is "nothing silly please", which asserts that the system behaves in a way that the user *could* have achieved by non-exceptional commands.

This analysis of exceptions will lead naturally on to considering systems where the input language is for some reason restricted. This is not the normal "grammar" of the dialogue, which describes what the system *expects* of the user, but expresses some form of *physical* limitation of the user's input. The obvious example is a bank teller machine which covers the keypad with a perspex screen until the customer inserts a cash-card. The user is not free to perform all possible action sequences but, on the other hand, it may not be wise to assume that "illegal" sequences will not occur. The various principles need slight reformulations in the light of these properties.

2.2.5 Relationships between PIEs

We will briefly examine the ways in which we can relate PIE models to one another. This forms a basis for understanding the way that we can have several models of the same system, perhaps with different scopes or at different levels of abstraction. This will be taken up again in Chapter 7.

2.2.6 Health warning

The rest of this chapter will be quite "heavy", so as a reminder to the reluctant formalist, there are easier waters in subsequent chapters.

2.3 PIEs – formal definition

There are several ways we could use to define PIEs. The one we shall regard as basic, and probably the simplest, is as a triple $< P, I, E >$, as follows:

P – The set of sequences of commands from C (P stands for programs). More generally P can be a semi-group, which has little relevance for the user but can be useful occasionally in constructions (see §2.8).

E – The effects space, the set of all possible effects the system can have on the user. This may be thought of in different ways, and at different levels. For example, it may be regarded as the actual display seen by the user, or as the entire text of a document being edited, perhaps even the entire store of information available to the user.

I – The interpretation function, $P \to E$, representing all the computation done by the system.

We can represent a PIE as a diagram:

$$P \xrightarrow{I} E$$

Later we will use the following class of functions. For any $p \in P$, we define I_p by:

$$I_p(q) = I(pq)$$

That is, I_p gives the functionality of the system as if when you get to it the command sequence p has already been entered. Obviously this will have relevance to the "gone away for a cup of tea" problem.

Alternative, equivalent formulations of the PIE model will be useful in different circumstances.

By state description and doit functions

Sometimes it is easier to describe a system as a state with transitions on each command. We will call the transition function *doit* and the set of states S:

$$doit : S \times C \to S$$

There will be, of course, some initial state $s_0 \in S$.

In general, the state will contain more information than we want to consider as the effect. In these case we will require an abstraction function *proj* yielding the effect:

$$proj : S \rightarrow E$$

That is, we need a sextuple $< C, S, E, s_0, doit, proj >$ in place of the triple!

We can then define a corresponding PIE with interpretation function I_{doit} as follows:

$$I_{doit}(p) = proj(doit^*(s_0, p))$$

where $doit^*$ is the iterate of $doit$:

$$\begin{aligned} doit^*(s, null) &= s \\ doit^*(s, p::c) &= doit(doit^*(s, p), c) \end{aligned}$$

and "::" is the concatenation operator.

It is possible to reverse this and obtain a state representation for any PIE by simply taking $S = P$, and defining s_0, $doit$ and $proj$ thus:

$$\begin{aligned} s_0 &= null \\ doit(s, c) &= s::c \\ proj &= I \end{aligned}$$

However, taking the complete command history as state is a little excessive. In fact this is a maximal state representation: no other state representation can distinguish more reachable states. We will later, as promised, define a minimal (and much more useful) state representation, the *monotone closure*.

The fact that systems defined by *doit* functions can easily be related to PIEs, and hence have the various principles we will define over PIEs applied to them, is very important. The actual systems that I have specified have used *doit* functions as these are often the simplest way to describe interaction. In particular, the specification described in Chapter 11 uses just this mechanism. Also, in later chapters, we will use whichever representation is most convenient for the particular circumstances, and may even switch fluidly between the two representations.

By equivalence relations on P

A mathematically very simple (although not necessarily very meaningful to the user) way of defining an interactive system is by equivalence relations on P. Essentially, two command histories are equivalent if they have the same effect. That is, for any PIE $< P, I, E >$ we can define a relation $\equiv_I$:

$$p \equiv_I q \quad \hat{=} \quad I(p) = I(q)$$

Similarly, given such an equivalence, it defines (in some way) an interactive system and we can obtain a PIE $<P, I_{\equiv}, E_{\equiv}>$ from it:

$$E_{\equiv} \quad \hat{=} \quad P \,/\equiv$$
$$I_{\equiv}(p) \quad \hat{=} \quad [\,p\,]_{\equiv}$$

where $[\,p\,]_{\equiv}$ is the equivalence class of p in $P \,/\equiv$. These are inverses of one another, in that $\equiv_{I_{\equiv}}$ is the same as $\equiv$ and $I_{\equiv_I}$ is the same as I up to a one-to-one map on E. To prove the former:

$$\begin{aligned} p \equiv_{I_{\equiv}} q \quad &\hat{=} \quad I_{\equiv}(p) = I_{\equiv}(q) \\ &= \quad [\,p\,]_{\equiv} = [\,q\,]_{\equiv} \\ &= \quad p \equiv q \end{aligned}$$

The other way round is similar, except it is relative to the one-to-one map between $P \,/\equiv_I$ and E given by $[\,p\,]_{\equiv_I} \rightarrow I(p)$.

We will define other relations on P later which will be useful for defining observability properties.

By complete effect histories

We have given the basic definition of a PIE using an interpretation function relating complete command histories to the single effect they yield. On grounds of symmetry (and in analogy with transfer functions), we could have chosen to relate complete histories of commands to complete histories of effects. For example, for any interpretation I we can define a new interpretation I^* yielding the complete history of effects given by I:

$$I^* : P \rightarrow E^+$$

where E^+ is the set of non-empty sequences of effects from E, such that:

$$\begin{aligned} &I^*(c_0 c_1 c_2 ... c_n) \quad = \\ &\qquad (I(\mathit{null}), I(c_0), I(c_0 c_1), I(c_0 c_1 c_2), ..., I(c_0 c_1 c_2 ... c_n)) \end{aligned}$$

This form of definition is very similar to the use of streams to define interactive systems in lazy functional programming.

Any function generated thus satisfies some simple properties:

$$\forall p, q \in P$$
(i) $p \leq q \;\Rightarrow\; I^*(p) \leq I^*(q)$
(ii) $\mathit{length}(I(p)) = \mathit{length}(p) + 1$

Contrariwise, any such stream function can be turned into a PIE by taking the last element. Examining these conditions leads one to imagine generalisations of the PIE model based on such stream functions, but with a relaxation of one or other of these conditions:

(i) This is a necessary condition asserting that the function is temporally well behaved. If we drop this function we have systems that change their mind about effects already produced!

(ii) This is not so obviously necessary and it could fail to hold for one of two reasons (or both):

$$(ii_a) \quad \exists\, p, c \;\; \textbf{st} \;\; I^*(pc) = I^*(p)$$
$$(ii_b) \quad \exists\, p, c \;\; \textbf{st} \;\; length(I^*(pc)) > length(I^*(p)) + 1$$

(ii_a) This says that the new command has no additional effect. Note, this is not the same as saying the current effect is the same as the last one, which would be $I^*(pc) = I^*(p) :: last(I^*(p))$. How would you distinguish these at the interface? This would only seem useful as an abstraction from the true interface.

(ii_b) This has a more reasonable interpretation, namely that the system has some sort of dynamic behaviour after the command c. This (and ii_a) could be captured by using a more expressive effect space which represents within it the elements of dynamism: however, in doing this one might loose naturalness of expression.

Elements of this generalisation can be found in several places in this book. It could be seen as a half-way house between the PIEs and the fully temporal model considered in Chapter 5. Although we do not use the generalised form of stream function, we use the stream representation of PIEs when considering non-determinism in Chapter 6. Finally, there is some similarity to the model developed to describe status input in Chapter 10.

2.4 Some examples of PIEs

Although almost all interactive systems can be cast into the PIE framework, choosing appropriate examples is a little difficult. "Toy" examples are useful for getting to grips with nitty-gritty properties, but are of course unrealistic. (In fact, the toy examples may be parts or abstractions of more substantial systems.) If we look at realistic systems, we cannot expect to express the interpretation function very concisely or fully (if we did we would have a complete specification of the entire system). Bearing this tension in mind, we will look at a few examples.

2.4.1 Simple calculator

The first example we shall look at is a calculator that adds up single digits:

$$C = \{ 0,...,9 \}$$
$$E = \mathbb{N} \quad - \quad \text{the natural numbers } \{ 0, 1, ..., 57, 58, ... \}$$

$$I(\,null\,) = 0$$
$$I(\,pc\,) = I(\,p\,) + c$$

The interpretation basically says:

- We start off with a running sum of zero.
- If at any stage we enter a new number (c) it gets added to the current running sum ($I(\,p\,)$).

To make the example a little more interesting we could add a clear key #. We augment C appropriately:

$$C_{\#} = \{ 0, ..., 9, \# \}$$

The effect is the running total as before. The interpretation is built up recursively in a similar manner to I above:

$$I_{\#}(\,null\,) = 0$$
$$I_{\#}(\,pc\,) = I(\,p\,) + c \qquad c \neq \#$$
$$I_{\#}(\,p\#\,) = 0$$

The only difference is the obvious one, that if a clear (#) is entered the running total is zeroed.

2.4.2 Calculator with memory

Now we will add a memory to the calculator. We will call the new commands MS and MR (memory store and memory recall). The command set is obvious:

$$C_m = \{ 0, ..., 9, \#, MS, MR \}$$

Again the effect space is simply the natural numbers. The interpretation function is as before, but must "scan back" into the command history to find the last memory store, every time a memory recall is used:

$$I_m(\,null\,) = 0$$
$$I_m(\,pc\,) = I(\,p\,) + c \qquad c \notin \{ del, MS, MR \}$$
$$I_m(\,p\#\,) = 0$$
$$I_m(\,p\ MS\,) = I(\,p\,)$$
$$I_m(\,p\ MR\,) = 0 \qquad MS \notin p$$
$$I_m(\,p\ MS\ q\ MR\,) = I(\,p\,) \qquad MS \notin q$$

Notice that the scanning back process makes the interpretation relatively complex. In fact, if readers are not familiar with such tricks they may well want to check it a few times to see that it really does what they would expect. The reason for the complexity is that the effect does not hold all the state information necessary for interpreting the next command. The appropriate information has to be gleaned from the command history.

We can produce an identical system using a state representation that explicitly includes the calculator's memory. We model the state as a pair of numbers, the first being the running total and the second the memory contents. The initial state is with both zero:

$$\begin{aligned} S &= \mathbb{N} \times \mathbb{N} \\ s_0 &= (0, 0) \end{aligned}$$

The projection function *proj* that extracts the effect is simply the first component:

$$proj((e, m)) = e$$

The *doit* function just does the obvious adding and swapping around between memory and running total:

$$\begin{aligned} doit((e,m), c) &= (e + c, m) \qquad c \notin \{\#, MS, MR\} \\ doit((e,m), \#) &= (0, m) \\ doit((e,m), MS) &= (e, e) \\ doit((e,m), MR) &= (m, m) \end{aligned}$$

2.4.3 Typewriter

From numbers to text. To simplify descriptions we will just consider simple typewriters and editors with no line-oriented commands. The whole text will be on a single line. Adding multiple lines just makes the descriptions rather longer. First the simple typewriter:

$$\begin{aligned} C &= Chars = \{\text{a,b,c,d,e, ...,z,A,B, ..., etc.}\} \\ E &= P = Chars^* \end{aligned}$$

That is, the commands are the printable keys on the keyboard and the effect is a single sequence of these characters.

The effect of any key sequence is simply the sequence of keys hit:

$$I(p) = p$$

or equivalently:

$$I(\,null\,) = null$$
$$I(\,pc\,) = I(\,p\,)::c$$

The second recursive description can be used to extend the system, first to include a delete key. We will use ∇ for this:

$$C_{\nabla} = Chars + \nabla$$

The effect is as before:

$$E_{\nabla} = Chars^{*}$$

The effect is built up in a similar manner to the simple typewriter except that the delete key removes the last character:

2.4.4 Editor with cursor movement

We can now add a cursor position with left and right movement. The command set has these additional two commands:

$$C_{cursor} = Chars + \nabla + LEFT + RIGHT$$

The effect we can regard as two sequences, those *before* and those *after* the cursor:

$$E = Chars^{*} \times Chars^{*}$$

This method of describing a cursor position is due to Sufrin (1982). Typing a character C from *Chars* simply adds text to the *before* half:

$$I_{cursor}(\,null\,) = \{\ null, null\ \}$$
$$I_{cursor}(\,pc\,) = \{\ before::c\,, after\ \}$$

where

$$\{\ before,\ after\ \} = I_{cursor}(\,p\,)$$

The delete ∇ operates on the *before* half in a similar manner to the typewriter above, so we will skip it and move on to the cursor commands:

$$I_{cursor}(\,p\ LEFT\,) = \{\ before,\ c:after\ \}$$

where

$$\{\ before::c\,,\ after\ \} = I_{cursor}(\,p\,)$$

and

$$I_{cursor}(\,p\ RIGHT\,) \;=\; \{\ before::c\,,\,after\ \}$$

where

$$\{\ before,\ c\!:\!after\ \} \;=\; I_{cursor}(\,p\,)$$

That is, the *LEFT* key moves characters from *before* to *after*, and *RIGHT* *vice versa*.

2.4.5 Editor with MARK

Finally (well almost finally), we have an editor with a marked position to which we can jump:

$$C_{mk} \;=\; Chars\ +\ \nabla\ +\ LEFT\ +\ RIGHT\ +\ MARK\ +\ JUMP$$

The effect space must be extended so that we can see where the mark is. We do this by just having the mark as a special character in the effect too:

$$E_{mk} \;=\; (\,Chars + MARK\,)^{*}\ \times\ (\,Chars + MARK\,)^{*}$$

This allows any number of marks, but we will arrange it so as there is only ever one. The interpretation of all the existing commands is just as before, and *MARK* is treated almost like any other character. We let delete remove a mark if it is the last "character" before it, and cursor movement go back and forth over it. However, to ensure that there is only one mark the adding of it must be slightly different:

$$I_{mk}(\,p\ MARK\,) \;=\; \{\ strip(\,before\,)::MARK,\ strip(\,after\,)\ \}$$

where $\{\ before,\ after\ \} \;=\; I_{mk}(\,p\,)$. We assume *strip* is a function that gets rid of any marks.

That only leaves us with the *JUMP* command:

$$\textbf{let}\ \ \{\ before,\ after\ \} \;=\; I_{mk}(\,p\,)$$

$$\begin{aligned}
I_{mk}(\,p\ JUMP\,) &= \{\ b_1\ MARK\,,\ b_2\,after\ \} \\
&\quad \textbf{if}\ \ before \;=\; b_1\ MARK\ b_2 \\
I_{mk}(\,p\ JUMP\,) &= \{\ before\ a_1\ MARK\,,\ a_2\ \} \\
&\quad \textbf{if}\ \ after \;=\; a_1\ MARK\ a_2 \\
I_{mk}(\,p\ JUMP\,) &= \{\ before,\ after\ \} \\
&\quad \textbf{otherwise}
\end{aligned}$$

These definitions simply move the cursor to just after the mark if there is one, or leave it where it is if there is none.

In Chapter 8 we will look more closely at positional information such as cursors and markers, and give a cleaner and more uniform treatment. However, these more detailed descriptions will still be able to be rendered as PIEs like the above.

2.4.6 A full wordprocessor

The last example will be the word processor I am using to write this. Unfortunately, if described in full it would take up the rest of the chapter, so we shall elide a few of the details.

First, let's say what are the commands and effect space:

$$C \;=\; Keys \quad - \text{ the keys on my PC keyboard}$$
$$E \;=\; \{1...25\} \times \{1...80\} \;\rightarrow\; Glyph$$

Here *Glyph* is the set of characters, spaces and symbols that can appear on my screen. Basically, E is the set of all possible screen displays.

The interpretation function is quite simple:

$$I(p) \;=\; \text{what I see when I have typed the keystrokes } p$$

2.5 Predictability and monotone closure

Now we come to the "gone away for a cup of tea" problem. Can we predict the future effect of commands from the current effect? It is likely that two different sequences of commands may yield the same effect, but if more commands are entered some difference comes to light. We then say that the original effect is *ambiguous*, as more than one internal state is possible:

$$ambiguous(e) \;\hat{=}\; \exists\, p,q,r \in P \;\; \textbf{st} \;\; I(p) = e = I(q) \textbf{ and } I(pr) \neq I(qr)$$

If we look at the examples in the previous section, the calculator with memory is always ambiguous, since from the effect, we only can tell the running total. Thus, if we have a current effect of "57" and then use the memory recall command, we may get any effect depending on the precise history that led to the running total of "57". Similarly, the word processor I am using is ambiguous (when the effect is regarded as the screen image), as a scroll up key will reveal text that cannot be inferred from the current screen.

The other examples, such as the simple calculator and all the simple editors, have no ambiguous effects at all. If no effect is ambiguous then things that look the same are the same. We call a PIE that has this property *monotone:*

monotone:

$$\forall\, p,q,r \in P \quad I(p) = I(q) \;\Rightarrow\; I(pr) = I(qr)$$

Or alternatively, in terms of I_p:

$$\forall p,q \in P \qquad I_p(\, null\,) = I_q(\, null\,) \;\Rightarrow\; I_p = I_q$$

Is this a useful property, and does it say what we wanted it to say? What it says is that the entire future effect can be predicted from the present one. In other words, the current effect could be regarded as a state, and in fact, any monotone PIE can be represented with a *doit* without a corresponding *proj* function.

If we think about an actual computer system (like my word processor) and take the effect to be the actual screen display, this monotone property is far too strong: we know that the screen could not possibly hold sufficient information for that. We don't even want it to. We want to view only relevant parts of what we are manipulating. Clearly, if this property is to be useful it should be applied to an effect which is not the underlying machine state (when it becomes tautologous) and not the display itself, but something "just underneath" the display, a widening of the display view. In the next section we will look at just such a widening.

We can give a very simple statement of monotonicity using the monotone equivalence $\equiv^\dagger$ defined by:

$$p \equiv^\dagger q \quad \hat{=} \quad \forall r \in P \;\; I(\,pr\,) = I(\,qr\,)$$

That is, p and q are monotone equivalent if they have the same interpretation and whatever commands are entered afterwards they both yield the same effects. Monotonicity then becomes:

monotone:

$$\forall p,q \in P \quad p \equiv q \;\Rightarrow\; p \equiv^\dagger q$$

We can use $\equiv^\dagger$ to define a PIE using the construction defined in §2.3. This PIE is clearly monotone and can therefore be described completely using state transitions. Further, we can define a projection function from $E_{\equiv^\dagger}$ to E by $[\,p\,]_{\equiv^\dagger} \to I(\,p\,)$. This is a function, as clearly $p \equiv^\dagger q \;\Rightarrow\; I(\,p\,) = I(\,q\,)$.

This interpretation and projection together give us the *monotone closure*, the minimal state representation of the PIE. We will call the resulting interpretation $I^\dagger$, the effect space $E^\dagger$ and the projection $proj^\dagger$.

If it is preferred, we can give an (up to isomorphism) equivalent definition without the use of relations:

$$E^\dagger \subset P \to E$$

$$I^\dagger(\,p\,) = I_p$$

The monotone closure is minimal in that given any monotone (state-based) PIE $< P , I', E' >$ and projection $proj'$ such that $I \;=\; proj^{\dagger} \circ I^{\dagger}$, we can define a projection $proj''$ from E' to $E^{\dagger}$ so that the following diagram commutes:

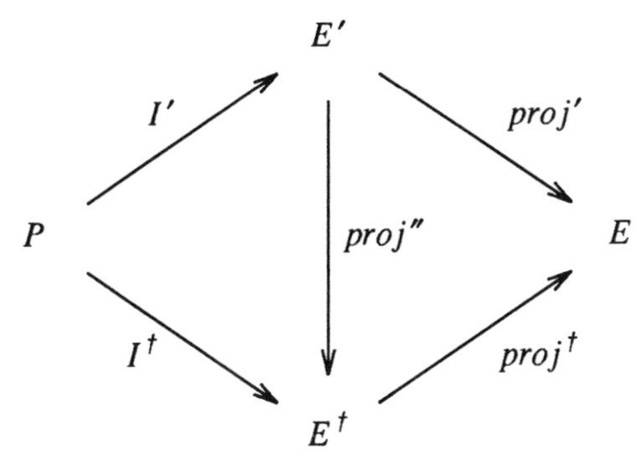

Another way we can view this minimality is that any distinctions that are made between states in $E^{\dagger}$ are implied directly by the possible observations. In that sense it has no implementation bias (Jones 1980). This is important because it means that we can freely assume the existence of a state-transition implementation of a PIE without sacrificing our surface philosophy, as the states within PIE† are directly observable from the interface.

It should be noted that this is intended as a theoretical construction, useful for the statement of principles and for defining terms, but not for the implementation of systems. The minimal state representation need not be realisable in practice, as even when I is computable and we can thus for any p represent the function $I^{\dagger}(p)$, we cannot in general decide whether two such functions are identical. Having said that, however, the construction can be useful for implementation.

If the PIE represents a finite automaton then $I^{\dagger}$ is computable and this gives us a construction for the finite state automaton, with minimal number of states satisfying the behavioural specification given by the PIE.

Even when the PIE is not a finite automaton the monotone equivalence is sometimes useful. In many cases it turns out to be computable and it is possible to label the equivalence classes giving a constructed minimal state representation. Alternatively, the attempt to compute the monotone equivalence may suggest a non-minimal, but still good, state representation.

As an example, if we look at the calculator with memory, the monotone closure of the interpretation function yields precisely the state representation that we gave. The monotone closure of my wordprocessor is much more complex, including at least all the text off-display, an invisible marked location, find/replace buffers, a copy of the last deleted item, and a ruler with tab positions.

2.6 Observability and user strategy

In the last section, we said that we will often want the effect itself to be something that is not predictable, yet there is another wider effect just beneath the surface of what is seen directly and which can be inferred from the immediate defect. For example, when using a display editor, there is (say) a 25-line window which is visible on the screen, yet one conceptually "sees" an arbitrarily large text just beneath this. Because the process of discovering this underlying text from the display is relatively easy, it will for most purposes be safe to ask for the PIE defining the underlying text to be predictable.

How do we define this wider view? One answer has already been suggested in the introductory chapter, namely as the information contained in all possible subviews. However, we also saw there that such a definition is probably too strong: aliasing prevents it from working. A different way to approach the problem is to examine what a typical user might do if he wanted to know the entire text of a document. The scenario might be a bit like this:

1. Note where I am.
2. Use the TOP button to get to the top of the document.
3. Use the SCROLL DOWN button to move down page by page.
4. When I get to the end, use the SCROLL UP button to get back to where I started.

Depending on the functionality of the editor there may be more complex "short cuts" to speed up this process, but the essential feature is that the user has a sort of program or algorithm which he follows which "covers" the entire text. Further, when bringing these snippets of information together, the relative positions are known because of the strategy followed. So in the example in the Introduction, one could tell "»1011«" from "»1101«" because in the former the "11" window would come after the "10" window and in the latter it would precede it. We will use this idea to extend the concept of observability. We will call such a user algorithm a *strategy*, and the resulting wider view the *observable effect*.

One possible formulation of the requirement for strategies is as follows. If E^+ is the space of non-empty sequences of elements of E, we define a strategy as a function $s : E^+ \rightarrow P$. To use this function starting at a given initial effect e after a sequence of commands p, we define $q_1 \ldots$, $p_0 \ldots$ and $e_0 \ldots$ as follows:

$$\begin{aligned} p_0 &= p \\ p_{n+1} &= p_n\, q_{n+1} \\ e_n &= I(p_n) \\ q_{n+1} &= s(e_0, e_1, e_2, \ldots, e_n) \end{aligned}$$

That is, from all the effects so far in our use of the strategy we work out a new command sequence q_n using s which is then issued resulting in a new set of commands to update p_n and a new effect e_n. We would demand that the strategy s actually comes to an end sometime and this is represented by $q_n = null$, the empty sequence.

We can distinguish between two states of the system generated by p and p' if for some i, $e_i \neq e'_i$ and the observable effect – "what you see" in the wide sense – is precisely the equivalence classes generated by this. That is, we can define an equivalence relation using the strategy $\equiv_s$ by:

$$p \equiv_s p' \quad \hat{=} \quad \forall i \quad e_i = e'_i$$

The observable effect O_s is then precisely $P/\equiv_s$ and we have the corresponding interpretation function I_s, derived in the standard manner. Note also that $e_0 = I(p_0) = I(p)$: thus $p \equiv_s q$ implies $p \equiv q$, and so $\equiv_s$ is stronger than $\equiv$. This means that there is a natural projection $proj_s$ from O_s to E factoring I, ($I = proj_s \circ I_s$):

$$P \xrightarrow{I_s} O_s \xrightarrow{proj_s} E$$

On the other hand:

$$\begin{aligned} p \equiv^{\dagger} p' &\Rightarrow \quad \forall q \;\; p \leq q \;\Rightarrow\; I(pq) = I(p'q) \\ &\Rightarrow \quad \forall i \;\; I(p_i) = I(p'_i) \\ &\Rightarrow \quad \forall i \;\; e_i = e'_i \\ &\Rightarrow \quad p \equiv_s p' \end{aligned}$$

That is, $\equiv^{\dagger}$ is stronger than $\equiv_s$, and thus there is a projection $projo^{\dagger}$ from the monotone closure to O_s factoring I_s, ($I_s = projo^{\dagger} \circ I^{\dagger}$):

$$P \xrightarrow{I^{\dagger}} E^{\dagger} \xrightarrow{projo^{\dagger}} O_s \xrightarrow{proj_s} E$$

It may be that there is a strategy such that the projection from the monotone closure is one-to-one, or in other words so that I_s is monotone. In this case the strategy has revealed sufficient information to predict all future effects. We will say that such a strategy *tames* I and if such a strategy exists, we will call the original PIE *tameable.*

Later we shall see a PIE which is not tameable: that is, how ever cleverly one devised a strategy, there will always be states that are indistinguishable using the strategy, but which eventually may differ given certain commands. We will prove this by using the following constructions and lemma.

For any p, p' and set $Q \subset P$ we say Q distinguishes p and p', if adjoining some sequence to p and p' yields a different effect:

$$Q \text{ distinguishes } p, p' \ \hat{=} \ \exists\, q \in Q \ \textbf{ st } \ I(pq) \neq I(p'q)$$

We will say that a PIE is *grotty* if there is some p such that given any finite Q there is some p' which cannot be distinguished from p and yet is not equivalent to it using the monotone equivalence That is:

grotty:
$$\exists\, p \in P \ \textbf{ st } \ \forall\, Q \subset P \ \ Q \textit{ finite } \Rightarrow$$
$$\exists\, p' \ \textbf{ st } \ \forall\, q \in Q \ \ I(pq) = I(p'q) \ \textbf{and not} \ p \equiv^{\dagger} p'$$

Note especially the order of the quantifiers: the indistinguishable element p' depends on the distinguishing set Q.

The lemma is that any grotty PIE cannot be tameable.

LEMMA: grotty $\Rightarrow$ **not** tameable.

We shall prove this by showing that any strategy we may choose does not tame the PIE.

PROOF:

Let p be the element which has the indistinguishability property, and assume the strategy is s. Define Q as follows, using the sequence q_i defined above.

$$Q \ \hat{=} \ \{ \textit{null}, q_1, q_1q_2, q_1q_2q_3, \ldots \}$$

This set is finite because the strategy terminates, and thus the q_is are eventually null. Therefore, because the PIE is grotty, there must be some p' not monotone equivalent to p which cannot be distinguished using Q, that is:

$$\forall\, i \in \{0, \ldots\} \ \ I(p\, q_1q_2 \ldots q_i) = I(p'\, q_1q_2 \ldots q_i)$$

But this means that $e'_0 = e_0$ and hence $q'_1 = q_1$, and then by induction $e'_n = e_n$ and $q'_n = q_n$ for all n. Thus:

$$\forall\, i \in \{0, \ldots\} \ \ I(p\, q_1q_2 \ldots q_i) = I(p'\, q'_1q'_2 \ldots q'_i)$$

That is (by definition) $p \equiv_s p'$.

Hence we have a pair p and p' which are not monotone equivalent and yet $p \equiv_s p'$, so that the projection $proj_O{}^{\dagger}$ is not one-to-one. However, this argument was for an arbitrary strategy and hence there is no strategy that tames I and it is not tameable.

All the arguments and definitions in this section have been for an arbitrary strategy s, but often the strategy of interest is clear by context, and in this case we will drop the suffix s, for instance using O instead of O_s.

2.7 Reachability

Reachability properties are about what can be done with a system, and whether there are states one can get to from which other states are inaccessible. The simplest reachability condition is to demand that I is surjective. If the set E describes the intended set of effects then it will often be a basic requirement of the system that all such effects can be obtained by some sequence of commands. The surjectivity of I ensures this. We will call this condition *simple reachability*:

simple reachability:
I is surjective

Of course this is not enough: although we may originally be able to reach any given effect, we may well be able to get ourselves up a "blind alley", forever after being unable to obtain a desired effect. To make a stronger statement we must look at the situation when we have entered some sequence of commands p; I_p (defined in §2.3) must still be surjective. We will call this stronger condition *strong reachability*:

strong reachability:
$\forall\ p \quad I_p$ is surjective

That is, "you can get anywhere from anywhere".

There are equivalent definitions using I or $\equiv$ alone and assuming simple reachability:

$$\forall\ p, q \in P \quad \exists\ r \in P \ \ \textbf{st}\ \ I(pr) = I(q)$$

$$\forall\ p, q \in P \quad \exists\ r \in P \ \ \textbf{st}\ \ pr \equiv q$$

There is a third, yet stronger, reachability condition we may want to impose. Strong reachability says that any desired effect can be obtained after any initial command history: however, this ignores the "hidden" state that may manifest itself latter. We may want to say that we can get to anywhere in the stronger sense that we cannot distinguish the way we got there by later observation. This obviously refers to the monotone closure and we can express it as:

megareachability:
$\forall\ p \quad I^{\dagger}_{p}$ is surjective

Again there are equivalent definitions using $I^{\dagger}$ or $\equiv^{\dagger}$ alone, assuming simple reachability:

$$\forall\ p, q \in P \quad \exists\ r \in P \ \text{ st } \ I^{\dagger}(pr) = I^{\dagger}(q)$$

$$\forall\ p, q \in P \quad \exists\ r \in P \ \text{ st } \ pr \equiv^{\dagger} q$$

2.8 Undoing errors

2.8.1 Importance and problems of undo mechanisms

The ability to spot errors quickly is one of the advantages of highly interactive computing. But this ability is only appealing if the correction is just as easy! Many systems supply some form of undo facility, either by direct command or as a side effect of the form of the system. Shneiderman (1982) suggests that the ability to easily undo one's actions incrementally is one of the hallmarks of a direct manipulation system.

Undo mechanisms can get quite sophisticated, ranging from the simple restoration of delete buffers to the keeping of entire session history trees (Vitter 1984). It is important to realise that there are fundamental incompatibilities between the various possible refinements. In practice, by ignoring these incompatibilities, designers may fall into several traps.

Any editor that retains deleted line buffers or complete copies of the last editor state is too inflexible to deal with the more general types of mistake, only being able to recover back one or at most a few steps. But any editor that retains a complete command and editor state history for the purposes of recovery has further problems:

- *Observability* – The behaviour of the editor is obviously determined by the state of the current history. So if we want to preserve monotonicity this history must be part of the observable effect: that is, there must be commands for perusing this structure (in fact, these commands are often included in the undo procedure itself).
- *Reachability* – Supposing the command and state history exists in some form and is observable, if the editor as a whole satisfies the reachability properties, there must be some way for the user to edit this structure since

it is part of the editor's state. This is usually the point at which an undo system would call a halt, only asserting reachability for the unadorned system. On the other hand, if this editing is allowed...

- *Undoability* – Are the relevant commands themselves undoable? If so we start quickly to chase our tail!

2.8.2 Examples

To illustrate some of these problems, we consider two simple undo editors. Both are based on the simple typewriter but could easily have been defined generically over any PIE. They both seem quite reasonable at first glance, but they are lacking in either functionality or in observability.

First, a basic one-step undo:

$$C_1 = \{ a, b, c, ..., z, \# \}$$
$$E_1 = \{ a, b, c, ..., z \}$$

$$I_1(null) = null$$
$$I_1(\#) = null$$
$$I_1(p::c) = I_1(p)::c \qquad c \neq \#$$
$$I_1(p::c::\#) = I_1(p) \qquad \forall\, c$$

This is the sort of undo mechanism that is used in editors like vi: it essentially retains two "states" and flips between them. It is not strong reachable, and requires an additional delete mechanism to be so. Neither is it predictable, as it is not possible to tell from the current display what a future undo (#) will result in:

$$I_1(null) = null = I_1(a\#)$$

but

$$I_1(\#) = null \neq a = I_1(a\#\#)$$

It is, however, tameable, using the strategy "type '#' twice". We can prove this by defining a state transition function *doit* with associated projection *proj* equivalent to I_1 and then showing that the strategy can observe this state:

$$S = E_1 \times E_1$$

$$s_0 = (null, null)$$

$$doit(c, (e_1, e_2)) = (e_2, e_2::c) \qquad c \neq \#$$
$$doit(\#, (e_1, e_2)) = (e_2, e_1)$$

$$proj((e_1, e_2)) = e_2$$

One can quickly verify that $proj \circ doit(\,.\,, s_0)$ satisfies the conditions for I_1.

Further, it is clear from it that the sequence "# then # again" followed no matter what the effect is, will give rise to the two effects e_1 and e_2.

One might be tempted to design a more elaborate and powerful undo system, where an indefinite number of commands could be undone:

$$C_2 = \{ a, b, c, ..., z, \#_1, \#_2, ... \}$$
$$E_2 = \{ a, b, c, ..., z \}$$

$$\begin{aligned} I_2(null) &= null \\ I_2(\#_n) &= null \\ I_2(p::c) &= I_2(p)::c \qquad c \neq \# \\ I_2(p::c::\#_n) &= I_2(p::\#_{n-1}) \qquad \forall\, c \end{aligned}$$

That is, $\#_n$ is the n-step undo which gets one back to the effect one had n commands ago; however, this is not exactly the same state as measured by the monotone closure. This new undo editor is in fact strong reachable, but not mega reachable. It is also not only, not predictable, but grotty!

The minimal state representation is in fact huge:

$$S \subset E_2^{\infty-}$$

where $E_2^{\infty-}$ is the set of sequences from E_2 semi-infinite to the left (in fact, the sequences generated will always have only a finite number of non-null entries) and:

$$s_0 = null^{\infty-} = (..., null, null, null, null, null, null)$$

$$\begin{aligned} doit(c, s::e) &= s :: e :: (e::c) \qquad c \neq \# \\ doit(\#_n, s) &= s : s[-n] \end{aligned}$$

$$proj(s::e) = e$$

where $s[-n]$ represents the n th element of s from its end.

This state representation is clearly minimal, since any two states s_1 and s_2 can be distinguished by repeating the command "$\#_{len}$", *len* times, where *len* is the number of elements from the ends of the two sequences we need to go before all preceding elements are null. (This must be finite by simple examination of *doit*.)

This makes the non-megareachability obvious. If we entered a then b, we could never get from the latter state to the former, as the *doit* function only adds to the end of the state sequence and thus the state would always have ab as its first two non-null elements.

We can also now prove that I_2 is grotty. Take the command sequence a and any finite set of command sequences Q. Each command in Q may have some ordinary commands and some commands of the form $\#_n$. Let m be the maximum of all these ns over all the elements of Q. Finally, take p' to be the command history "$a\#_1\#_2...\#_m a$" and p to be simply "a". p and p' will yield the states:

$$null^{\infty-}\, a \;=\; (\,..\, null\,, null\,, null\,, a\,)$$

and

$$null^{\infty-}\, a\; null^m\; a$$

The *doit* function always adds to the end of the state, and the commands within any element of Q can at most look m from the end of the state on which they act, and hence at most m from the original state: hence the two states generated by p and p' will yield exactly the same effects for each element of Q. Yet p and p' are not monotone equivalent by the minimality of S. Thus we have shown that for any finite Q we can construct a p' such that Q does not distinguish them, but which is not monotone equivalent to p ($\#_{m+2}$ distinguishes them). Hence I_2 is grotty.

2.8.3 A simple definition of undo

As quite reasonable and simple undo mechanisms seem to have quite complex semantic problems, can we progress further by framing some sort of formal definition of undo? A weak form of error correction has already been described in the reachability conditions. These tell us that we can get anywhere from anywhere, so in particular, if we do something wrong we can get back to the position before the mistake. However, the definitions of the reachability conditions involve the entire program. It is more attractive if error recovery is incremental, depending only on mistakenly entered commands, because typically these are most recent and are short. Imagine a user telephoning an advisor with the request "I've just done so-and-so and now I seem to be stuck; how can I get out of it?": a helpful reply should exist and not depend on any other information. Formally we could write this:

$$\begin{array}{l} \exists \;\; undo \in P \rightarrow P \\ \quad \textbf{st} \;\; \forall \; p, r, s \in P \quad I(\,r\,p\; undo(\,p\,)\,s\,) = I(\,r\,s\,) \end{array}$$

This appears to be not as strong as one might want: for instance, one might really want the undo to always be the same command. However, it turns out to be far more restrictive than is apparent. To see this we will define the *strong equivalence,* and restate the simple undo condition using it.

2.8.4 Strong equivalence

We have already used a fairly strong equivalence $\equiv^t$. Regarding P as a semigroup, one would call such an equivalence a *right congruence*, as equivalence is preserved by adding commands to the right. This instantly suggests investigating the full congruence, where equivalence is preserved in all contexts. We will call this strong equivalence ($\sim$):

$$p \sim q \quad \hat{=} \quad \forall\, r, s \in P \quad rps \equiv rqs$$

That is, two command sequences are strong equivalent if they have exactly the same effect, no matter the context in which they are used (ignoring the temporary effects while they are being invoked). If the PIE represents a finite-state automaton, then the equivalences $\equiv^t$ and $\sim$ are called Nerode and Myhill equivalence respectively (Arbib 1969).

We can construct the semigroup $P^{\sim}$ as the quotient of P by the congruence $\sim$, $P^{\sim} = P/\!\sim$. We can then define the map $I^{\sim} : P^{\sim} \rightarrow E$ by $I^{\sim}([p]) = I(p)$. This is a well-defined function because $p \sim q \Rightarrow I(p) = I(q)$. We can then factor I using the canonical semigroup homomorphism $parse^{\sim} : P \rightarrow P/\!\sim$:

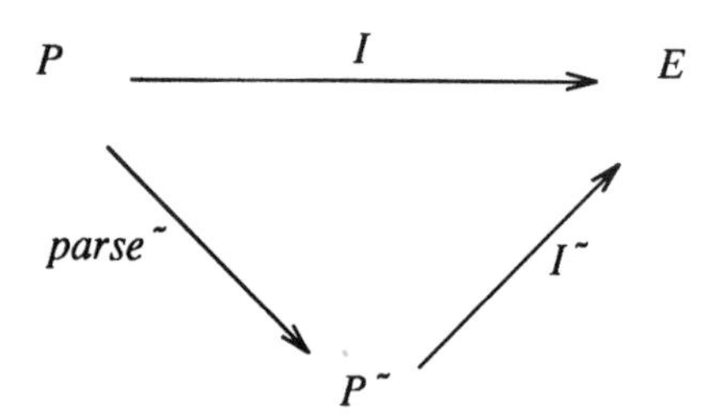

This semigroup is in fact minimal for expressing $<P, I, E>$ in the sense that for any $<P', I', E'>$ and semigroup homomorphism $parse' : P \rightarrow P'$ such that $I = I' \circ parse'$, we can construct $parse'' : P' \rightarrow P^{\sim}$ so that the following diagram commutes:

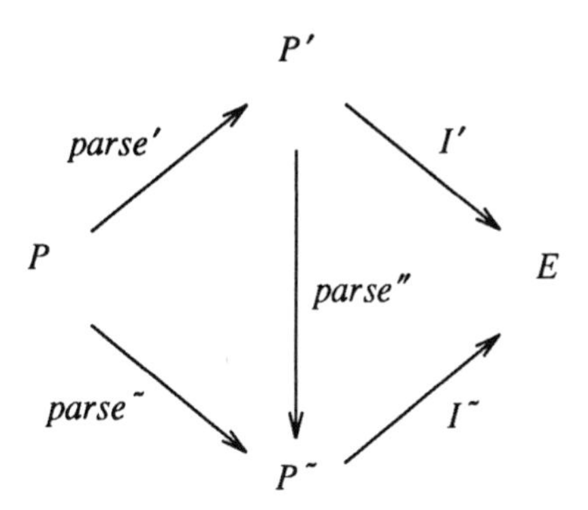

Combining this with the monotone closure, we have the following decomposition for any PIE:

$$P \xrightarrow{parse^{\sim}} P^{\sim} \xrightarrow{I^{\sim\dagger}} E^{\dagger} \xrightarrow{proj^{\dagger}} E$$

where $P^{\sim}$ is "as far to the right" as possible for a semigroup, and $E^{\dagger}$ is "as far to the left" as possible for a state.

Again, we have no guarantee that $P^{\sim}$ is computable, as $\sim$ is in general not decidable.

2.8.5 Undo and group properties

Using this strong equivalence, we can restate the undo condition as:

simple undo condition:
$\exists \; undo \in P \rightarrow P$
st $\forall \; p \in P \quad p \; undo(p) \sim null$

This is in fact a way of saying that $P/\sim$ is a group. The resultant class of objects has been studied in the theory of formal languages (Ansimov 1975). One consequence of this is that not only would there exist an undo for each command, but that they would all be different! Thus knowing only that there exists a way of undoing any action is of little comfort; we must also ask how easy it is to remember and to perform.

2.8.6 Undo for a stratified command set

A more realistic requirement is an editor with both ordinary commands, and a small distinguished set of undo commands. Ordinary commands obey much more stringent rules than undo commands, under the assumption that the latter are used more cogently. This distinction could be expanded to other classes of special commands.

Intuitively the class of ordinary commands includes some commands which have undos (e.g. letters being undone with delete), some commands that do behave like subgroups (e.g. cursor movement), and others like delete itself with no undo; the special undo commands would then be added. However, it is worth noting that in general even if some commands like cursor right/left have inverses relative to the ordinary commands, they do not do so once the new undo commands are added.

Formally, we have two alphabets C and U. We then have P, the sequences from C, and P_u, the set of sequences from $C+U$, and the following equivalent conditions:

$$\exists \; undo \in P \to P_u$$
$$\textbf{st} \; \forall \; p,q,r \in P \quad I(p\,q\,undo(q)\,r) = I(p\,r)$$

$$\exists \; undo \in C \to C+U$$
$$\textbf{st} \; \forall \; p,r \in P, c \in C \quad I(p\,c\,undo(c)\,r) = I(p\,r)$$

2.9 Exceptions and language

When designing a concrete system, we often begin by designing an idealised system, such as an editor with an unbounded text length or a graphics device with arbitrary resolution. However, when turning this ideal into a running system various boundary conditions arise. The user of a system may also have an idealised model of the system; for instance, "UP-LINE followed by DOWN-LINE gets me back where I started" (*cf.* undo above). This too may fail at boundaries. For a system designed with user engineering in the foreground the two ideal views should correspond. Given that some properties of the ideal system fail in the actual system (if nothing else, the commands that hit the boundary condition would behave inconsistently), we would expect that when an exception does arise there is some sort of consistent system action. That the system warn the user of such an exception by some means – bell, flashing light, etc. – goes without saying!

2.9.1 Modelling exceptions

Suppose then that we have two PIEs $<P, I, E>$, the ideal, and $<P, I_{ex}, E_{ex}>$, the exception PIE. They are related by a map *proj* from E_{ex} to E. There is also a boolean function *ex* on the elements of P, that satisfies $ex(p) => ex(pq)$ for all q. This means exceptions don't go away – not as bad as it seems! The following diagram commutes on the set $Ok = \{ p \in P \mid \textbf{not } ex(p) \}$:

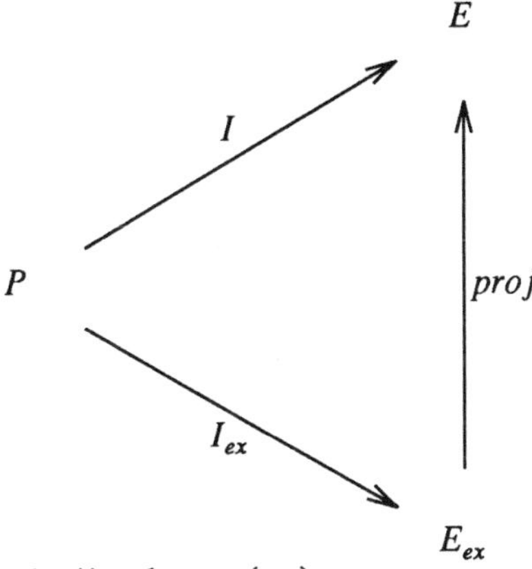

That is, $I(p) = proj(I_{ex}(p))$ unless $ex(p)$.

Note that in general the map *proj* need be neither surjective nor injective. The exception effects may have additions such as error status lines, which leads to *proj* taking several members of E_{ex} to one of E. Also there may be some effects in E that can never be reached by $proj \circ I_{ex}$ because exceptions block all paths. For instance, if the exception represents a bounded version of a text editor, then elements of the effect corresponding to large texts may never occur.

2.9.2 Detecting exceptions

Clearly we should be able both to detect when an exception has occurred, and ideally be able to predict whether an exception will occur before submitting the command that raises the exception. We can state this formally by requiring two user decision functions, *is_ex?* and *will_be_ex?*. The former can tell from an effect whether an exception has occurred, and the latter can tell from the current effect whether a particular command will cause an exception:

$$is\_ex? : E \rightarrow Bool$$
$$will\_be\_ex? : C \times E \rightarrow Bool$$

$$is\_ex? \circ I_{ex} = ex$$
$$\forall p \in P, c \in C \qquad ex(pc) = will\_be\_ex(c, I_{ex}(p))$$

We could equally well have been more positive and defined the "ok" operators *is_ok?* = **not** *is_ex?* and *will_be_ok?* = **not** *will_be_ex?*: the two formulations would be logically (but perhaps not psychologically) identical. Also one should note that although these are expressed as user decision functions their importance lies also in the constraints they put on the system.

The latter, prediction, information may be hard to achieve, and in case of doubt one should of course aim for safety, warning the user against exceptional conditions if one is unsure, especially if the consequences are major. This corresponds to a weakening of the condition for this to an implication:

$$\forall p \in P, c \in C \qquad ex(pc) \Rightarrow will\_be\_ex(c, I_{ex}(p))$$

In Chapter 5, we use similar decision functions for detecting steady state and when commands will be ignored. These can be thought of as specific cases for particular exception conditions.

2.9.3 Exception recovery principles

Although we've said that the exception effect may, and should, contain error indicators, we will assume that we are dealing with an abstraction of the full effect which does not include this extra information for the statement of the following principles. These are concerned with the possible error recovery rules after an exception has occurred.

One likely design principle for exceptions is "no guesses please". That is, when a command causes an exception to be raised, subsequent commands behave as if the command had never been issued. Formally:

$$\begin{array}{l} \forall\ p, q \in P, c \in C \\ \quad \textbf{not}\ ex(p)\ \textbf{and}\ ex(pc) \;\Rightarrow\; I_{ex}{}^{\dagger}(pc) = I_{ex}{}^{\dagger}(p) \end{array}$$

Or there is a weaker condition, "nothing silly please":

$$\begin{array}{l} \forall\ p, q \in P, c \in C \\ \quad \textbf{not}\ ex(p)\ \textbf{and}\ ex(pc) \;\Rightarrow \\ \qquad \exists\ p'\ \textbf{st}\ \ \textbf{not}\ ex(p')\ \textbf{and}\ I_{ex}{}^{\dagger}(pc) = I_{ex}{}^{\dagger}(p') \end{array}$$

This just says that when an exception is raised, at least the position the user is left in is one that could have been reached by legitimate means.

The reason for not including the error signalling in these principles is that clearly the effect after an exception would be different, as the error message would be there. One can clearly extend the rules to cover the more complex case: for instance, one can ask that all but the immediate effect is identical when considering the additional information. This would correspond to asking that all error messages be cleared after the next (correct) response, rather than leaving them there until the next error. There are many permutations of this ilk, and the above simplifications give the flavour of possible exception conditions.

A tentative basis on which to choose between the alternative requirements is that "nothing silly" is appropriate where the exception is associated with a limit of the system, but "no guesses" should apply where the user has made an error. For example, line break algorithms for most editors follow "nothing silly", but cuts without previous marks follow "no guesses" – not deleting the entire document up to the cut point, say! We can also reformulate our request for undos, only asking for an undo when an exception has not occurred. (This is safe if the "no guesses" condition is used.) For instance, up-line/down-line undo one another *except* at the top/bottom of the document.

2.9.4 Languages

An exception occurs when a user does something that we wish he hadn't. There is the stronger restriction on input arising from what is *possible*. That is, at the level we are designing the PIE it may be impossible for certain command sequences to be generated. This may occur in two ways:

- There may be some *physical* constraint. For instance, a cash dispenser may have a screen across it until the user puts the card in. The screen closes down when cash is removed. Similarly, the "cash removed" event cannot occur until the cash is presented to the user.
- There may be some *logical* constraint imposed by surrounding software. For instance, we may be considering an abstraction of the real system which is enclosed by a layer performing some syntactic checking.

In either case we may represent the set of possible actions by a subset L of P. However, this subset must satisfy a simple temporal condition. If some command sequence is possible, then all initial subsequences must also be possible, else it wouldn't have been possible to produce the longer sequence. That is:

$$\forall q \in P, p \in L \quad q \leq p \quad \Rightarrow \quad q \in L$$

If we have an exception PIE then the persistence condition on ex is exactly equivalent to this condition on the exception-free language:

$$L_{ex} = Ok = \{ p \mid \textbf{not } ex(p) \}$$

Note that this condition is as weak as possible, as the temporal condition is a minimal one to make the language sensible. Often there will be some more complex condition. For instance, the possible command histories may be initial subsequences of a context-free grammar, as is the case in Anderson's work (Anderson 1985).

We may require that it is possible to predict whether a command will be part of the language from the effect, as we have for exceptions, but usually the constraints will be evident if they are physical, or given by the enveloping system if they are logical. Thus in general such a requirement will not be necessary; if it is, then the exception conditions give a reasonable template.

Of greater importance is the way this affects the functionality of the system. For instance, if we design cash dispenser software that asks for the customer's PIN number, then reads in the cash-card, then gives the money, it will appear to have sufficient functionality when considered without the language constraints. However, when we take into account the language, we see that the customer cannot type in the PIN number until the card has been read and the cover raised. The machine turns out to be rather tight-fisted. On the other hand, it is rarely safe to assume when one is designing a system that the input will conform to what one expects to be inviolable constraints: the cash machine's window that covers the keyboard may be broken!

The latter problem means that one of the exception rules should be used for inputs that are not part of the language, and because this is a serious problem the "no guesses" rule would be appropriate. Thus we can assume that any command history a system has is effectively from the language, even if not actually so. The restriction in the input set must, however, be reflected in the formal statement of our principles. For instance, strong reachability for language PIEs becomes:

strong reachability with language:

$$\forall p, q \in L \quad \exists r \in P \quad \textbf{st} \quad pr \in L \ \textbf{and} \ I(pr) = I(q)$$

The monotone closure and hence megareachability must also be modified, as two states cannot be said to be equivalent based only on their effect: they must also have the same language extensions. Thus we redefine monotone equivalence for languages:

$$\forall p, q \in L \quad p \equiv^{t} q \quad \hat{=}$$
$$\forall r \in P \quad pr \in L \Leftrightarrow qr \in L \ \textbf{and} \ pr \in L \Rightarrow I(pr) = I(qr)$$

If the language does have warning information like *is_ex?* then this is no different from the standard definition, as the effects being the same ensures the same language extension.

2.10 Relations between PIEs

As a mathematician, one of the first things one does when faced with a new class of objects is to look at the relationships between those objects. This section will examine the relationships beween pairs and collections of PIEs. There are several ways one can relate PIEs at an abstract level, and it will be seen how these correspond to different ways of describing interactive systems. On the whole, the tenor of this section will be rather abstract, but it forms a sound basis for the more concrete discussion of layered systems in Chapter 7.

This examining of the relationship between things is the very stuff of mathematics, but is not just abstract speculation. It is fairly obvious that there is some insight to be gained by, for instance, looking at the way the cursor behaves in a word processor in isolation. For the purposes of thinking about the cursor, our keyboard could effectively have just a single "x" key for typing and the normal complement of cursor control keys. By thinking about the cursor on its own, we lose the distraction of other aspects of the system, and are able to better understand the specific problems related to cursor movement. We then happily relate this abstract view back to the full system, enabling us to extend our understanding of the part to the whole. This process of relating different models of a single system happens all the time in our informal analysis, often without our ever noticing. The different relationships in this chapter attempt to capture some of this informal reasoning in a formal framework.

2.10.1 Isomorphism

We saw in §2.3 that the $<P, I/\equiv_I, P/\equiv_I>$ was equivalent to $<P, I, E>$ in the sense that the following diagram commuted, where *proj* is a one-to-one function:

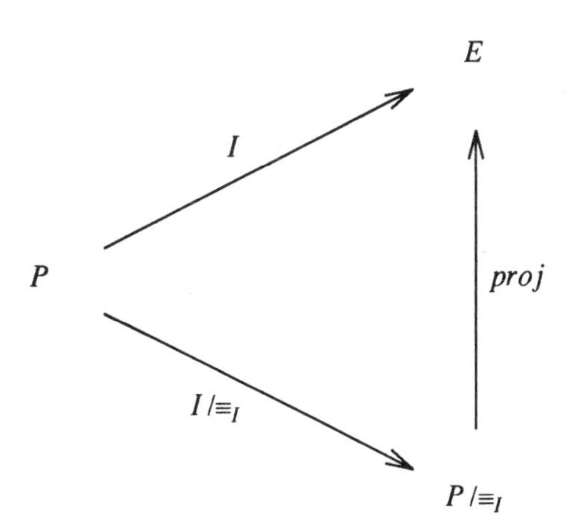

This is, of course, a special case of isomorphism of algebras. The general form of an isomorphism between $<P, I, E>$ and $<P', I', E'>$ requires two one-to-one relations, *parse* between P and P' and *proj* between E and E', such that the following diagram commutes:

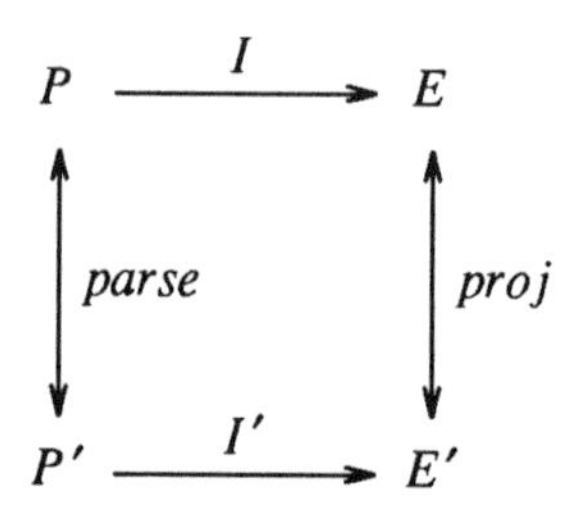

In fact, as we are interested in P as a semigroup, we also require that *parse* is a semigroup isomorphism. That is:

$$parse(p,p') \textbf{ and } parse(q,q') \quad \Rightarrow \quad parse(pq,p'q')$$

For the normal case, where P is freely generated over the command set C and likewise P' from C', this means that *parse* is generated from a one-to-one relation between C and C'.

Of course, although this seems like a sensible isomorphism when considering the PIEs as mathematical objects, it is not necessarily sensible for the user. For instance, a projection that involved reversing the letters in the alphabet ("abc" to "zyx", etc.) would be perfectly acceptable mathematically, and in regard to its information content, but no user would regard them as being the same. Similarly, when we say that E and $P/\equiv_I$ are equivalent up to isomorphism, that does not mean that a display editor and an equivalence class of command histories are equally useful for the user!

2.10.2 Other relations

Clearly, the most general relationship between PIEs would be obtained by letting *parse* and *proj* be general relations, perhaps with some restrictions on *parse*. However, we will only deal with a few special cases, in all of which we assume that the relations are in fact functions. There are clearly two major cases to consider:

- *1-morphisms* – both *parse* and *proj* go in the same direction. Without loss of generality, say $parse: P \to P'$ and $proj: E \to E'$.
- *2-morphisms* – *parse* and *proj* go in opposite directions. Say $parse: P \to P'$ and $proj: E' \to E$.

In addition, there is the case when one or other of *parse* or *proj* is one-to-one, and hence 1-morphisms and 2-morphisms coincide. We will call this special case:

- *0-morphism* – either *parse* or *proj* one-to-one.

Further, we will consider only two special cases of 1-morphisms: when both *parse* and *proj* are injective, which we shall call *extension,* and when they are both surjective, which we will call *abstraction.*

We will deal with each of these cases in the succeeding sections in the order 0-morphism, extension, abstraction and finally 2-morphism. But first we will consider the restrictions we may want to place on *parse*.

2.10.3 Restrictions on *parse*

If we are interested in P and P' as general subgroups then it is reasonable to consider the meaning of requiring *parse* to be a subgroup homomorphism. That is:

$$\forall p, q \in P \quad parse(p;q) = parse(p);parse(q)$$

In the normal case, when P is generated from C, this corresponds to simple macro expansion of commands from C into sequences of commands from P'.

Often this will prove a too stringent a condition, and in these cases we will often find that the temporal well-ordering condition introduced in §2.3 will be useful:

$$\forall p, q \in P \quad q \leq p \quad \Rightarrow \quad parse(q) \leq parse(p)$$

This, of course, only has meaning if the semigroup is in fact free (that is, P is the sequences from C): however, this is the most usual case, the only exception we have dealt with being the construction of $P/\sim$. Note also that it is strictly weaker than the semigroup homomorphism condition.

This condition does seem to be quite general; the only reason for breaking it is when we want to "backtrack" on some decision of *parse*. Not surprisingly, the relationships that do not obey this condition, are those generated by undo mechanisms. We will therefore assume that the temporal relation holds for all the morphisms we consider, but only ask for the semigroup condition when necessary.

Why didn't we ask for this rather than semigroup isomorphism when considering PIE isomorphism above? It would indeed have been possible, and indeed the sort of bisimulation condition needed between the handle space models of window managers would obey the weaker condition only. However, for simple PIEs, the informal idea of two systems being isomorphic must surely be that they differ in a one-to-one way between keypress and effects (perhaps also in any manifest effect structure). There is, of course, no right answer: one uses the definition that is most appropriate to the circumstances.

Before we move on, the temporal condition does suggest the possible generalisation of PIEs to where P is simply a well-ordered set. However, this has no obvious meaning at the user interface, and I have found no useful construction yielding such an object.

2.10.4 0-morphisms

We have already dealt with several 0-morphisms: in particular, the relation between monotone closure and its original PIE is a 0-morphism where the P spaces coincide:

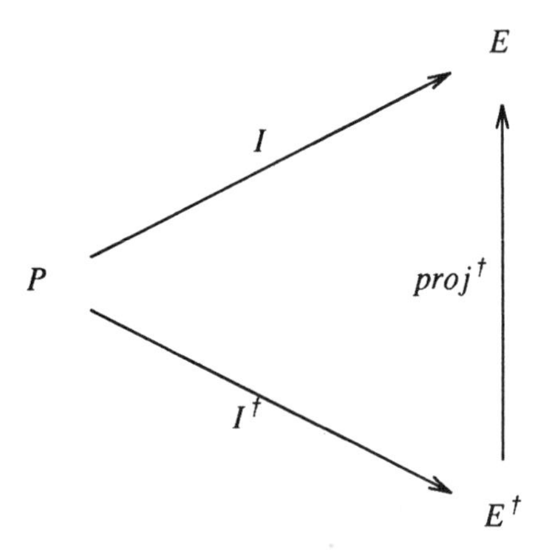

The relation between P and $P^{\sim}$ also forms a 0-morphism, except here it is the effect spaces which coincide:

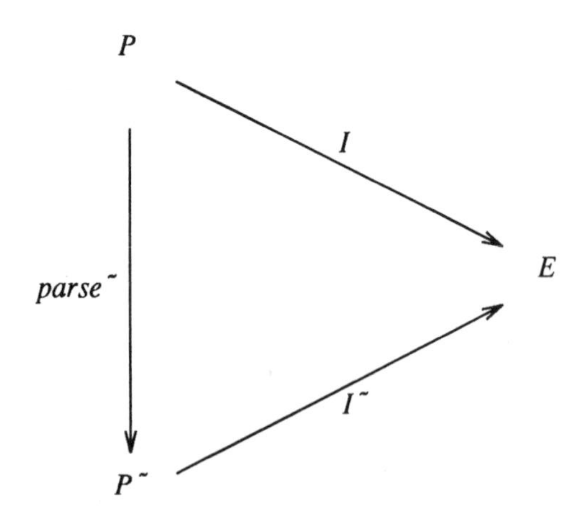

In this case we have a general semigroup, and the semigroup homomorphism condition does in fact hold. The temporal condition is of course inapplicable to general semigroups.

The exception relation (§2.9) is a generalisation of the 0-morphism. Two further examples will come in the next chapter when we consider red-PIEs: the relations between I and I_r and between I and I_d will both be 0-morphisms.

2.10.5 Extension and restriction

This is the special case of a 1-morphism where both *parse* and *proj* are injective. A 1-morphism, we recall, is when *parse* and *proj* both go in the same direction:

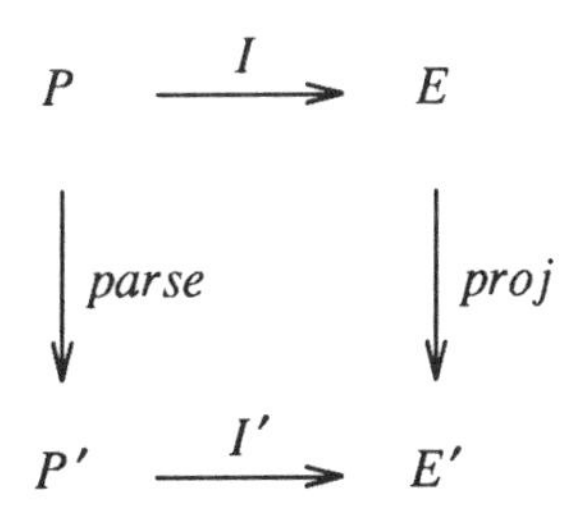

This would arise if, having an existing system $<P, I, E>$, we added extra commands to it to increase its functionality. Alternatively, we might start with $<P', I', E'>$ and *restrict* it by cutting down the commands available (perhaps to meet size or performance constraints).

In §2.5 when we considered simple predictability, we suggested that it would be a suitable condition to apply to certain parts of the system. These parts could be specified formally as a restriction of the system. For instance, we may restrict the command set of an editor to cursor movement and direct typing, and apply the condition to the restricted PIE so obtained.

Another example would be the editor with simple cursor movement of §2.4. We extended it by adding commands to set and jump to a mark. In these examples the command and effects of the latter were simple extensions of the former. So long as we never used the mark, the two interpretation functions matched exactly.

In both these examples the *parse* function of the extension is a semigroup homomorphism, and I cannot think of any reason to have it otherwise in the case of extensions in general. If we do assume this, then it is fairly simple to prove that monotonicity of PIE′ implies monotonicity of PIE. So, if we have a predictable system, restricting it will not spoil it. On the other hand, if we have an unpredictable system we cannot make it predictable by extending it. This is a case of "adding commands to a bad system doesn't make a good system".

It must be emphasised, however, that this only applies to the definition of predictability given by monotonicity. If we look at strategies and observable effects, and define predictability in terms of these, we see that restricting the command set may remove commands necessary for the strategy and hence destroy predictability; contrariwise, adding commands may make an effective strategy possible.

The notion of extension is not really that useful, as although the extended system behaves exactly like the original when *no* other commands are entered, it tells us nothing about whether there is any sort of behaviour consistent with the original when other commands have been used. We would normally want to say something a bit stronger. For instance, in an editor with cut/paste, we want to say that the effect of using just the cursor keys and normal typing is "the same" even when some other commands have been typed. The start state will of course be different. This can be partially captured by a "bundle" of extensions, one for each member of P'. Explicitly, for each p' from P' there exists p and $proj_{p'}$ as follows:

$$p \in P$$
$$proj_{p'} : E \rightarrow E'$$
$$proj_{p'} \text{ is injective}$$
$$I'_{p'} \circ parse = proj_{p'} \circ I_p$$

That is, the following is an extension:

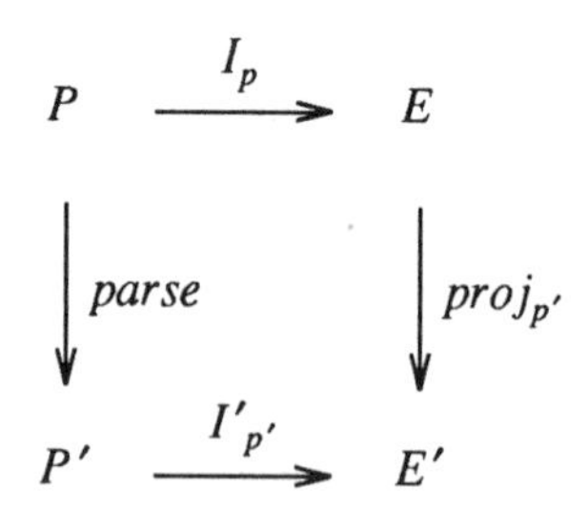

The *parse* functions are all the same, as we would expect the connection between commands to be the same. The projections need to be different, however, as the results of the "extra" commands in p' may change the connection between the effects.

As an example of such a bundle, let us look at the extension of the simple typewriter to the editor with cursor movement. The *parse* function is the identity, and for any sequence of commands p' from P', the characters with cursor movement, we simply take p to be *before* where:

$$\{ \ before, after \ \} \quad = \quad I'(p')$$

The projection then merely maps an effect in E into the first component of E':

$$proj_{p'}(e) \quad = \quad \{ \ e, after \ \}$$

This says that for any period when the cursor commands are not used, the text before the cursor acts as if generated by a simple typewriter.

Returning to the general case of a bundle, the fact that there are different projections does unfortunately allow anomolies. It is quite possible for one projection to take the text "abc" to "abc" as one would expect, but for another to take the same text to "def": we would obviously like them to agree on the "unextended" part of the effect. In the section on 2-morphisms we will introduce a further condition that will enforce such consistency.

The bundle also does not help with more complex extensions like the MARK/JUMP editor. This is because even if no more MARKs or JUMPs are entered, the projection function cannot keep track of the position of the MARK.

2.10.6 Abstraction

The case of the 1-morphism where *parse* and *proj* are both surjective, we call abstraction. It is probably the most important of the PIE relations. It is called abstraction because the two PIEs have identical behaviour except that PIE distinguishes more commands and effects than PIE′. Again we could look at it the other way round, and say that PIE refines PIE′.

Consider the typewriter with delete from §2.4.3, $<P_\nabla, I_\nabla, E_\nabla>$, and the simple counter $<P', I', E'>$ defined by:

$$\begin{aligned} C' &= \{ -1, +1 \} \\ E' &= \{ 0, 1, 2, 3, \dots \} \end{aligned}$$

$$\begin{aligned} I'(null) &= 0 & \\ I'(p::+1) &= I'(p) + 1 & \\ I'(p::-1) &= I'(p) - 1 & \quad I'(p) > 0 \\ &= 0 & \quad I'(p) = 0 \end{aligned}$$

The counter PIE is an abstraction of the typewriter when they are related by the maps:

$$\begin{aligned} parse(c) &= +1 & \quad c \neq \nabla \\ parse(\nabla) &= -1 & \end{aligned}$$

(the rest of *parse* is the semigroup homomorphism generated from these)

$$proj(e) = length(e)$$

This is the abstraction whereby the counter just keeps a tally of the number of characters typed.

In various places in the book we will mention the idea of something being an abstraction of another, referring to the general idea of having one or more abstraction functions. However, some instances deal with models other than the PIE model and thus cannot be handled explicitly by the above. For example, in Chapter 5, when we consider time-dependent systems, we find that systems may have some facet such as a clock which would never reach steady state, but we can look at some abstraction of the system which would stabilise. We could formalise this abstraction as desiring the existence of a *proj* map between the actual display (with clock) and the abstracted display (without). We can then apply all the analysis which we develop concerning steady-state behaviour to this abstracted temporal model.

We can investigate some of the properties of a PIE by looking at the simplified PIE′. One result that is useful here is that if *parse* is a semigroup homomorphism and PIE is strong (mega) reachable then so is PIE′. This would tend to be useful in the negative: if we were to find that an abstraction of our system is not reachable then we would know that some modification of the original system is necessary. The relevant modifications may become obvious when considering the abstraction. Abstraction is also useful when considering how one system is built in layers upon another, a point we will return to in Chapter 7.

If we allow abstractions to have exception conditions as in §2.9, we can have bundles of abstractions in a similar way to bundles of extensions. This would enable us to handle the relation between the MARK/JUMP and the plain cursor editor. The *parse* function simply throwing away *additional* MARK commands, with an exception being raised for JUMPs. The projection then just strips the MARKs from the effect. That is, for any p in P_{mk}, we choose p' in P_{cursor} to be *before* and define *parse* and $proj_p$ to be:

$$
\begin{aligned}
parse:\ & P_{mk} \rightarrow P_{cursor} \\
& parse(p) = strip(p) \\
proj_p:\ & E_{mk} \rightarrow E_{cursor} \\
& proj((b, a)) = (strip(b), strip(a))
\end{aligned}
$$

where *strip* removes any MARKs or JUMPs. These form an abstraction with exceptions for each p. We notice that the exception condition obeys the "nothing silly" condition, since every time we have an exception, the cursor editor can be seen as being in the same state as if the relevant p' had been entered.

2.10.7 2-morphisms, implementation

In a 2-morphism the two functions *parse* and *proj* go in opposite directions:

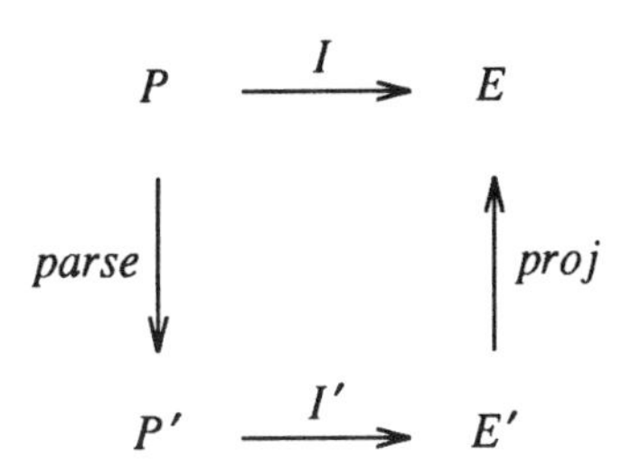

The commutativity condition is therefore:

$$I \;=\; proj \circ I' \circ parse$$

We have seen an example of this already, namely the relation between any PIE and its associated PIE generated from $E^{\dagger}$ and $P^{\sim}$:

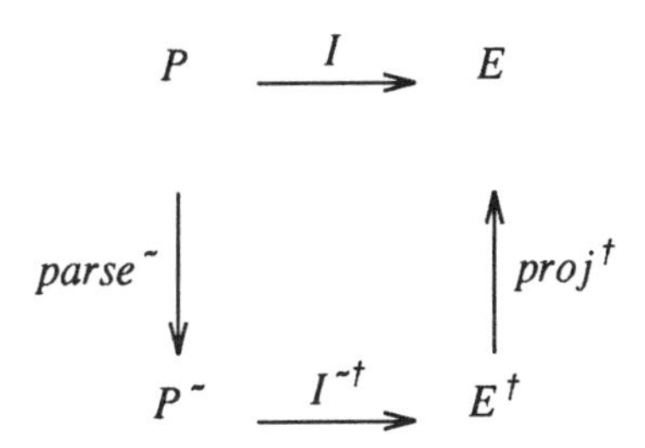

We will call the 2-morphism an *implementation* because if we had coded $< P', I', E' >$ already, one way to implement $< P, I, E >$ would be using the functions *parse* and *proj*. For example, if we had implemented the calculator with memory for PIE′, we could implement the simple calculator by making both *parse* and *proj* identity maps. Although this seems a very useful construction it does not live up to its promise. This will become apparent when we consider such constructions in detail in Chapter 7. But now we will show how a 2-morphism *can* be useful in combination with an extension.

We recall that in order to have a better concept of a restricted system, we introduced a bundle of extensions parametrised over the extended command sequences P'. Each extension had a separate projection function $proj_{p'}$. These needed to be different, as the "extended" part of the effects might differ, but we wanted a way of enforcing consistency on the "unextended" part. We can ensure this by having an additional function *proj* from E' to E, the same for all the PIEs in the bundle, which makes all the PIEs in the bundle into 2-morphisms when taken with the original (constant) *parse*:

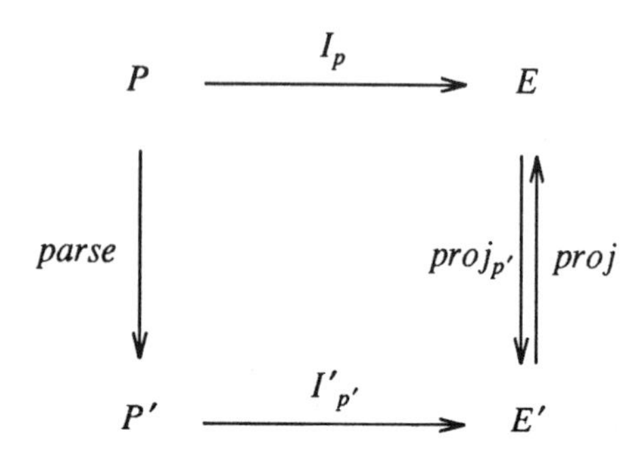

This extra implementation projection function ensures the consistency between the different extension projections. To be precise, the separate commutativity conditions imply that:

$$proj \circ proj_{p'} = identity$$

For example, consider extending the simple editor with cursor movement. Recall for any p' from the extended commands we chose p to be *before*, and defined $proj_{p'}$, where:

$$\textbf{let} \ \{ \ before, after \ \} = I'(p')$$
$$proj_{p'}(e) = \{ \ e, after \ \}$$

The projection *proj* is then simply the function that extracts the "before" half of the effect:

$$proj((b, a)) = b$$

2.10.8 What we can and cannot capture in PIE-morphisms

In this short analysis we can see some success, but also some problems in describing the relationship between systems. We will see in Chapter 7 that the *abstraction* relation is particularly important and has wide generality when decomposing systems. Difficulties in relating formally systems that we believe to be similar can arise for two reasons.

First, the problem may be a strictly technical. Our models miss out the significant features that would enable us to express the relationship. If this is the only barrier, a more extensive model will allow us to proceed. For instance, the *pointer space* model described in Chapter 8 gives us just such enhanced expressive power. The price we may pay for such power will typically be a more complex model, and therefore a model which is more difficult to analyse (although which is capable of more extensive analysis). However, if we are careful the model may not necessarily be more complex, just more appropriate for the relationship we wish to express.

The second, more fundamental problem is that the similarity we perceive informally may be in some way unformalisable. Its expressiveness may be related to a deep understanding of different situations that is lost if we try to tease out the similarity too precisely. This is the contrast between metaphor on the one hand, and allegory on the other. When we formalise a relationship, we effectively allegorise it; this feature in the one situation corresponds to that feature in the other, etc. If we are unaware of this problem attempts to formalise a metaphor may be counterproductive, capturing the peripheral parts whilst missing the central issues entirely. However, if we keep these limitations in mind it can be useful to find just how far a similarity can be captured formally, and thus highlight where we do rely on intuition.

2.11 PIEs – discussion

Even using the very simple PIE model, we have been able to investigate some non-trivial properties of interactive systems. We have introduced the principles of predictability, observability and reachability, and the important notion of monotone closure. We have also examined problems of undo, and dealing with exceptions in interactive systems. Finally, we have seen how we can relate different systems to one another. This was done at a largely theoretical level, but gives us the machinery with which to discuss layered design and implementation in Chapter 7.

CHAPTER 3

Red-PIEs – result and display

3.1 Introduction

In the previous chapter, we considered a very simple input–output model of an interactive system. We noted that the *effect* part of the model could correspond to various levels of system responses. This rather extreme level of abstraction allowed us to discuss properties which we can then apply to various facets of an actual system.

Perhaps the most well-quoted interface property is *what you see is what you get* (WYSIWYG). In the context of a word processor this refers to the close relationship between the displayed and printed version of the document. In order to discuss properties of this nature we will introduce just such a distinction. We will distinguish those parts of the effect which are to do with the final end-product of the system, the *result*, and those that are more ephemeral, the *display*. So in the case of the word processor, the final result is the printed, formatted output, and the display is the moment-by-moment view the user has on the screen:

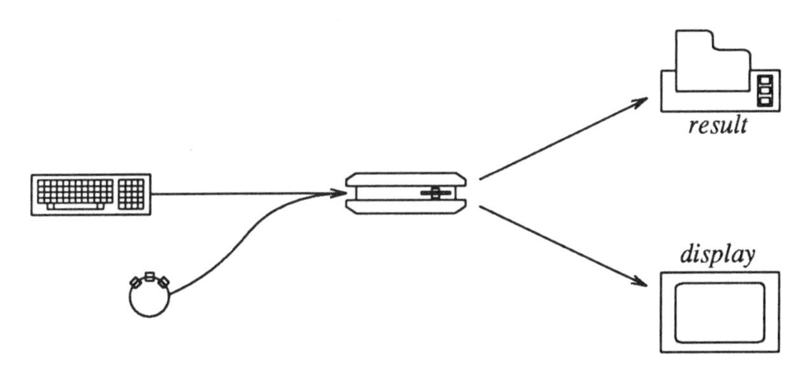

At a physical level, this distinction can be associated with the actual screen and printed output. However, similar distinctions appear at a more abstract level. For instance, when we think about the word processor we may be interested in the *document* as an entity in its own right, which may be formatted in different ways. It may thus be the document itself which is regarded as the result rather than a particular printing of it. Similarly, on the display side, we may want to ignore the limited size of the screen and think of the display as the entire view the user can get by scrolling up and down. Regarding the result and display in this light makes some properties (such as the exact visual appearance) less important, but may highlight other issues. For instance, spaces at the end of lines, which may be invisible on screen and in a particular printout, may cause the document to be badly formatted at some future stage. A similar problem, and now a familiar sight, is where hyphen- ation intro- duced by a word processor gets "left behind" in the docu- ment when it is re-formatted to a diff- erent line width.

In more complex systems such as databases or integrated CAD–CAM systems it is more clear that the significant result of an interaction is not a single hard-copy. On the other hand, we do not wish simply to identify parts of the system's representation and call these the result. Ideally, the result should be identified with a user-oriented model of the end-product of the system, and identifying this is a major part of the task analysis relating the user to the formal model.

To a large degree, we can regard the result as being the element of the system which is there because of the task or goal, and the display as the elements required for interaction. This can be a useful parallel to help us appropriately define the result of a system: for instance, we would not expect the cursor position in the result. Sometimes, especially when considering a system at several levels of abstraction, we may want to include interaction elements in the result. For instance, when looking at a word processor as a whole, we may only want to include the document itself in the result. However, when we look at a subdialogue, it may be that the result of that dialogue is a style sheet. The style sheet will affect the formatting of the document, but is of itself an artifact of the system, not the task.

Our main interest in this chapter will be on these *observability* properties, asking what the user can infer about the final result by examining the display. The properties will be information rather than presentation oriented. So, for example, we will not consider issues such as whether the visual appearance of text on screen and paper is identical. Instead, we would ask merely whether we can *work out* what the printed form will look like from the displayed form.

In addition, the introduction of the result and display will enable us to take a first look at what it means for a command to be *global,* that is, affecting the whole result without regard to the current display, or *local,* that is, all the changes are visible from the display. An example of the former would be a global search/replace command and of the latter, typical character-by-character typing or direct manipulation.

3.2 The red-PIE model

3.2.1 Definition

We will model the distinction between result and display as a refinement of the PIE model. To do this we assume that from our effect space E there are two maps, *display* and *result*, to sets D and R respectively. These correspond in the obvious manner to the immediately visible display and the product that would be obtained if the interactive process were finished immediately. Diagrammatically we have the following:

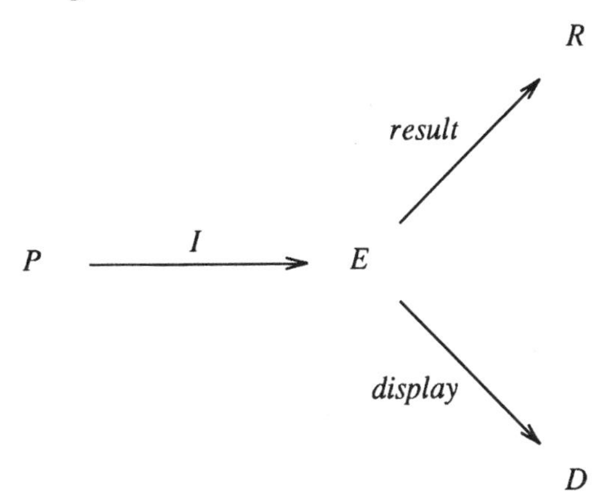

For obvious reasons we refer to this structure as a red-PIE. We assume that the effect space E can be regarded as a system state and also that it is minimal in order to give the display and result the desired behaviour.

In the last chapter we expressed concern about using state representations, as these may introduce additional complexity which has no effects at the user level. However, we also saw that we can derive minimal states, using the monotone closure operator. This means that we will be able to assume that E has just, but only just, enough in it. We will make this more precise in the following sections.

3.2.2 Resolution (product) of result and display

The definition of a red-PIE given implies the existence of two additional interpretation functions:

$$I_r = result \circ I$$
$$I_d = display \circ I$$

giving rise to two PIEs sharing a common command set, $< P , I_r , R >$ and $< P , I_d , D >$. One could argue that these two PIEs are a more basic configuration than the red-PIE above, and that we should instead have ignored E and just asked for two such PIEs:

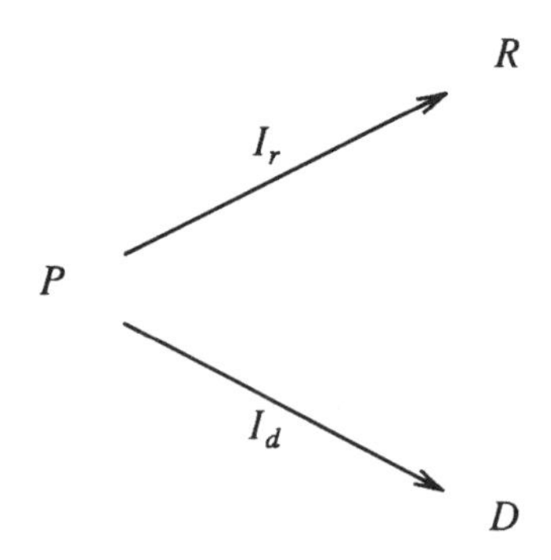

However, given any pair of PIEs with a common command set we can define a new PIE which factors both. We will call this PIE the *resolution* of the two PIEs. We define the resolution using an equivalence on P. Define the resolution equivalence $\equiv_{res}$ by:

$$p \equiv_{res} q \;\hat{=}\; I_r(p) = I_r(q) \;\; \textbf{and} \;\; I_d(p) = I_d(q)$$

That is, two command histories are resolution equivalent if they yield both the same result and display. Equivalently, we could use the equivalences defined by I_r and I_d and define $\equiv_{res}$ as:

$$p \equiv_{res} q \;\hat{=}\; p \equiv_r q \;\; \textbf{and} \;\; p \equiv_d q$$

That is, $\equiv_{res}$ is the weakest equivalence stronger than both $\equiv_r$ and $\equiv_d$. We can define E_{res} as $P/\equiv_{res}$ and I_{res} as the canonical map in the standard way. Because $\equiv_{res}$ is stronger than $\equiv_r$ and $\equiv_d$, we can factor I_r and I_d with I_{res} and projection maps $proj_r$ and $proj_d$, so that the following commutes:

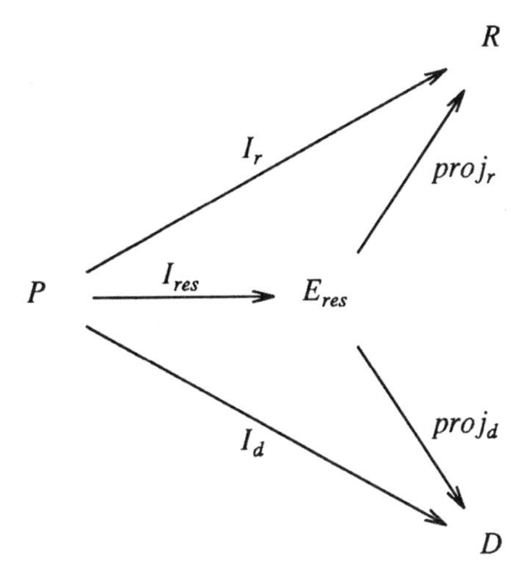

The resolution is in fact minimal, in the sense that given any PIE $< P, I', E' >$, with projections $proj_r'$ and $proj_d'$ which factor I_r and I_d respectively, e.g. $I_d = proj_d' \circ I'$, then there exists another projection $proj_{res}'$ from E' to E_{res} such that:

$$\begin{aligned} I_{res} &= proj_{res}' \circ I' \\ proj_r' &= proj_r \circ proj_{res}' \\ proj_d' &= proj_d \circ proj_{res}' \end{aligned}$$

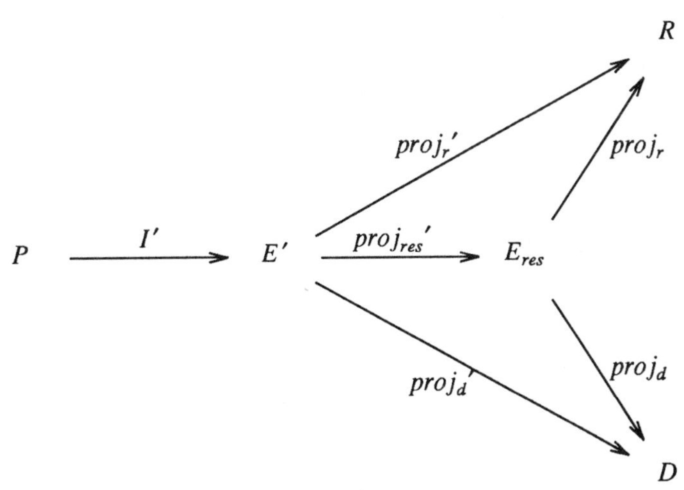

To prove this, we simply construct $proj_{res}'$ such that:

$$\forall\, e' \in E' \quad proj_{res}'(e') = I_{res}(p)$$

where

$$I'(p) = e'$$

This is a well-defined function since:

$$I'(p) = I'(q) \quad \Rightarrow \quad I_r(p) = I_r(q) \ \textbf{and} \ I_d(p) = I_d(q)$$

(because I' factors I_r and I_d)

$$\Rightarrow \quad I_{res}(p) = I_{res}(q)$$

and it clearly satisfies the required conditions.

We are almost at the original definition, only we also asked that the effect be monotone. We can easily achieve this by factoring I_{res} by its monotone closure, $(I_{res})^\dagger$:

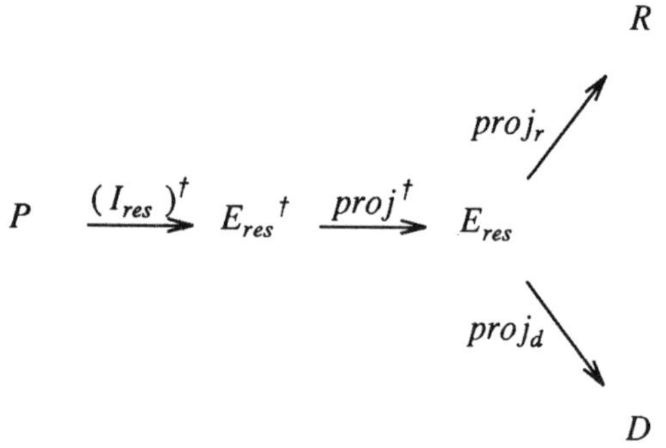

$$P \xrightarrow{(I_{res})^\dagger} E_{res}{}^\dagger \xrightarrow{proj^\dagger} E_{res} \qquad E_{res} \xrightarrow{proj_r} R \qquad E_{res} \xrightarrow{proj_d} D$$

We can then define I, *result* and *display* by:

$$\begin{aligned} I &= (I_{res})^\dagger \\ result &= proj_r \circ proj^\dagger \\ display &= proj_d \circ proj^\dagger \end{aligned}$$

Which is minimal by the minimality of the constructions for resolution and monotone closure. To be precise, any other monotone PIE with relevant projections will sit to the left of E_{res} by the minimality of resolution, and thence must also sit to the left of $(E_{res})^\dagger$ by the minimality of monotone closure.

Thus we have obtained the original definition. That is, although the definition using two PIEs appears more basic, we can, without loss of generality, assume the existence of a single monotone effect as in the first definition. This makes the statement of principles of observability and predictability much simpler.

3.2.3 Commutativity of resolution and monotone closure

Before moving on it is worthwhile noting the equivalence of an alternative derivation of this state E. Given the desire for a monotone effect factoring both the result and display, we might have chosen to move towards monotonicity first, and then later using the resolution. That is, we might have constructed $I_r{}^\dagger$ and $I_d{}^\dagger$ and then looked at their resolution, $(I^\dagger)_{res}$ say. It turns out that the resolution of two monotone PIEs is itself monotone:

LEMMA: $(I^\dagger)_{res}$ is monotone.

$$\begin{aligned} p\ (\equiv^\dagger)_{res}\ q \ &\hat{=}\ p \equiv_r{}^\dagger q \ \textbf{ and } \ p \equiv_d{}^\dagger q \\ &\Rightarrow\ (\ \forall s \in P \ \ ps \equiv_r{}^\dagger qs\ ) \\ &\qquad \textbf{and} \ \ (\ \forall s \in P \ \ ps \equiv_d{}^\dagger qs\ ) \end{aligned}$$

(by monotonicity of $I_r{}^\dagger$ and $I_d{}^\dagger$),

$$\begin{aligned} &\Rightarrow\ \forall s \in P \ \ (\ ps \equiv_r{}^\dagger qs \ \ \textbf{and} \ \ ps \equiv_d{}^\dagger qs\ ) \\ &\Rightarrow\ \forall s \in P \ \ (ps\ (\equiv^\dagger)_{res}\ qs\ ) \end{aligned}$$

Thus $(E^\dagger)_{res}$ would have been a good candidate for the state effect E. Happily these two constructions yield the same effect. That is:

THEOREM: commutativity of resolution and monotone closure.

$$(I_{res})^\dagger \ = \ (I^\dagger)_{res}$$

We can prove this by showing the equality of the two equivalences $(\equiv_{res})^\dagger$ and $(\equiv^\dagger)_{res}$. We have already shown that $(I_{res})^\dagger$ is minimal: hence we already know that:

$$p\ (\equiv_{res})^\dagger\ q \ \ \Rightarrow \ \ p\ (\equiv^\dagger)_{res}\ q$$

Thus we only need to prove the reverse implication.

PROOF:

$$\begin{aligned} p\ (\equiv_{res})^\dagger\ q \ &\Rightarrow\ \forall s \in P \ \ (ps \equiv_{res} qs\ ) \\ &\Rightarrow\ \forall s \in P \ \ (ps \equiv_r qs \ \ \textbf{and} \ \ ps \equiv_d qs\ ) \\ &\Rightarrow\ (\ \forall s \in P \ \ ps \equiv_r qs\ ) \\ &\qquad \textbf{and} \ \ (\ \forall s \in P \ \ ps \equiv_d qs\ ) \\ &\Rightarrow\ p \equiv_r{}^\dagger q \ \ \textbf{and} \ \ p \equiv_d{}^\dagger q \\ &\Rightarrow\ p\ (\equiv^\dagger)_{res}\ q \end{aligned}$$

(The astute reader might have noticed that all the implications were in fact equalities, and we needn't therefore have used the minimality of $(I_{res})^\dagger$; however, it seems nice to do so!)

Having proved the equivalence of these different definitions we can drop the distinction and refer to $E_{res}{}^{\dagger}$ as E, etc. We have, of course, from the second definition projections from E to $R^{\dagger}$ and $D^{\dagger}$ factoring $I_r{}^{\dagger}$ and $I_d{}^{\dagger}$; thus we have the complete picture:

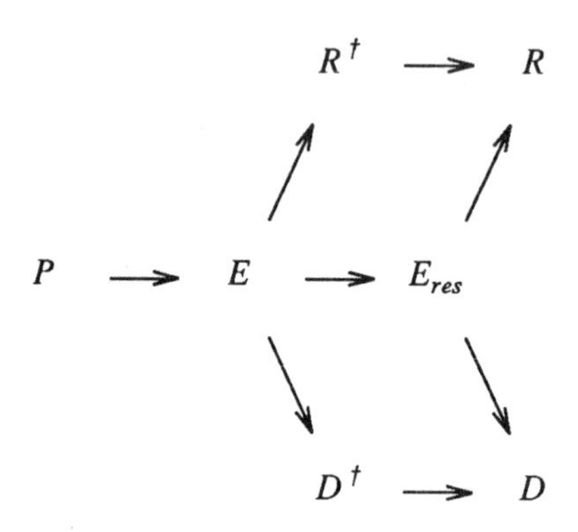

Note that often one will start by specifying $P \rightarrow R^{\dagger} \rightarrow R$ first. That is, the task or application-level objects and operations are defined giving a functional core. Later the display component may be added. The diagram above shows that it is an *abstraction* from the state of the system as whole. In Chapter 7 we will see how important it is *not* to try to build a system from this as a *component*.

3.3 Observability and predictability

When dealing with the simple PIE model, the notions of predictability and observability were slightly unclear as we were uncommitted as to what exactly the effect represented. We now have a richer effect space and we can give better definitions of these concepts. We will phrase each predictability principle as a requirement for the existence of some function between sets in the diagram, such that the resulting diagram with the function arrow added would still commute. In each case this will imply that certain arrows in the diagram must represent one-to-one mappings and thus the diagram will collapse a little.

One of the reasons for introducing strategies in §2.6 was for use at this stage; however, before wheeling in this extra complexity, we will look to see if any of the existing interpretation functions we have deserves to be monotone.

3.3.1 Simple predictability

In (almost) all editors we would expect some sort of additional editing "context" in addition to the resulting object (e.g. a cursor). Thus it will not be useful for the result interpretation to be monotone. Similarly, the display (as we've said before) contains only a part of the required information, and this is its role. So again we don't want I_d monotone. E is the result of a monotone interpretation by definition, as of course are $R^\dagger$ and $D^\dagger$, so this leaves E_{res}. What does it mean for this to be monotone?

Essentially E_{res} contains the sum of the information from R and D. (In fact, we could have defined it as the range of $I_r \times I_d$ in $R \times D$.) If I_{res} were monotone, then this would mean there was sufficient information in R and D together to predict the future behaviour of the system. For instance, if we were editing a program then it would mean that an up-to-date listing of the program together with the current display would be sufficient for all prediction. We can state this formally then as:

simple predictability:
I_{res} is monotone

In terms of the diagram, this means that E and E_{res} are no longer distinct (a monotone effect is equal to its own monotone closure).

As we promised, we can also define this in terms of a prediction function $predict_{simple}$:

$$\exists\ predict_{simple} : R \times D \rightarrow E$$
$$\text{st}\ \ \forall p \in P\ \ predict_{simple}(result(I(p)), display(I(p))) = I(p)$$

Is this a reasonable demand on an interactive system? Some very simple editors would satisfy it, as would some simple graphical systems. However, it forbids any sort of off-screen memory for objects to do with the editing task that are not actually part of the finished product. So for instance find/replace strings would have to permanently visible and one would not allow any buffers for temporary pasting to and from unless these could also fit on screen. Thus although useful for simple systems it appears too restrictive for systems with greater functionality. It would, however, be sensible to require it of certain parts of the total system. For example, it would be true of a cut/paste editor when one ignored the cut/paste actions.

3.3.2 Observability from D and strategies

Simple predictability, as defined above, depends on knowing the current result. The result is of course potential, and unless one has just collected an up-to-date listing of the document (or produced an example of the product with a CAD–CAM system), this information is not available. What is available is the display, and a primary observability aim for a system would be to view the result using the display. Clearly the current display alone will not be sufficient in general, and we will require a widening of it. That is, we want a strategy s on the display PIE $<P, I_d, D>$ which will give rise to an observable effect O which will fit into the diagram thus:

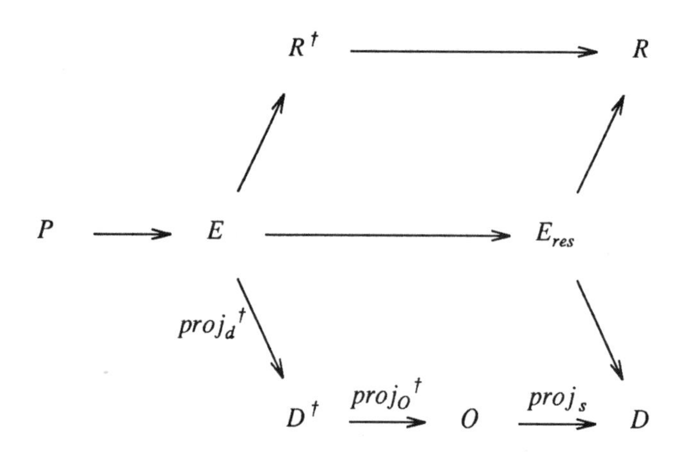

We will call the composite map from E to O *observe*, that is, $observe = proj_O^\dagger \circ proj_d^\dagger$. The rest of our observability principles will be based on maps between O and other sets on the diagram.

The first requirement, which we will term *result observability*, is that we should be able to observe the result via this observable effect. That is, we want a prediction function $predict_R$:

result observability:

$$\exists \; predict_R : O \rightarrow R$$
$$\textbf{st} \quad predict_R \circ observe = result$$

As O already factors the map to D and it also factors the map to R, it must be equal to E_{res} by the minimality of the resolution. Thus the diagram is simplified by identifying E_{res} and O. Further, since I_r is factored through O it is also factored through $D^\dagger$. Thus $D^\dagger$ is monotone, and factors both I_d (by definition) and I_r. It must therefore be equal to E, by the minimality of E. That is, our diagram is simplified again, by the unification of $D^\dagger$ and E:

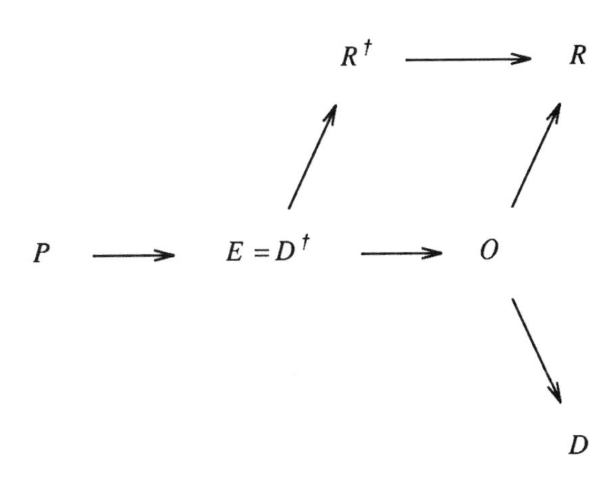

Observing the result is, of course, useful especially if we also have simple predictability. In this case the result and display together tell us everything about the system; thus, if the observable effect can tell us everything about both the result and the display, it is sufficient for all predictions about the system. However, as we have said this applies only to simple systems. If the system is not simple predictable, we will want conditions to hold on O which are stronger than simply result observability. In particular, as well as wanting to know the current value of the result, we may well want to predict what will happen to the result in future. $R^\dagger$ contains sufficient information for this by its definition, containing all information about the system pertaining to its finite state, but ignoring purely ephemeral details. So for example, it would contain the current cursor position. If there were commands based on the screen coordinates, such as a "move to the top of the screen" command, then it would also have to include the screen framing information. It would *not* have to contain information such as what error messages were currently displayed, or, if there were no explicit use of screen coordinates, the screen framing. The ability to view this information underlies a lot of what might be termed informally the predictability of the system, and thus we define *result predictability* as the ability to observe this information from the observable effect:

result predictability:
$\exists \; predict_{R^\dagger} : \; O \to R^\dagger$
st $\; predict_{R^\dagger} \circ observe \; = \; result^\dagger$

That is, $I_r{}^\dagger$ is now factored through O:

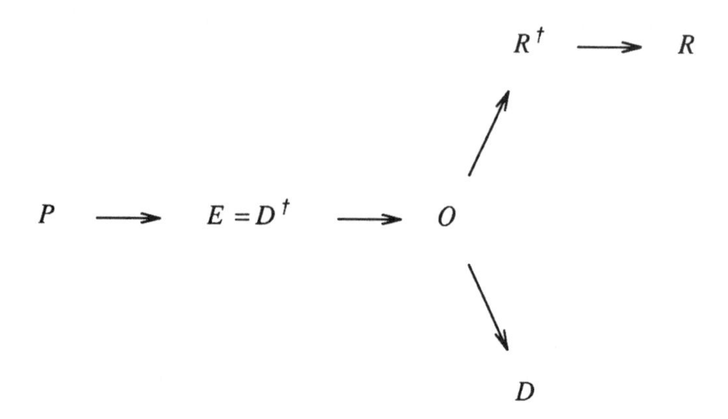

We can, of course, also define *display predictability* as the ability to observe $D^{\dagger}$ through O, and this is equivalent to the PIE $< P, I_d, D >$ being tameable. This together with result predictability would mean that one could predict not only the persistent effects of one's action, but also the ephemeral ones. On its own it seems less useful, and we won't dwell further on its implications.

Result and display predictability together we will call *full predictability*, and this can be given a definition in terms of a single prediction function which we will call simply *predict*:

full predictability:
$\exists$ *predict* : $O \rightarrow E^{\dagger}$
st *predict* $\circ$ *observe* = *identity*

In terms of the sets of the diagram, this means that O and E (and thus $D^{\dagger}$) will coincide, giving:

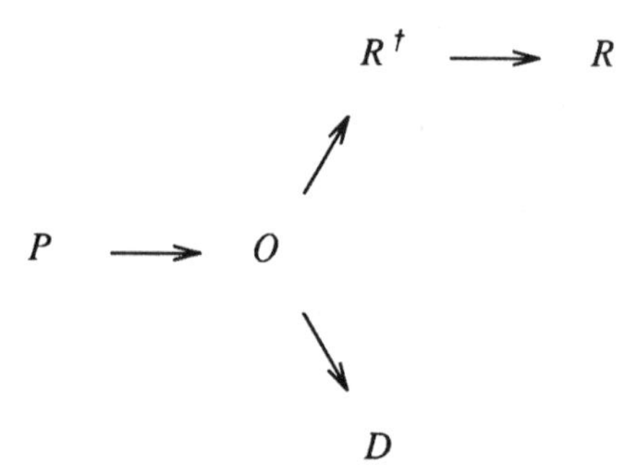

This is nearly as simple as the diagram we started with!

3.3.3 Different strategies and the limitations of observability principles

For simplicity, we have considered a single strategy. In practice, of course, the strategy used would depend on the purpose we intend. For instance, there would be a simpler strategy providing result observability than for result predictability (the latter would have to examine all buffers, search/replace strings, perhaps macro definitions, etc.)

It would be unacceptable to ask the user to use the whole strategy all of the time, and a real system would additionally need partial strategies for obtaining partial information. Further, the various observability principles state merely that the information is available, not how easy it is to use. For instance, a screen that displayed all its characters upside down would be as acceptable as a standard screen. Equally the strategy may be unusably complex.

We could address these problems of ease of use in various ways:

- We could accept their limitations and use informal reasoning to extend the formal concepts.
- We could produce richer models with principles that capture more of the notion of ease of use.
- We could produce formal measures of complexity for functions and algorithms the user is expected to use.

The last course could be followed by using programmable user models following Young *et al.* (1989) and Runciman and Hammond (1986) or using the sort of complexity analysis of Kiss and Pinder (1986). Alternatively, one could follow a naive approach such as demanding that the strategy can be described using a finite-state machine with a small (7?) number of states and simple recognition functions (like "am I at the top of the document?"). All these approaches have to be applied as they now stand during the production of a particular system and do not lend themselves well to reasoning at the abstract level of the interaction model. It is to be hoped that this will change as these methods mature.

The second option, that is, richer models and principles, will be taken up later. Chapter 5 gives specific predictability principles for temporal systems. Chapter 8 uses pointer spaces to address the issue of fidelity between display and result.

It is my belief, however, that these formal techniques will (in the foreseeable future) always require some sort of informal psychological analysis in addition to the formal statement of principles. Thus even when more specific temporal prediction principles are defined in Chapter 5, there will be an informal discussion as to the application of these.

There is one condition on the strategies that we use for observation of the general red-PIE, that can be given some formal statement, and this we examine next.

3.3.4 Passivity

The definition of a strategy that we have given so far allows some very silly ones. For instance, it would include the strategy for a simple word processor: "use the DELETE UP key until you get to the top of the document, then use DELETE DOWN till there's no document left". It is clearly sensible that any strategy designed to observe the result should not affect that result in so doing. There are several formulations of this informal principle, depending on how strict we want to be, but we shall call the general concept *passivity*.

First we will define what it means for a command or sequence of commands to be passive, then apply this to strategies. We will say a command sequence is passive, if the result before and after is the same, that is:

$$p \text{ is } passive \;\hat{=}\; \forall q \in P \quad I_r(qp) = I_r(q)$$

This still leaves open the possibility that intermediate results are different (e.g. delete everything then type it in again!). If we don't want this we have a stronger condition, *strong passivity*:

$$p \text{ is } strong\ passive \;\hat{=}\; \forall c \in p \quad c \text{ is } passive$$

That is, a sequence of commands is strong passive, if every command in the sequence is passive.

Moving back to our strategies, we recall that the use of a strategy leads to a sequence of command sequences being invoked, q_i,... These will terminate sometime, meaning that after some n all the q_i will be null. We will call the complete sequence up to termination q:

$$q = q_1 q_2 \ldots q_n$$

We will obviously want the whole sequence to be passive:

$$\{ \; passive\ strategy \; \} \quad q \text{ is } passive$$

Further though, we want to stop silly strategies, like "delete and re-type", so we want each of the q_i to be passive and probably strong passive:

$$\{ \; strong\ passive\ strategy \; \} \quad q \text{ is } strong\ passive$$

Even this is not enough: although we may have the same result, it is no good if the rest of the system state has been mangled. Thus we will also want the whole strategy to be passive with respect to $result^{\dagger}$, or even better to return completely to the original state:

$$\{ \textit{strategy returns} \} \quad I(pq) = I(p)$$

History and undo mechanisms may make this difficult to achieve, as they complicate all reachability principles, and this would probably have to be applied to the functionality disregarding such facilities.

Such a strong condition would also be unacceptable for the individual commands of the strategy, as they must change the state if they are to achieve anything. Even passivity with respect to $R^{\dagger}$ is likely to be too strong. For instance, the strategy that includes scrolling up and down in a word processor seems quite reasonable; however, this would almost certainly involve moving the cursor, which will be part of $R^{\dagger}$. To achieve strong passivity with respect to $R^{\dagger}$, we would need something like a separate browsing window in addition to the editing one.

We would like to say, for example, that the strategy may move the cursor, but doesn't mangle anything useful like buffers, even temporarily. To do this it is necessary to layer the design, and separate out those features regarded as inviolate functionality, and those that are acceptable for browsing.

3.4 Globality and locality

The observability and predictability principles are concerned with the static state of the system: although the strategy itself is dynamic, it has been reduced to a static state, the observable effect. The concepts are all to do with what can be observed of what is there, rather than acting on the objects of interest. When considering simple PIEs it was the reachability conditions that were of importance with regard to these considerations of functionality. These can be applied directly to the red-PIE.

The result space R is most critical, as this represents the final output of the system, and we would clearly want I_r to be at least strong reachable, and probably megareachable too. The display is more dodgy. It may include start-up banners and the like that are not intended to be returned to; on the other hand, it is useful if one can set up the display exactly as it is most pleasing and useful, since it is through the display that one interacts with the system. In short, the reachability conditions are applied but with the willingness to make exceptions, or abstract from the full display as necessary.

If the system is highly interactive then one is also interested in the observation of the effects of commands. In particular, one wants the effects of many commands to be immediately visible from the display. Informally, it may be said that a command is *local* if its effects are contained entirely within the current display. The opposite case is a *global* command, which takes no notice of the display whatsoever. For example, typing a character would be a local command since one can see the character being inserted on the screen, and this is the only effect it has. On the other hand, "replace all words 'fred' with 'Fred'" would be a global command, as it makes no reference to the current display.

It is easier to attempt a formal definition of globality, and we can say that a command c is global if its effect can be modelled by a function on R, that is:

$$\begin{aligned} & c \text{ is } global \quad \triangleq \\ & \quad \exists\, r_c : R \to R \quad \textbf{st} \\ & \quad\quad \forall p \in P \quad I_r(pc) = r_c(I_r(p)) \end{aligned}$$

Or graphically, regarding c as the state transition function from E to E:

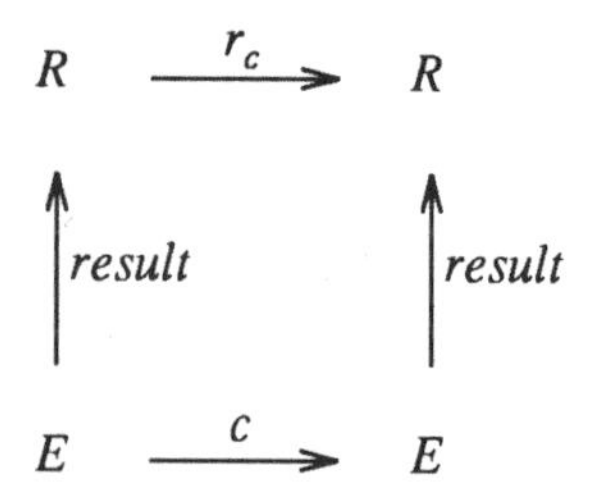

Notice that using this definition, a passive command is a special case of a global command! This is not exactly as intended, and we really would like to say "and the rest of the state doesn't change". Even if we did state this formally, it would not be right, as things like the cursor position and the display, would in fact change, but they would follow "in line" with the result. In Chapter 8 when we consider pointer spaces we will consider one way of expressing this requirement formally.

Locality is more complex; we could say:

$$\begin{aligned} & \exists\, d_c : D \to D \quad \textbf{st} \\ & \quad \forall p \in P \quad I_d(pc) = d_c(I_d(p)) \end{aligned}$$

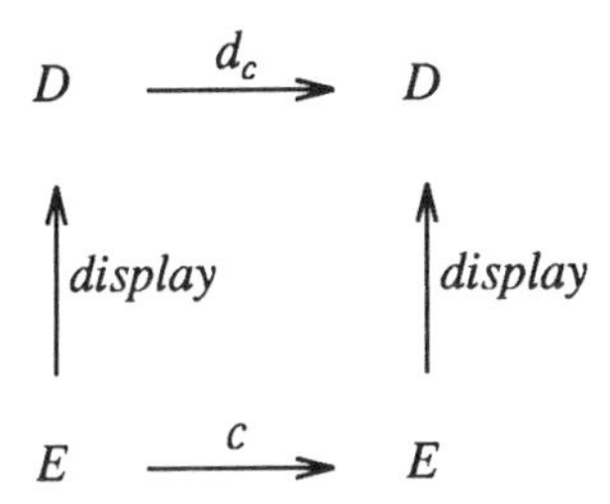

This would make many commands on graphical systems, such as Mac-paint, local; but on text systems like word processors, many commands which are local given the informal definition, are not using this definition. For instance, "delete line" would mean that an extra line of text would be brought into the display at the top or the bottom. Where this line would appear might be predictable from the current display; what it would contain would not be.

Both of these functions r_c and d_c are what I will term *complementary functions* to the command c, and Chapter 9 discusses the properties of such functions and includes a more complex framework in which globality and locality can be defined. However, for the time being we will leave the above two definitions as a flavour of what the true definitions should be.

Even without an exact definition of local commands, we can state some of the properties we would expect from them, in particular the ability to spot and correct mistakes easily and using only the current and previous display. Thus we could assert the following (L being the local commands):

mistakes in local commands are obvious:

$$\forall\ c, c' \in L, p \in P$$
$$I(pc) \neq I(pc') \quad \Rightarrow \quad display \circ I(pc) \neq display \circ I(pc')$$
and
$$I(pc) \neq I(p) \quad \Rightarrow \quad display \circ I(pc) \neq display \circ I(p)$$

So if we try something (c) and we accidently type something else (c'), if it makes any difference at all then we can see that difference in the display. Additionally, we can tell the difference between the effect of a local command and typing nothing at all.

Note that these conditions only tell us that if there is a change (in the effect) we will be able to see some change (in the display). It does not tell us whether the changes we see are the only changes. More sophisticated approaches are required to ensure locality in this strong sense. In particular, Chapter 8 develops stronger statements of this type of property.

One stronger property that we can state using the existing model is the ability to easily work out how to reverse the effect of local commands:

local commands suggest their inverses:

$$\exists \quad invert : D \times D \rightarrow P \quad \textbf{st}$$
$$\forall \; p \in P, c \in L \quad I(pc \; invert(d, d')) = I(p)$$

where

$$d = display(p) \quad \textbf{and} \quad d' = display(pc)$$

That is, from the previous display and the current one, alone, it is possible to recover back to the previous state. This is perhaps more the sort of undo principle to which Shneiderman is referring for direct manipulation systems (Shneiderman 1982).

3.5 Limitations of the PIE and red-PIE models

Abstract models will, by their nature, serve to specify only a subset of properties of interest. For any particular property, we may want a different model, or a refinement of the model. We cannot predict all possible such refinements and alternative models, but the succeeding chapters attempt to fill some of the more obvious holes.

The PIE model is clearly a single-user single-machine model; it does not cater well for the concept of multiple input streams. In the next chapter we consider a slightly different model that is better suited to such descriptions.

The PIE model expresses well the sequentiality of user input and machine output. It does not attempt to describe the interleaving and exact timing of these input/output events. Chapter 5 uses a model very similar to PIEs in order to describe some of the characteristics of this real-time behaviour, and to factor it out of the rest of the design process.

That leads on to the use of PIEs as the design of a system continues. The more detailed models in the succeeding chapters, can be seen as refinements of PIEs, although we will see in Chapter 6 that to do so requires a generalisation of the PIE. In Chapter 7, we consider how we can model parts of the internal specification of an interactive system using relations between PIEs. This represents, in a sense, development *within* the PIE framework.

When we considered globality and locality, the definitions were crude, because it was difficult to express the way changes in one projection of the effect should be reflected "naturally" or "pull along" the rest of the effect. We can think of this as "the spider's legs problem."† We have a spider and pull one of its legs. What do the rest do? They must move somewhat as they are attached to the body, but where exactly? Chapters 8 and 9 deal with this problem from different perspectives. Chapter 8 on pointer spaces considers the "editing the object" perspective, whereas Chapter 9 on views takes the opposing "edit the display" perspective. In fact, these two are not as far apart as it seems, as the more natural the map between object and display the more blurred the distinction, and both approaches try to make the map as natural as possible from different directions. These two chapters are also, in a way, less abstract than the earlier models, being concerned with particular models of construction rather than entirely behavioural models. Both also, being more constructive, can be useful tools for the detailed specification and implementation of particular system. This could be seen as a useful property, or as excessive implementation bias.

We expect interaction models to be used within the framework of any formal process and of course the PIE model can be coded into whatever specification language one is using: thence the various principles become system requirements in addition to the application-specific requirements. We discuss this further in Chapter 11.

The PIE model attempts to be as abstract as possible, and one would think that this would make it unsuitable for use nearer the implementation end. However, work by Runciman and Toyn (1987) has shown that expressing PIEs in an executable functional language can be a useful implementation technique.

---

† Arachnophiles should skip the rest of this paragraph.

CHAPTER 4

Sharing and interference in window managers

4.1 Introduction

Multi-windowed systems allow the user to compare several related displays on the same task, or to intersperse activities on more than one task. Both add to the complexity of the tasks that can be performed and to the complexity of the interface. It is the latter on which we will particularly focus in this chapter. In particular, we will be interested in the way that activities in different windows may interfere with one another and hence how the tasks associated with these windows may interfere.

First of all, to put our discussion in context, we will consider different types of window manager and their purposes. We will compare the MAC-type window manager with the sorts of windows typical in programming environments, and for the latter discuss the problems they are designed to solve. Section §4.3 will then discuss informal concepts associated with sharing, concepts such as level, granularity and perspective. We will reject models based around the computer's data model and instead opt for a behavioural view of sharing.

We will then get on to define a model of window managers suitable for expressing properties of dynamism and sharing, the *handle space*. Previously this has been called a view space (Dix and Harrison 1986) but the less evocative name is used here to avoid confusion with the "views" of Chapter 9. In section §4.5, we will use this model to give definitions of two types of sharing: *result independence* and *display independence*. In addition we will discuss general properties of these and other dependency relations. A formal definition of independence could be used during design to stop a windowed system having any sharing at all; however, in practice this will not be a totally desirable goal (we want some sharing). Because of this, the next section goes on to discuss the various ways the dependency structure of the windows can be used to improve the user interface and reduce the disorienting effect of unexpected sharing. If this sharing information is to be used, there must be effective procedures for detecting it. Section §4.7 discusses the various ways this can be achieved for different types of sharing. It also mentions the far more difficult issue of detecting interference induced by the user's own model.

4.2 Windowed systems

4.2.1 Windowing styles

There are two major genera of windowed systems, the first aimed primarily at office automation, Xerox Star, Apple Lisa and the ubiquitous Macintosh. The other major strand comes from programming environments, principally Smalltalk, Cedar and InterLisp. Most traditional Unix-based windowed systems lean towards the latter style, but a few, such as Torch Opentop and X-Desktop, are aimed towards the former (reflecting the expected markets). Comparing the MAC-style interface with these UNIX-based systems uncovers some important distinctions in the purpose and paradigm of mult-window environments.

MAC style

The metaphor that the MAC (and its predecessors) follow is physical, the desktop. Windows and icons are related to manipulable objects. You select objects and you do things to *them*. That is, the screen model reflects the *computer data* model. The purpose of the interface is to open up the guts of the computer and remove as much hidden functionality as possible.

A consequence of espousing a physical metaphor is that users will infer that it obeys physical laws. This is exploited when, for instance, selecting an object and dragging it about is very similar to the corresponding physical action. This physical metaphor does have important repercussions for sharing. In the physical world, when you do something to one object, you do not expect other objects to be affected unless they are physically connected: action at a distance, magnetism or radio controls have an almost magical air to them and are "not really" physical. This would cause problems if several windows were allowed onto a single database. If you were allowed to "peel off" several database windows, then these physically separate windows would be secretly linked. Alternatively, if you opened the database as single window, and within this application had several subwindows, these subwindows would have to be considered as different in kind from the main windows (even if they share the same physical appearance and controls). Whichever way the problem is tackled the user is involved in a context switch when changing from normal physical windowed work to the database.

UNIX style

At the simplest level this can be regarded as the multi-tty style: the windows are virtual terminals, some graphic, some character-based and these are multiplexed onto the physical screen. At this level the window manager is a *screen resource manager*. Each window corresponds to an application or a standard terminal session, and looking at it this way we see that the screen model most closely reflects the *computer process* model.

One reason for having several parallel terminal sessions, is in order to engage in several tasks and considered thus the screen model to some extent reflects the *user task* model. Viewed as such we could say that the window manager multiplexes the user, or is a *user resource manager*. We will examine this aspect and why it is necessary in the next section.

4.2.2 Handling multiple user tasks

To see why multiple windows are necessary for handling multiple user tasks, we consider a typical terminal session:

```
% type old_prog.c
    ....
% edit new_prog.c
    ....
    ....
% compile -Debug new_prog.c &
% edit letter
    ....
    ....
% typeset letter
% run new_prog
    ....
    ....
```

First of all the user looks at an old program; she wants to write a new but similar program. Having examined the old program, she then proceeds to create the new program using the editor. When she has finished, she compiles her program as a background process (indicated by the &). To fill in the time while it is compiling she types in a letter and then gets it typeset ready for mailing. By this time the program has compiled and she runs it to see if it has worked as expected.

Looking at this example we can see first of all that there are two main tasks, writing the new program and writing the letter. The user will experience several problems when trying to perform these two tasks:

- *Granularity of interleaving* – The user switches between the tasks but the granularity of interleaving is gross, at the level of complete invocations of applications. Whereas the computer when it is compiling concurrently with running the editor will interleave these at the level of arbitrary groups of machine instructions, the user cannot easily switch back and forth between editing the program and the letter.
- *Loss of context between tasks* – When the user compiled the program she specified a debug option. By the time she comes to run the program, the editor will have cleared the screen and the compile command will no longer be visible for her to refer back to.
- *Loss of context within a task* – Even when looking just at the programming task we see that the listing of the old program which is being copied from will be cleared from the screen by the editor, making it difficult to refer back. Many editors allow the user to execute other commands "shell out" while in the middle of an edit, partly relieving this sort of situation, but of course this is still a very gross and unnatural level of interaction.

Programming environments' window systems are an attempt to deal with these problems. Separate tasks are run in separate windows, so the interleaving can be as fine as the user wants and is limited only by quirks in the system (like dialogue boxes or menus that take precedence over normal windowing) and the cost of popping the required windows (Card *et al.* 1984). Because the tasks are in different windows, when a task is resumed its window is as it was when it was left, thus dealing with loss of context between tasks. In a similar way applications that use up a lot of screen space, such as editors, are usually run in separate windows to avoid context loss within tasks.

This example also gives us examples of two of the ways sharing can arise in windowed (and non-windowed) environments:

(i) We said that the user would know that the compilation had finished. Quite likely she would know because it would have displayed diagnostics all over the screen while she was in the middle of editing her letter! The two tasks seemed totally independent with no shared data, but they interfere with one another because they share the screen as a common resource. Having the two tasks running in different windows would not remove this problem completely, as the user might well have started preparing a data file for the test run while the program compiled. This is often particularly disconcerting, as having entered the command invoking the background process, the user will reach closure and forget about it.

(ii) The second example is due to shared data. If several windows belong to one task, as would happen if editing the program in the example produced a separate window, then the two windows would both be acting on `new_prog.c` and the user may accidentally invoke the compiler without writing the edited file. Again this is a closure problem, as the user reaches closure when the new program is completed, but before it is written. The non-windowed version did not have this problem, as it was impossible there to start the compilation until the edit was complete; however, it would be possible to edit the program while the compilation was proceeding in the background, a common error. On balance though, these closure errors do seem more prevalent when the user has several concurrent activities.

4.2.3 Importance of simplicity in windowing systems

It is especially important that users can use a windowed system without regard to the complexity of the windows themselves, as they are likely to be using the windows to offset mental load.

Two analogies of user–computer interaction can help us to understand this process.

(i) Card *et al.* (1984) suggest that we can draw an analogy between the user's use of windows and a computer's use of virtual memory. Essentially the screen corresponds to the primary storage, which is not big enough for all the memory requirements (the windows) and therefore some information is consigned to secondary storage (hidden windows).

(ii) Young *et al.* (1989) and Runciman and Hammond (1986) suggest the use of programmable user models to estimate the cognitive load of a system. This essentially draws the analogy between the user's process of using the system and a program to use it. Limitations to the ability to perform a task would include its time/space complexity, especially as regards short-term memory. If using a system exceeds these limits the user's performance will either degrade or they will have to off-load some of the memory demands, perhaps using windows to store information that would otherwise have to be remembered. We can liken this very small short-term memory to the registers in a CPU, and we get the following combined model:

user's short-term memory	—	visible windows	—	hidden windows
	(i)		(ii)	
CPU's registers	—	primary storage	—	secondary storage

Typically a user will gather information onto the visible windows to perform some task; this process may possibly use information from hidden windows. Because people tend to drive their abilities to the limit, and because of the cost of transferring information between memory media, the user will probably only use each method of off-loading information when they are close to their cognitive capacity. Therefore any additional load caused by the volatility of the storage media (lack of display independence), can be disastrous in terms of lost efficiency. In particular, events that require movement between visible and hidden windows will cause the user to forget information relevant to the present task in this "system" task of paging.

In an attempt to reduce the cost and complexity, Card and Henderson (1987) designed a windowed system around the idea of rooms, wherein one performs specific tasks, and in which are collected the various windows connected with a particular task. Additional methods to reduce the need for paging might include mechanisms by which the user can view critical parts of hidden windows on the visible screen; these include the use of extensive folding to minimise window size, or active icons to display status information such as compilation progress.

As well as the direct cost of paging it is necessary to consider the complexity of the interaction between different windows: any unexpected interaction will add mental load to a system which should be reducing it.

4.2.4 Summary

There are two major genera of window manager, one shaped around the computer's own data model, and the other around the support of multiple user tasks. Problems of interference between tasks are very important, especially if the windows are used to hold contextual information when task switching. This chapter is aimed mainly at the task-based view of window managers, although much of its analysis will be applicable to the MAC-style as well.

4.3 Sharing, informal concepts

4.3.1 Levels of sharing – actors

As we have seen, problems of sharing can occur in both windowed and non-windowed systems; however, the separateness of windows tends to emphasise independence and thus interference can be more disconcerting. We can classify the types of sharing that can occur at three levels:

- *Several independent actors* – This is what is normally understood as sharing, in database applications for instance, where several different users may access the same piece of data or possibly a user may access data at the same time as a system process.
- *One controller – several actors* – This is where the user sets off one or more background processes and these interact with each other or with the user's foreground activities. As all the processes are under the user's

control she should in principle be able to predict the sharing. However, especially if one of the processes is invoked long before another, she may not. Further, the form of interference may not be obvious, as an application with a window of its own may still print certain error messages in its parent window, which the user may no longer associate with it. As we have seen, this is not a problem merely of windowed systems.

- *One actor – several personae* – This is a particular problem of windowed systems: two windows may be accessing shared information and, for instance when examining data from several perspectives, perhaps necessarily so. As the user is consciously acting on each window, one might argue that she had little excuse for errors arising from this type of sharing; however, if the user is context switching between several tasks she is in effect acting as several personae and should not be expected to carry over information from one to another.

When thinking of the particular problems of windowed systems it is this last type of sharing which we will have predominantly in view. It is not confined solely to windowed systems: for instance, a clerk may be entering sales figures into a database but, when he has spare time, trying to produce sales totals, and the two tasks may easily be performed on a traditional terminal system in the same way as the two tasks in the programming example. The reason for saying that this is particularly important in windowed environments is the frequency, pace and granularity of context switching which is rarely encountered elsewhere. Having said that it is this last type of sharing that is of interest, most of what follows could be applied to all of them.

4.3.2 Granularity of sharing

Although we have been talking about sharing, it is not obvious what precisely is meant: the three levels given apply to who or what experiences the sharing, but there are also obviously different types of sharing, for instance sharing of physical resources such as the screen, or information resources such as data-bases. If we are to understand sharing and perhaps detect and warn the user of it we must define it more precisely.

The most obvious option is to look at some sort of data model such as files, relations or records, then say that two windows or processes interfere if they have access to a common resource. This approach has two major drawbacks:

- *Granularity* – What level of resource granularity ought we to use? Should we say two windows share if they have access to the same file system, file, record or field? If we opt for the smallest level possible, then we might say that if two windows are editing opposite ends of the same document they are not sharing, is this acceptable?

- *Choice of perspective* – From what perspective ought we build our data model? Do we say windows share if they have access to the same record or the same disk block? At a very concrete level, if a process uses up all the free space on a disk, any other processes accessing that disk will be interfered with. Logically equivalent database designs may have very different expressions in terms of file and record structure, giving rise to different definitions of sharing for the same logical design.

When considering these points we realise that a definition based around a specific data model is not sufficiently abstract; further, it may be both too strong and too weak in its definition. It may be too strong in that it gives rise to spurious sharing, for instance if two processes access the same record but are using logically distinct parts of it. It may be too weak in that it may ignore less obvious paths of interference. The most obvious case concerns consistency relations on databases, such as not being able to delete a non-empty directory. Further, by concentrating on the data model one ignores other links in the system, such as interprocess communication. Lastly and most difficult to detect are indirect links, where one process does something which causes an intermediate process to behave differently, thus affecting a third process. We cannot deal with this last problem simply by taking the transitive closure of our dependency relation as this would almost invariably be far too strong: in most operating systems, all processes interact with the kernel and would hence all be said to be interdependent.

We can avoid these problems by taking a *behavioural* or *implicit* definition of sharing as opposed to the constructive or explicit definition implied by a data model. This is (of course!) also more in keeping with our surface philosophy for describing interface issues, and we will define sharing in terms of the observed effect at the interface.

4.4 Modelling windowed systems

4.4.1 Fundamental features of windows

In order to build a model within which we can give an implicit definition of sharing, we will have to decide what are the features of windowed systems that are relevant and that should be included in the model. The first thing that we will abstract away from is spatial positioning of windows and the method of selecting them. These features are relevant for the user, but in this chapter we consider only the computer's implicit data model. In essence, we assume that we are looking beyond the physical screen and have unrestricted access to the entire array of virtual displays, in the same way as we might abstract beyond the physical display of a word processor and consider the document being edited.

Having so abstracted there seem to be two remaining essential features:

- *Content* – This concerns what is actually displayed at the window, whether graphical or textual.
- *Identity* – We assume that the window can be regarded as an object in its own right, and can be identified and manipulated as such. The identity of windows is important when considering change in time: as a window's content change it still remains the same window.

Some user actions may be addressed to a single window: for instance, we might set compiler options by opening up the compiler icon then typing them into slots in a form. Other actions may require two or more windows: for instance, we might compile a file by dropping the file icon onto the compiler icon. Some actions may require no windows at all: for instance, a global undo as a menu option. All these cases can be managed quite easily, but to avoid complexity we will confine ourselves to unary commands, limiting consideration to UNIX-style rather than MAC-style windows. The extension to the n-ary case is straightforward but increases the combinatorics with little gain in understanding.

4.4.2 Aliasing

It was said above that we were going to ignore presentation issues. However, there is one important issue that we should discuss before continuing. Several authors have produced specifications of window and graphical presentation managers. These are at the level of specific windowing policies and are aimed at defining the graphical nature of these systems. It is also useful to tie in the underlying applications to the presentation component at a more generic level. Previously (Dix 1987b), I have experimented with models of windowing based on collections of PIEs (as the underlying applications). The presentation component was left rather vague: the model assumed only that there was a selection method to choose the current window for input and that the currently selected window was always visible. This allows many different kinds of windowing policy, tiled or overlapping, and different kinds of selection mechanism. The model was also very uncommitted with regard to the applications, as the PIE model covers nearly all simple interactive systems. However, the very fact that it was a collection of different PIEs meant that the windows were necessarily independent and without sharing.

The intention was to prove that if the component systems had various predictability and reachability properties, then so would the collected system. This sort of modular proof is of course very powerful in building systems, especially because of the wide applicability of the models concerned. In fact, the experiment was largely a success, limited only by the simplicity of the data model (which is addressed in this chapter). The proofs all fell out as expected. In order to prove predictability properties (not surprisingly), I had to assume that there was some way of telling from the display not only *what* was in the current window, but also *which* was the selected window. Now by this I mean not only which of several displayed windows was the selected one (although this is part of the problem), but which underlying application was associated with the current window. To paraphrase in the terms introduced a short while ago, we need to know not only the *content* of the currently selected window, but also its *identity*.

What does this "proof problem" mean in practice? The necessity in the proof arises because in some circumstances two applications may be different but may temporarily have identical displays. Thus knowing what is displayed in the current window does not tell us which application is selected. We recall that a similar problem arose when we considered a simple model to express WYSIWYG (§1.5). There we could not tell where in a document a part of it belonged because different sections could look the same. So this is another instance of *aliasing* as applied to window managers.

Now the language we have just developed for windowing applies reasonably well for editing systems. We need to know not only the content of the visible part of the document (or diagram) but also where in the whole it is (the identity). We could therefore define the problem of aliasing as: *content does not determine identity.*

When we discussed aliasing before, we discussed solutions such as scroll bars. We can look for similar potential solutions for windowed systems. Most window managers allow title bars naming the windows. Assuming the naming is unique, this would (at least at a formal level) solve the aliasing problem. Of course, people may not notice the banners, so this may not be sufficient. Also in many systems the name bars are heavily under-used (most of the windows on my screen are called "shell" or "spy"). Screen positioning is also a very powerful cue to users, as are window shapes, colours, etc. Most probably, window appearance would be determined by the function being performed (editing, compiling, mail, etc.), but it is most likely that aliasing problems would arise between different instantiations of the same function. Perhaps appearance and naming should be associated more with topic than function?

So we have seen aliasing raise its head again. For the rest of this chapter we will assume that aliasing has been dealt with at the presentation level, and that we can look beneath this. That is, we assume that there is a collection of windows and that it is possible to associate the content of windows with their identity, and to address commands to any window (by identity) at any time.

4.4.3 A model of windowed systems – handle spaces

We now need to flesh out the assumptions and frameworks of the previous sections at a formal level.

We will call the set of machine states E and the set of commands C, to emphasise similarity to the red-PIE. We also have a result space (R) and a result map:

$$result : E \rightarrow R$$

The difference here is the set of *handles* (Λ) used to embody the idea of identity. These handles are not meant to represent anything meaningful to the user and are purely a device to represent the fact that the user can identify the display of a particular window and address commands to it. At any time only a subset of handles will be meaningful; we will call this set the *valid handles* and we will represent it by a map:

$$valid\_handles : E \rightarrow \mathbb{P}\,\Lambda$$

The result is defined without respect to a particular handle, and would probably be the file system or something similar. The display, on the other hand, depends on the window, and this is represented by including a handle as parameter:

$$display : E \times \Lambda \rightarrow D$$

This is a partial map, however, only defined for valid handles:

$$e, \lambda \in dom\ display \quad \Leftrightarrow \quad \lambda \in valid\_handles(e)$$

In terms of windows, we cannot ask for the display of a window that doesn't currently exist! In a similar way, commands are directed to a particular window, and hence the state transition function *doit* also includes a handle as a parameter and a similar condition:

$$doit : E \times C \times \Lambda \rightarrow E$$

$$e, c, \lambda \in dom\ doit \quad \Leftrightarrow \quad \lambda \in valid\_handles(e)$$

As we've said, we will restrict ourselves to analysing the case of unary commands, and the model has therefore exactly one handle parameter for update. For different purposes it is easy to develop variants allowing allowing n-ary and 0-ary handle parameters.

Many different kinds of system as well as window managers can be described using this model. For instance, it may also refer to the internal process model (assuming this is not in one-to-one correspondence with the windows) where each handle is a process, or to the file system where each handle is a file name.

In UNIX files are referred to at a deeper level by numbers called i-nodes and these are handles at a different level of abstraction, so both the file system with file names and i-nodes and the windows and processes can be regarded as handle spaces at different levels of abstraction. With the file system, we get at least two different handle spaces depending on whether we take the commands available to the programmer, or the commands available to the user. With the latter view, we would say that typing into a file was a unary operation, copying from one file to another was binary (requiring the more complex handle space model), and an editor which can work simultaneously on several files would be n-ary. A 0-ary operation might be file creation. The file system with file names is probably easiest to conceptualise but is not perfect as a handle space, as the presumption is that the handles themselves have no meaning whereas this is not usually true. However, it applies at a certain level of abstraction.†

In each of these example we can identify phases that are *static* or *dynamic* in terms of the handles available for operations. In the file system, editing a file is static in that the set of available files remains constant, whereas file creation and deletion are dynamic. Similarly, window creation and deletion are dynamic events.

The same command may be static or dynamic in different circumstances. For instance, looking at the process-level abstraction, if an editor is in insert mode "x" may type as itself, whereas if it were in command mode it might exit and hence destroy the process. We must therefore define whether a command is static in relation to a particular handle in a particular state:

$$static(e, c, \lambda) \equiv valid\_handles(doit(e, c, \lambda)) = valid\_handles(e)$$

Having said all that, however, in most systems many commands are always static, so we can define a predicate to apply to commands:

$$static(c) \equiv \forall\ e, c, \lambda \in dom\ doit \quad static(e, c, \lambda)$$

For most of this chapter, we will restrict ourselves to looking at the sharing properties of static commands, to avoid dealing with two types of complexity at once! We will return to dynamic commands in §4.8, when we will consider the possible principles for them.

---

† This is similar to the problem in denotational semantics where the denotation of a fragment of program will abstract away from the textual form and hence lose connotation:
`year_average = year_total/12` and `week_average = week_total/12`
are not equally valid.

4.5 Definitions of sharing

As we've said earlier, we are looking for some sort of implicit or behavioural definition of sharing. We will consider two ways of measuring independence of commands issued to windows: using display or using result. By comparing their properties we are led to consider several attributes that will distinguish general dependency structures.

4.5.1 Result independence

The result map is defined to be the final result of the system as a whole and is thus, in a sense, more important than the ephemeral displays. In the simple model of a window manager where each process had its own private display, result and state, it was easy to see that each process was completely independent, because they didn't change each other's internal state. However, even when processes use shared resources, file systems, databases, etc. they may still be accessing "independent" parts of the resource and at least over a subset of their available commands be "independent" themselves. We need to characterise this independence externally to a detailed breakdown of particular shared states which may hide some interdependencies and imply others which don't exist. We can do this using the handle space model and the notion of commutativity. When we are interested only in the final state of something this is the natural method of definition and is used, for instance, in deciding what database updates can be executed, locked and backtracked over independently. Essentially, two actions A and B commute if the combined effect of A followed by B is the same as that of B followed by A. So we say that two static commands c_1 and c_2 issued to handles λ_1 and λ_2 in state e are *result independent* if:

$$\begin{aligned} & \mathit{result}(\,\mathit{doit}(\,\mathit{doit}(\,e\,, c_1, \lambda_1\,), c_2, \lambda_2\,)) \\ & \qquad = \mathit{result}(\,\mathit{doit}(\,\mathit{doit}(\,e\,, c_2, \lambda_2\,), c_1, \lambda_1\,)) \end{aligned}$$

Essentially, this is drawing an analogy between the user with multiple windows and a scheduler with several processes. The scheduler has an easier job if there are no hazards caused by interference, and by analogy if those hazards are not present or minimised between windows the user's job will be eased.

We may want to strengthen this definition so that not only is the current result the same but all future results are the same no matter what *doit* s follow. To do this we need to define result congruence ($\equiv_r$) on the state space E. This is defined inductively by:

$$e \equiv_r e' \mathrel{\hat{=}} result(e) = result(e') \\ \textbf{and } valid\_handles(e) = valid\_handles(e') \\ \textbf{and } \forall c \in C, \lambda \in valid\_handles(e) \\ doit(e, c, \lambda) \equiv_r doit(e', c, \lambda)$$

We can then say that c_1 and c_2 issued to λ_1 and λ_2 in state e are *strong result independent* if:

$$doit(doit(e, c_1, \lambda_1), c_2, \lambda_2) \equiv_r doit(doit(e, c_2, \lambda_2), c_1, \lambda_1)$$

We may also want to say that two commands are independent only if they commute in all contexts. However, because the same command given to different handles may have completely different interpretations, we will almost certainly either have to limit ourselves to a specific pair of handles or introduce a type structure for handles. If we take the first case we need to restrict ourselves to the set of contexts that preserve the nominated handles (λ_1,λ_2, say). That is, we would demand the commands to commute when issued to λ_1,λ_2 in any of a set of states (E') defined recursively by:

$$\forall e \in E', c \in C, \lambda \in valid\_handles(e) \\ \lambda_1, \lambda_2 \in valid\_handles(e') \Rightarrow e' \in E'$$

where

$$e' = doit(e, c, \lambda)$$

The second option, of typing the handles with a type set T and a typing function $type : E \times \Lambda \rightarrow T$, would lead to a definition of *typed result independence* for commands c_1, c_2 issued to handles of type t_1, t_2:

$$\forall e \in E, \lambda_1, \lambda_2 \in valid\_handles(e) \\ type(\lambda_1) = t_1 \textbf{ and } type(\lambda_2) = t_2 \\ \Rightarrow doit(doit(e, c_1, \lambda_1), c_2, \lambda_2) \equiv_r doit(doit(e, c_2, \lambda_2), c_1, \lambda_1)$$

4.5.2 Display independence

When discussing result independence we said that the result was what was really important. However, if we take human factors into account we must modify this position somewhat, firstly because we are interested not only in the functionality of the system but also in its impact on users, and secondly because what the user *does* depends on what he *sees* and thus the result is affected indirectly by changes in the display. For instance, if a message is sent to a window in which there is a screen editor, if the user does not notice the message arriving then he may enter erroneous commands based on the corrupted screen.

We can state a simple form of display independence easily by saying command c issued to λ is display independent of λ' in state e if:

$$display(e, \lambda') = display(doit(e, c, \lambda), \lambda')$$

That is, the command issued via window λ has no effect on the display in window λ'. We can extend this definition in similar ways to that for result independence. There are, however, some major differences between the two:

- Result independence (and its negation, result dependency) is symmetric. Display independence, on the other hand, is directed: we can talk of one window/process affecting another without the opposite being true. This is especially true where the second window is a read-only view of a data structure being manipulated in the first: for instance, a formatted view of a text.
- If a set of commands and handles are pairwise (strong) result independent, then they may may be composed in any order without affecting the result. Pairwise display independence, on the other hand, does not in itself imply any group property.
- As formulated, display dependency is an event at particular instant: this command to this handle changes that display *now*. Result dependence is harder to pin down, as it involves two commands, and we need to ask which command actually causes the sharing.
- If the handle space corresponds to an underlying data model then display dependency can be inferred "from the outside". Thus a window manager would be able to detect and report such dependency. Result independence is much more difficult to infer "on the fly", and probably requires some extra information to be passed from the underlying data to the window manager.

4.5.3 Observability

We recall the example given above of the message arriving on a screen editor window and causing command errors. It has its impact because the user was expecting a correspondence between the display and the object being edited, a correspondence that was disrupted. Thus the lack of display independence led to a lack of ability to observe the object and predict the outcome of actions. This effect is exacerbated if the individual processes are well behaved, being themselves observable and predictable. (Dix and Runciman 1985) In this case the user may have inferred these principles and will rely on them. The scenario presented is not unlikely, as, in many window managers, utilities frequently report errors back to the parent window, a relationship that may well be forgotten by the time an error occurs.

One reason for the problem is that many utilities are defined in terms of the changes they make to the screen, because this is how they are implemented, and the window manager's job is to ensure that those changes are carried out. If the contents of the window are not as expected then the changes may be meaningless. If instead the window manager were defined in terms of preserving a correspondence between the screen contents and some description of it in the individual process, then this would not occur so easily. (Note that this does not mean that the implementation cannot make use of change information supplied by the process but merely that the *definition* has this property; the separation of definition and implementation implied by formal specification techniques facilitates this). Gosling and Rosenthal (1985) noted that when the style of the window manager implementation requires the individual processes to be able to regenerate the screen contents, this imposes a certain beneficial discipline on the programmer.

Even if the user is aware of the change to the current display, there may still be a significant increase in the cognitive load as the user tries to reconstruct the desired display. This load is especially onerous as the user is likely to be using the window manager precisely to off-load such effort.

4.5.4 General dependency

The above definitions are not the only structure on the processes presented by a window manager. As we have noted before, there will be user model structure and explicit data model structure as well. The importance of the above definitions in comparison to the explicit data model is that they refer to *implicit* dependency that might otherwise be missed. This is especially important as we would expect the user to be more aware of the explicit structure anyway and thus find non-explicit interference more difficult to assimilate. This could be an argument for demanding of the designer of the underlying data abstraction, that all implicit interdependence be contained within the explicit model: this has been a major concern of database theorists in the definition of the relational model and onwards. If the window manager could be assured of this then its job would be much easier. The relation to the user model is more complex and is discussed later in this chapter.

As we have noted these definitions are by no means the last word, either in implicit or explicit structure; however, in comparing them we noted some important differences and these can be used to classify dependency structures whether arising from implicit or explicit data structures, or from the user model:

- *Symmetry* – Is the relation between objects symmetric or is there a particular direction of dependence? This will appear in the explicit data structure, as for instance the difference between the "sibling" and "parent" relations for families, or between the "in the same module as" and "compiled version of" for program sources and objects. It also may appear in the user model: for instance, windows displaying documentation, running a spreadsheet and running a mail utility may all be in the symmetric relation of being part of the task of discovering and reporting a discrepancy between the documented and actual behaviour of the spreadsheet. On the other hand, if I am reading the documentation in order to run a program then I could also regard the information flow as being more one way. As this last example demonstrates, the distinction is by no means as clean as it might be.

- *Group properties* – Can group properties be inferred from pairwise properties? This is especially important, since properties of the form "A doesn't affect C and B doesn't affect C, but A and B together do" are especially difficult to assimilate.
- *State and event properties* – Is the property to do with a pair or group of handles in general, or can the dependency be isolated to a particular event? Often a definition that is framed in one context can be moved to the other. For instance, as we noted, display dependence is essentially an event; however, we could change it to a relation between handles by saying that λ_1 affects the display of λ_2 if there is any command that can be issued to λ_1 that is not display independent of λ_2. Similarly, we could take the much more diffuse property of result independence and take a subset of commands for two handles that are all result independent and say that the event of result dependency occurs on the first command not in one of these sets. Alternatively, we may want to partition the result space by projections for each handle signifying the part they are "interested" in, and assign the event of result dependence in an identical manner to display dependence.
- *Inferring structure* – Can the property be inferred by the window manager, or does it have to ask someone (the data model or the user)? This is most important for the efficiency of the window manager. If the dependency relation is difficult for the window manager to detect, it probably means that it is difficult for the user to assimilate also and thus an efficiency issue becomes a cognitive human factors issue.

4.6 Using dependency information

Assuming there is some sort of interdependency between windows, and that it has been detected by the window manager, how are we to use this information?

We consider two situations. First, we consider ways that the computer can make the dependency information available to the user. Then we look at the way that sharing properties influence our choice of undo strategy.

4.6.1 System response to sharing

The appropriate response to make in order to inform or warn the user of sharing will depend on the type of sharing that has occurred. Also we have earlier intimated that it depends on the user's task model.

Display dependence, we recall, is an event. Such event-based dependencies lend themselves well to some sort of *post hoc* warning, perhaps a bell or a flashing window border. As the focus of the user's attention is not on the changed window, one could argue for an audible signal in order to make the change obvious but, on the other hand, the problems arising from display dependence are to do with the user feedback loop and thus the signal could wait until the user actually focuses on the changed window. The latter argument would push towards a strategy such as altering the border of the changed window, possibly supplemented by a more explicit warning when it is next selected. A further advantage of this approach is that it does not interrupt the flow of work on the selected window. By the same token, however, by the time the changed window is selected the context in which the interference took place may be forgotten. A window manager could address this latter problem by allowing the user to examine the previous window contents. In a system with an existing history mechanism the user could even be allowed to find out which window caused the changes, browsing the context in which it occurred. Almost certainly the closer two windows are in the user's task space, the more subtle the form of signalling that is required. In the extreme, where two windows are just different views of the same object, no signalling is required at all.

Result dependence is a relation between windows. The obvious way to make it apparent to the user is by some correspondence with the spatial and visual relationship of the windows. If the relation is explicit in the data model then this will also be a correspondence between the data model and the screen model in the tenor of the MAC-style interface. Specific options might include always displaying all interdependent windows together, having interdependent windows all subwindows of some grouping window, or linking together interdependent windows (with rope?). A more fundamental way of dealing with the problem is to structure the command set so that most of the commands never cause sharing, the remaining commands requiring more conscious and deliberate use in order to remind the user of the impact. The further from the user's focus of attention (as measured by the user's task) the sharing is, the more difficult and conscious the use should be.

4.6.2 Undo strategies

When we look at the problem of undo in a multi-window environment, we are instantly posed the question of whether we mean a global undo (rolling the clock back for all windows), or a local undo (undo for a single window). The former method has least semantic holes and is most easy to understand; however, it has the unfortunate consequence that by the time I discover my mistake in one window I may have engaged in some independent activity in another window, which will then have to be undone in order to rectify my mistake. This problem is not just one for window managers, as the same sequence of events could happen, for instance, in a word processor where edits to one paragraph are found to be erroneous after another has been edited: however, it will tend to be more recurrent in windowing environments because of the greater tendency to switch contexts, and also more major as the units of interaction may have greater effect.

Only allowing global undo would be unacceptable because of the implied independence of windows. We can thus look to the dependency structure for help in understanding the semantics of local undo. Of particular value is the notion of commutativity as used in the definition of result independence. If actions A and B are independent, in the sense that doing A followed by B is the same as B followed by A, then we can undo A even if it occurred before B because it *could* have been issued after it. That is, the local undo is equivalent to a global undo on a *possible* derivation. In terms of the dependency structure we can undo any process until we get to an action that affected another process.

The situation is not quite as simple as this, however, as in practice we may only have the implicit and explicit data structure, and not the user semantics of the operations. For instance, if information has been written to, and then read from a common cut/paste buffer, and we wish to undo beyond the write, is this allowable? If the semantics of the operations were "I want this text in the buffer, and then I want the text from the buffer copied out", then we are all right in doing the undo; if, however, the semantics are "I want the same text in document B as in document A and I will achieve this aim using cut/paste", then clearly the aim fails if thereafter the contents of that text in A are undone and those of B left alone. Propagating such undos, even if semantically feasible, would probably be almost impossible to predict by the user, so we would probably have two classes of dependency; one where the user absolutely cannot undo beyond without major effort, such as sending mail, and the other as in the above where undo is possible but the user is warned that corrective action may be necessary, and is made aware of the scope of the broken dependency.

When we consider user semantics, life will not even be as rosy as painted above. For instance, if I use information in one window to guide my decisions about actions in another window then I will set up a dependency that is not captured in the computer: therefore there will be no way I can be warned of such broken dependencies. We at least have the consolation here that we can expect the user to be aware of such explicit dependency when performing the undo – or can we?

4.7 Detection of sharing

We can use the dependency structure only if the window manager can detect it. Explicit data dependency such as sharing files is fairly easy to detect, and is well understood. We examine below the additional methods used to detect the implicit dependency obtained through the result and display, and also the even more difficult issue of detecting the user's own model of the system, and the interdependency that it implies.

4.7.1 Detecting result and display dependency

In order to make use of the definitions of sharing, we will need to decide when to detect it. If we are interested in forbiding sharing altogether, and we are designing a complete system, we can apply the relevant independence principle to the specification and use it to constrain design. More likely the situation will be more complex:

- The restrictions on sharing may not be absolute. Perhaps certain windows are expected to share and we want to signal this to the user.
- The entire system is not under our control. We are merely designing the windowing and system architecture around applications that already exist and for new applications written by third parties.

These considerations mean that in practice we are more likely to be designing a system that seeks to detect sharing at run time rather than prevent it all together. Depending on which type of sharing we are interested in, different parts of the system will be more appropriate to detect it.

Display dependence

This is simple for the window manager to detect. It merely has to signal any output to windows other than the one to which commands are addressed. It becomes more complex if we allow the possibility of background processes which may direct output to the current window without responding directly to commands from it.

Result dependence

The window manager clearly cannot perform experiments to determine commutativity, so detection of result dependence must rely on the underlying database manager and process architecture. This is not back-pedalling on our decision to opt away from data models, as any algorithm for the detection of sharing will be validated by the implicit definitions developed independently of the particular model. In other words, an explicit data model is fine for dealing with the specification and implementation of a specific system, but is dangerous for the initial definition of our terms and principles.

4.7.2 Capturing the user model

We have talked about the user's task model and using that to guide the use of the sharing information. We have also alluded to the way that the user may cause unseen sharing and interdependence. To see this we consider several updates to a database from two sources A and B:

Scenario 1

A reads $x = 10, y = 5$
B reads $x = 10, y = 5$
A writes $x \rightarrow 15$
B writes $y \rightarrow 10$

operations commute?

Scenario 2

A reads $x = 10, y = 5$
B reads $x = 10, y = 5$
A writes $x \rightarrow 15$
B writes $y \rightarrow x$

operations do not commute

If we just consider the updates then the operations in scenario 1 commute with one another whereas those in scenario 2 do not. However, if we consider the intentions and the information available the situation is different. From B's point of view, both scenarios could be attempting to set y to the value 10, in which case scenario 1 is acceptable and B just chose a poor way of expressing that intention in the second case. However, if B's intention is that y should have the value of x then this time scenario 1 is the bad choice of operation as it hides the interference due to the user's intention. If A and B were different users with different displays then they could be warned when the other user performed an update which invalidated the information shown on their screen. However, if A and B are the same user operating via different windows then the information available is not limited to the selected window. Further, the interdependence may be much more subtle in practice.

It is clear that we can never entirely capture this user model, and it is thus of prime importance that we make the implications of the user's actions as obvious as possible. However, while not neglecting that it is sensible to capture as much as possible in order to aid the user:

- *Automatic capture* – One option is to try to infer the user's model from behaviour, or to force the user to work in such a way that it becomes obvious. Direct manipulation systems already go a long way to providing this. Users of such systems often prefer to copy from window to window, even quite small pieces of text, rather than retype. This makes the user's cross-referencing obvious to the system. For instance, in the example above if b were to construct the $y \to 10$ request by copying from the display $x = 10$, the system could infer that the update was dependent on the value of x and warn the user of possible sharing in scenario 1. This is far from perfect, of course: the user may still choose to re-enter rather than copy, and it would miss more complex relationships that cannot be achieved by copying (although the calculator could be included). On the other hand, it might also produce spurious relationships where copying was for lexical rather than semantic reasons ($x = 10$ was just a good place to pick up a 10).

  Automatic capture could be a lot more hi-tech, perhaps attempting an on-the-fly working set analysis to determine interdependent windows.

- *Explicit capture* – A more conservative option is to encourage users *to explicitly tell the system* their task model. This explicit capture can be either in the data model or in the screen model (if these differ significantly). An example of the former would be a user having several roles: each window would be logged into a specific role and protection schemes would ensure that there was little sharing between roles. A similar, but more dynamic, example of the latter would be project workspaces, where the user would create projects and subproject windows on the fly within which the other tools windows would reside. If the separation between database and window manager is strong then one would imagine the more long-term relationships being logged in the former and the shorter-term subtasks in the latter, as the examples suggest.

  It is interesting to note that the "rooms" system of Card *et al.* (1987) would be suitable for these purposes. It was created in order to aid the user's context switching when working with several tasks each consisting of several windows. Although it has a different aim from the worries here about sharing, it has the same goal, namely to capture a certain amount of the user's task model.

4.8 Dynamic commands

All the discussion so far has been in the context of *static* commands, where no new windows are created or existing ones destroyed. These events are obviously very important, but are significantly more complex. We will just take an overview of some of the properties that we might want to consider in relation to dynamic commands without going into great detail. I have given this issue a slightly more formal treatment elsewhere (Dix 1987b) but even there the complexity of the issues means that a very limited treatment was given.

We have already given a formal distinction between *dynamic* and *static* commands. An obvious distinctions to make within the dynamic commands is between those that create and those that destroy windows. One requirement that we might wish to make about a system is that there should be a simple distinction for the user between these classes of commands. Thus the user can know whether a command is static but may change things, may create windows but do nothing else, or may destroy windows. This is unlikely to be true at the physical level as many different operations may be invoked by a mouse click, but if we look at a more abstract, logical level of commands this may be an appropriate distinction.

Whether such a split is desirable also depends on what we take as windows. If we include dialogue boxes then a large range of commands may be static in some circumstance and have an effect, and be dynamic in others when the dialogue box is required. Similarly, when the dialogue box is responded to, the command is likely to be both dynamic (because the dialogue box disappears) and have an effect (such as writing a file). This emphasises that the appropriate principles to apply to a system depend crucially on the level of abstraction we choose to adopt.

Another set of issues for dynamic commands are to do with setting limits on their dynamism. This is important since the appearance or disappearance of large numbers of windows is likely to be confusing. On the one side we may ask that windows are destroyed only when commands are addressed directly to them. On the other we may define the *fecundity* of a command to be the number of new windows created, and set strict limits on the acceptable fecundity.

We also need to extend the concepts of independence introduced earlier to the dynamic case. A natural extension would be to say that windows have some sort of dynamic independence if the way they create or destroy windows is independent of the order in which the commands are used. There is a conflict, however, between this sort of independence and the forms of static independence described previously. For instance, if we are opening windows onto a database the desire for static independence will require some form of locking to prevent the windows from opening onto the same part of the database. This locking will inevitably mean that the events of opening the two windows will interfere.

4.9 Discussion

We have produced a simple model of windowed systems, the *handle space*, suitable for the formal discussion of principles of sharing and creation and destruction of windows. Over this model we have identified two specific sharing properties, *result independence* and *display independence*. These can be used to warn the user of possible interference between windows, and to inform undo strategies of allowable rearrangements of the user's event history.

Display dependence can be detected by the window manager but result dependence requires aid from the database and operating system, the exact information that these supply being constrained by the *implicit* definitions given. Detecting dependency induced by the user's own model is more difficult; some suggestions have been given, but they leave many open questions. In particular, the information gleaned by implicit capture will tend to be partial, and it is not necessarily the case that some is better than none in this context. For this reason the explicit capture methods are probably to be preferred at the moment (but aren't as much fun.)

Dynamic properties have only been dealt with only briefly, but are clearly complex and need to be examined closely for any particular multi-windowed application.

CHAPTER 5

The myth of the infinitely fast machine

5.1 Introduction

Most specification and documentation of interactive systems covers only steady-state functionality. That is, the effect of each user command on the state and display of the system is described, but the effect of lags between the entry of these and when the system actually responds is ignored. Effectively, the system is seen as executing on an infinitely fast machine! The obvious exceptions to this are those systems which deal explicitly with real-time phenomena: response timing, games, simulations, etc. However, it is those systems whose real-time behaviour is incidental which this paper addresses.

There are good reasons for ignoring real-time behaviour. Specifications are deliberately aimed at some aspect of the functionality of a system, and from this abstraction they obtain their power of concise expression. It is therefore important that any considerations of non-steady-state behaviour do not clutter up these abstractions unduly, and can be considered independently. For this reason most of this book has deliberately ignored timing issues. Other formal techniques are equally silent on such issues, for example all the contributors to Harrison and Thimbleby's recent book on formal methods in HCI (Harrison and Thimbleby 1989).

Again, documentation is often complex enough anyway. It is comparatively easy to describe steady-state behaviour in terms of changes in the objects manipulated and snap-shots of screens, but it is both difficult and probably confusing to describe dynamic behaviour without on-line or video presentations. Further, the exact dynamic behaviour may depend on the run-time environment and machine, whereas manuals are often written in a machine-independent fashion.

Both these considerations prompt one to look for ways of dealing with real-time behaviour which are both simple and can be factored out from the main time-independent description of functionality. The next section examines various system compromises that are used to approximate the ideal machine (*buffering* and *intermittent* and *partial update*), and the problems to which these give rise. Section §5.3 introduces a simple formal model which includes temporal behaviour and uses it to give precise descriptions. This provides a framework within which to make precise the problems which arise and the possible solutions. In particular, a precise definition is given of the *steady-state functionality* of a system and of typical system behaviour, such as buffering, as they appear to the user, without regard to a particular model of implementation. This model is then used as a basis for proposing properties that a dynamic system must obey if its steady-state functionality is to be usable. Various concrete expressions of these properties are discussed in §5.5, both existing techniques and novel, including analysis of their user impact. These techniques put demands on the surrounding system software and hardware which are outlined in §5.6, which show where changes are needed in order to put these proposals into practice. Finally, these features are brought together in a design approach where the dynamic aspects of system behaviour are considered separately from the steady-state functionality.

5.2 Compromises and problems in existing systems

In reviewing the research into computer response times, Shneiderman (1984) found that, with a few exceptions, faster is better. This is certainly the general approach taken when implementing interactive systems. At first, care is taken to obtain general efficiency, concentrating on those areas where serious lags are expected. Then, where this proves insufficient, either because the computer is too slow, or because the output device cannot handle the rate of refresh, various additional strategies may be used.

The most common such strategy is for either the application or its surrounding environment (operating system, window manager, etc.) to provide *buffering* to avoid loss of user commands. So normal has buffering become that its absence can be quite unnerving. Slightly less frequently seen are *intermittent update* strategies where the application only occasionally updates the screen (Stallman 1981): for instance, after receiving 10 scroll-up-line commands the system may decide to display only one screen scrolled up 10 lines, ignoring the intermediate stages. Some versions of the word processor Wordstar augment this by adopting a *partial update* strategy, where the system attempts to keep the current cursor line permanently up to date but updates the rest of the screen only intermittently as it has the time. Both intermittent and partial update are particularly useful when the output device is the limiting factor. They also save some processor time. In Chapter 6, I will suggest *non-deterministic intermittent* update to reduce processor time further, but the analysis presented below does not cover this more complex case.

Not only can dynamic behaviour be confusing for the user, but it can also seriously undermine various desirable properties. A system that could be described as "what you see is what you have got" (Thimbleby 1983) on an ideal machine may well not be so on a real buffered machine, as the lag means that the current context as measured by the commands entered so far but still unprocessed, is not the context displayed. A case of "what you see is what you had"!

Problems arise particularly when applications are embedded within a surrounding environment; for instance, multi-processing often leads to long delays which would invalidate a partial update strategy. Further, the environment's strategies may conflict with those of the application, for example the operating system's buffering and the application's buffering may interact badly.

The worst problems occur when the user feedback loop is important, in other words, in the most interactive systems. Key-ahead is usually no problem so long as delays are not too long, but cursor movement and mouse positioning are far more critical. When using cursor keys, delays in cursor movement may lead to entry at the wrong point, and in the worst case to *cursor tracking*. This is the phenomenon where, because of display lag after using cursor keys, the cursor is moved too far and misses its target, then in moving back it again overshoots. Users develop strategies of their own for dealing with this, moving one character in the opposite or an orthogonal direction after a long sequence of moves, or even inserting a character just to see if the cursor has stopped. These strategies become so automatic that some users claim that cursor tracking is not a problem. On the other hand, for small moves, feedback is not so critical: correcting typing mistakes on the fly is relatively easy since the number of moves to make is known, and can be an almost automatic reaction.

Mouse-based systems pose more serious problems. Some systems obtain the mouse position independently of button clicks, and thus, for instance, in a drawing application, depressing a button and moving will result in a start point somewhere along the mouse's path. Users of such systems get used to a "click and hold" strategy when drawing or manipulating icons and menus. Because the meaning of a mouse action is so heavily dependent on the display context, some systems disallow mouse-ahead. Again this leads to a "click and wait for confirmation" strategy for the user. The task of accurately positioning the mouse cursor already places high cognitive demands on the user, which are increased if the user must be constantly vigilant to check that the various mouse actions have had the intended effect. This is stressful, makes it difficult for the user to think ahead, and prevents the user developing effective motor responses. In particular, the illusion that the mouse is an extension of the user (Runciman and Thimbleby 1986) is destroyed.

5.3 Modelling

In this section we develop an abstract formal model of interactive systems, taking into account the timing of input and output. We will use this to define explicitly what is meant by steady-state functionality, to describe the various departures from this ideal functionality and the various observability constraints that are required to make such a system usable. In line with the philosophy of this book, the model will as far as possible be a surface model, in the sense that it will try to describe the system from the outside. For example, the definition of a perfectly buffered system is not in terms of a physical representation, such as a block of memory in which keystrokes are stored, but in terms of the behaviour of the system. This approach both avoids odd hidden problems (e.g. the system may save keystrokes, then later ignore them) and, by staying outside the system, is a description of its interface only and has therefore a greater validity for HCI.

5.3.1 A simple temporal model

There are many possible formulations, and the one presented here is chosen because it is most similar to the previous models. We consider a sequence of discrete time intervals. At each time interval the user may enter a single command (from a set C) or do nothing (represented by τ – a tick). Also, with each time interval, we associate a current display (from a set D); however, we assume that this occurs just after the input for that interval, so that there is time for a fast computer response, but so that the gap is imperceptible to the user. Thus the display in the same interval can be seen as the instant response, but lags are represented by responses several time-steps later.

The display is obtained from the current machine state (from a set E), by a function $display : E \rightarrow D$. This state comprises all the computer's memory of the past. The relation of this state to the input history can be represented in two ways, either by considering the whole history at once, or by an incremental approach. The first method calls for an interpretation function (I) which gives the state due to any input history. The input history (from a set H) will consist of sequences of commands (from C) and ticks (τ):

$$I : H \rightarrow E$$

The alternative is to use a state transition function, which, given the input for any time step and the last state, gives the new current state:

$$doit : (C \cup \{ \tau \}) \times E \rightarrow E$$

These two representations are of course not independent and are linked by the equations:

$$I([\,]) = e_{init} \qquad \text{– the initial state}$$
$$I(h :: c) = doit(c, I(h))$$

That is the state obtained for a history is obtained by starting at the initial state and repeatedly applying *doit*. We could call this a temporal PIE or a τ-PIE.

Note that the results of Chapter 2 concerning monotone closure mean we do not have to worry about the full abstractness of assuming a state. In fact, we could proceed without the concept but the definitions would become more complex.

5.3.2 Defining steady-state functionality

To talk about steady-state functionality, we need an idea of when the system has stopped changing. This is clearly when further time intervals with no user input (i.e. τs) don't alter the state:

$$e \text{ stable} \quad \equiv \quad doit(\tau, e) = e$$

If the state does not change, then the display, which is obtained from the state, clearly doesn't change either. However, just because the display is stable, it does not follow that the state is. For instance, in a multi-processing environment, a background task may well be altering the file system but not produce any visual effect, or, when a large global edit is being performed, there may not be any indication of completion; in the most extreme case, consider a user response timer continually ticking away with no visible change until the user responds and the delay is recorded.

We will define steady-state functionality by associating it with the response obtained by a patient user who always waits for the system to stabilise before entering more commands. Unfortunately, from the last example, we see that there are in fact systems that never stabilise: another similar example would be an alarm clock. However, it is usually the case that either the system does always stabilise, or that the parts that do not, can be abstracted away. Making this assumption, we can associate with any sequence of inputs a longer sequence with additional ticks on the end, which gives rise to a stable state. These ticks can be thought of as the time a patient user would have to wait:

$$steady : H \rightarrow H$$
$$steady(h) = h :: w$$

where w is the shortest sequence of τs such that $I(h :: w)$ is stable.

If the machine were infinitely fast, it would reach a stable state instantly, and w would be empty. As there would be no time to type-ahead, all users would effectively be patient. For any sequence from H, we can obtain the sequence that a patient user would have entered:

$$patient : H \rightarrow H$$

$$\begin{aligned} patient([\,]) &= steady([\,]) \\ patient(h :: \tau) &= patient(h) \\ patient(h :: c) &= steady(patient(h) :: c) \quad \text{when } c \neq \tau \end{aligned}$$

We can now define the steady-state functionality to be exactly that which the patient user would obtain: that is, I composed with *patient*. This is precisely the functionality defined by most specifications and described in most documentation, and is therefore the behaviour that should be validated against them.

5.3.3 Buffering and update strategies

How does this differ from the actual functionality observed by a less patient user? To see this, we compare the steady state obtained from a sequence of commands (and ticks) h with that obtained if the user had been patient. That is:

$$I(steady(h)) \quad \text{compared to} \quad I(patient(h))$$

If these are equal, then the system obtains the same steady state no matter how fast the user enters commands. That is, equality in the above is the definition of a *perfectly buffered system*:

perfect buffering:
$$I(steady(h)) = I(patient(h))$$

Note that this definition is purely in terms of the behaviour of the system; it makes no reference to a buffer as an implementation device. Later on we shall consider departures from perfect buffering.

We can also examine total versus intermittent display update using this scheme. We say a system has a total update strategy if all steady-state displays occur in the actual display sequence:

total update:

$$\forall\ h \in H,\ h_0 \le h_1 \le h_2 \ldots \le h_n \le h$$
$$\exists\ s_0 \le s_1 \le s_2 \ldots \le s_n \le steady(\ h\ )$$
$$\textbf{st}\ \ \forall\ i\quad display(\ steady(\ h_i\ )\ ) = display(\ s_i\ )$$

A large proportion of systems are perfectly buffered and adopt a total update strategy: this is because they are effectively programmed for an ideal machine and buffering is supplied by the environment. The environment can supply degrees of automatic intermittent update in addition to buffering; for instance, TIP (Thimbleby 1987), a screen control package, attempts to suppress display update when there is input pending.

The decision whether or not to update is not just one of efficiency. Some intermediate displays convey information by their dynamic behaviour: for instance, seeing a display scroll up line by line gives the user more of an impression of direction than a sudden jump. On the other hand, if the user has typed ahead it is likely that any such feedback will be too late!

5.4 Dealing with display lag

As we have said in §5.3, the interpretation that the user gives to commands depends on the display context. The problem of display lag can be addressed in two ways:

- A computer response, trying to reinterpret or ignore commands.
- A user-based solution, supplying sufficient information for the user to make sensible decisions.

Typically a system will use a mixture of the two; in fact, we will see that responses of the first kind necessitate the provision of specific information for the user.

5.4.1 Computer response

In the first category, the most obvious example is disallowing any mouse-ahead. This amounts to:

$$h \ne steady(\ h\ )\ \textbf{and}\ m \in mouse\ commands$$
$$\Rightarrow\quad I(\ h :: m\ ) = I(\ h :: \tau\ )$$

This says that when we are not in steady state, any mouse command is treated just like a tick, an interval with no input. That is, mouse commands are ignored out of steady state. This may be too rigid, however, and mouse-ahead may well be acceptable if the displayed context when a command is issued is the same as the context when it is eventually applied. For instance, if there are two independent editing windows, mouse-ahead in one would be allowable even if there were commands outstanding for the other. Even if the contexts are different, the denotation in the context of invocation may still have meaning at the time of application. For instance, if there are still keyboard entries outstanding for a find/replace buffer and the user wants to initiate the action, then, if a simple function key was being used, we would expect most applications to queue it; it seems reasonable therefore that the lexical change to a find/replace button icon should behave similarly, allowing mouse-ahead.

A system that disallows some commands, but not others, can be represented by the existence of a function *noted*:

$$
\begin{aligned}
&\mathit{noted} : H \rightarrow H \\
&\forall\ h \in H \quad \textbf{let} \quad n = \mathit{noted}(h) \\
&\qquad\qquad n \text{ is a subsequence of } h \\
&\qquad\qquad \textbf{and} \quad I(\mathit{steady}(h)) = I(\mathit{patient}(n))
\end{aligned}
$$

This says that for any history, we can associate a shorter history of commands that have been taken note of. These noted commands are a subsequence of the original commands (dropping out the ignored ones), and the behaviour of the system is the same as if only the noted commands had been patiently entered. Note the similarity between this and the definition of perfect buffering in §5.3.3. This similarity is because systems that possess such a *noted* function will obey the exception rule "no guesses" (Chapter 2 §2.9), which suggests that when exceptions occur a system should not perform a special action, but instead merely ignore the command that raised the exception.

5.4.2 User solutions

The user should be able to discern what the correct context is. In particular, is this steady state? This decision must be possible from the contents of the current display, requiring the existence of a decision procedure for the user, *is_steady?*:

$$
\begin{aligned}
&\mathit{is\_steady?} : D \rightarrow \mathit{Bool} \\
&\textbf{st} \quad \forall\ e \in E : \ \mathit{is\_steady?}(\mathit{display}(e)) = \mathit{stable}(e)
\end{aligned}
$$

Although cast as a user decision function, its importance is of course the information it requires of the system. This information could be positive (e.g. a status light when the system is stable), or negative (a red cursor when it's not): formally both give equivalent information. Note, however, that the system response producing the information must be immediate: it is no use waiting for steady state for information to say the system wasn't in steady state when you typed the command! This is why the definition refers to the current state and its display. Some systems are always in steady state except when the screen is actually changing (e.g. cursor moving or text scrolling), and thus lack of change indicates steady state. We could modify the above condition to look at several displays; however, it is reasonable to maintain the definition and regard change as being an attribute of a single timeframe.

The user needs to determine whether the current context is steady, both to make valid decisions based on the display, and to predict dynamic response for cases like mouse-ahead. In the latter situation we see that there is interference between the computer's strategies for dealing with dynamic behaviour and the user's. For instance, if a user changes windows, the window manager might throw away key-ahead until the window has been displayed, to avoid accidental typing to the wrong window. This could be very annoying if the user does not know when type-ahead can begin again.

If the simple strategy of no mouse-ahead is used then the existence of *is_steady?* is sufficient. If, on the other hand, one of the context-sensitive mouse-ahead strategies is used a more complex observability criterion must be introduced. This will be related to the function *noted* introduced above. So that the user can know whether a command has been noted or not, we need a user predicate similar to *is_steady?*. We will call this *is_noted?*:

$$\begin{aligned} &is\_noted? : D \rightarrow Bool \\ &\textbf{st}\ \ \forall\, h :\ \ is\_noted?(\, display(\, I(\, h :: c\,)\,)\,) \\ &\qquad\qquad\qquad \Leftrightarrow\ \ noted(\, h{::}c\,) = noted(\, h\,) :: c \end{aligned}$$

That is, *is_noted?* tells us immediately whether or not our command will have an effect in steady state. An example of this would be a buzzer sounding whenever a mouse-click has been ignored. Perhaps even better would be to predict beforehand:

$$\begin{aligned} &will\_be\_noted? : D \times C \rightarrow Bool \\ &\textbf{st}\ \ \forall\, h :\ \ will\_be\_noted?(\, display(\, I(\, h\,)\,), c\,) \\ &\qquad\qquad\qquad \Leftrightarrow\ \ noted(\, h{::}c\,) = noted(\, h\,) :: c \end{aligned}$$

That is, *will_be_noted?* tells us before we enter a command whether or not that command will have an effect. An example of this would be hour-glass mouse icons (Goldberg 1984).

If the user is able to tell from the display whether or not a command has taken effect (but not of course whether the effect has been as expected, which requires steady state), then certainly the system needs (in some sense) to know. This disallows retroactive decisions, such as circular overwriting keyboard buffers (as in some UNIX systems). In practice this will mean being conservative about allowing mouse-ahead, only accepting it if we can be sure that the current context and the steady state context are in the right relation.

Note also that the information for these decision functions must be available *all* the time. In particular, it cannot be subject to intermittent or partial display strategies.

5.4.3 Summary of formal analysis

By considering the problem of real-time behaviour formally we have:

- Related precisely the steady-state functionality described in most specifications and documentation, to the actual temporal behaviour experienced by the user.
- Shown how the departures from this steady-state functionality can be described and recorded.
- Used the formal description of these departures to expose specific information that must be available to the user: *is_steady?*, *is_noted?* and *will_be_noted?*.

5.5 What would such systems look like?

Supplying sufficient information to ensure the existence of such decision procedures is not sufficient; the information must also be in a form comprehensible and noticeable to the user. We should also take account of whether we expect the user to be acting consciously or automatically, and where we expect the centre of attention to be. For example, a user having just asked for a program to be compiled would expect it to take a while, and would consciously look out for indications of completion (e.g. a new prompt); on the other hand, if the user were typing fast into a word processor, we could not expect the same degree of awareness. Again, if mouse-ahead is prohibited, flashing the mouse cursor would be an acceptable signal since attention is likely to be centred on the mouse, yet signifying a locked keyboard by a flashing LED on a key would be unacceptable for touch typists. Monk (1986) discussed this and similar issues when considering signalling modes to the user. In fact, the various conditions considered here (the existence of type-ahead and ignoring commands) could be thought of as a form of modiness.

We look first at information to represent the steady-state. Here there is a well-established use of hour-glass, busy bee or watch icons, substituted for the normal mouse icon, when the computer is busy. However, these tend to be used only when some significant pause, such as file transfer or global editing, is occurring. When used they tend to signify that either mouse input or possibly all input is being ignored. Thus they fall closer to the *will_be_noted?* than the *is_steady?* information. However, on such systems, it is often the case that mouse clicks can be ignored in other contexts also. The mouse icon could be used consistently in all non-steady-state situations, especially if we are using the rule of ignoring all such mouse input. This would be acceptable only if we could be sure that in all critical situations attention is focused on the mouse cursor, which does fortunately seem quite likely. The alternative centre of attention is the text entry point or current selection (the only one in non-mouse systems). Similar mechanisms can be used here; however, on more conventional terminals it is unlikely that there is much choice of cursor attributes, although the one choice that is often available is between flashing and steady cursor, and this could be quite appropriate.

The disadvantage of making steady-state information so prominent is that it could be a significant distraction: imagine the cursor changing icon for a fraction of a second on each keystroke! If we expect most uses of steady-state information (as opposed to explicit *will_be_noted?* information) to be in situations where users expect problems, then it could be consigned to a status line, either as a simple flag or more elaborately. Perhaps the most extreme form of such information would be the *munchman buffer*. As you type, the characters appear on the bottom line of the screen. As the application deals with each character, it is consumed in the bottom left corner (by the munchman!). Although slightly strange, such an interface might be useful in certain circumstances, such as command interfaces. Perhaps more likely, particularly where quite powerful personal computers are used as terminals, would be to have an icon associated with a window representing the state of the interaction. In its collapsed form it could have a small bar indicating the fullness of the buffer. When expanded, more information could be available; in particular, this could be combined with the status information for a terminal emulator. This would again be of particular value with command-based interfaces: it allows the user to determine steady-state, to make predictions about the context where commands will be interpreted, and gives feedback as to progress which can be very important if response is slow (Shneiderman 1984).

If we consider mouse-ahead acceptability, or keyboard locking, then changing mouse or cursor attributes again becomes by far the best solution. For the mouse these attributes will be context sensitive, so that the mouse may display a locked form when over a text window with outstanding updates, but be in its normal state when moved over a control panel where mouse input is acceptable. This effectively supplies us with the *will_be_noted?* information. However, even though it may be safe to assume the user's attention is focused on the mouse or cursor, merely visual cues may not be noticed if the user is acting semi-automatically. Thus it is insufficient to supply only *will_be_noted?* information: feedback (preferably aural) is required when the user makes a mistake. That is, we need to have the *is_noted?* information also, and in a different medium. The usual objections to such feedback (people don't like machines that beep at them) don't hold here, as there is sufficient visual warning before the fact: the self-aware user can use the visual *will_be_noted?* information and thus need never hear a thing.

5.6 System requirements

If any of the strategies proposed or any alternatives are adopted, and are only partly correct, then they may be worse than useless. Users may be lulled into a false sense of security and then make mistakes which would go unnoticed because of the reduced attention, or they may eventually come to ignore the feedback, considering it an unreliable and therefore irrelevant distraction.

Successful implementation depends on the mutual cooperation of the application, the environment (operating system, window manager, etc.) and the interface peripherals. We will see that one recalcitrant partner can spoil the system. Any of these parties, or a combination of them, can take the initiative in the process. Thus we will consider each of these (environment, application and peripheral) in turn.

5.6.1 Environment control

Except where communication is over large, slow networks, we can assume that the operating system has almost immediate control over its interface hardware. It may not be able to refresh an entire screen in an imperceptible interval, but at least it will be able to guarantee immediate control over some portion of the screen, enough to display some status information. Several problems may arise, however, if the application does not cooperate.

If the application buffers input for itself (e.g. in multi-user systems where system calls are expensive), the environment cannot know how much has been used and hence can give only coarse estimates to the user (i.e. no per character munchman buffers). In single-user systems (for instance CPM), system calls are usually on a per character basis and the system can thus keep a more exact track of input pending. In both cases if the operating system allows polling of input then there is the possibility of busy waits, not allowing the environment to detect steady-state. This will happen, in particular, when the application is attempting to perform screen update optimisations, as it will poll to decide whether or not to update: this is a case of antagonistic temporal strategies. Clearly the application must pass some message back, for instance guaranteeing to wait on input eventually. More problems arise when considering non-noted events. Unless there is a system-wide policy, such as ignoring *all* mouse-ahead *always*, only the application can know the allowable contexts. This is alleviated if the operating system includes a rudimentary user interface management system (Pfaff 1985).

The actual application would be perfectly buffered, and a separate lexical/syntactic module would determine acceptable events, this second module guaranteeing to run fast enough to be effectively instantaneous.

5.6.2 Application control

There are several reasons for investing temporal control in the application. It has most knowledge about the way it will be used and what sort of displays and feedback are appropriate and consistent with the rest of its operations. Further, the turnover of applications is far greater than that of operating systems and thus new ideas can be included more quickly. Finally, and more tentatively, similar to the desire for consistency across a system, is the desire for consistency of the same application on different systems.

The application is frequently prevented from exercising control by the primitives supplied by the environment. On the input side, only blocking input may be supplied, not allowing the application to decide whether there is any pending input. This is true, for example, of older Unix systems. In these cases intermittent update strategies are often possible by requesting large chunks of input and updating on each chunk; however, steady-state indicators are very difficult, even when scheduling can be guaranteed to be frequent.

Scheduling problems combine with the paucity of primitives to prevent proper synchronisation of input and output. Imagine we are about to output a screen that comprises a major context shift, and we wish to ignore all outstanding input. The operating system provides a *flush_os_buffers* call that does immediately flush all of its outstanding events. The following sequence is no good:

> *flush_os_buffers* ();
> *update* (*screen*);

because the application may be descheduled between the two lines and further input may be entered before the screen is displayed. Even if there is no delay, the update itself may take some time and unwanted input may still be accepted. However, if we reverse the order the situation is worse:

> *update* (*screen*);
> *flush_os_buffers* () ;

If there is descheduling, the user's input will be ignored, despite the fact that a new screen is displayed which, among other things, will imply that input is acceptable. On the other hand, if there is no delay, the call to *update* is likely to mean that the request has been placed in the appropriate queue, but not that it has completed, thus leading to the same problems as the previous order. These problems are particularly evident with bit-map displays where the refresh time can be significant.

Where events are time-stamped, we might try to compare these time-stamps against the time of last update; however, obtaining this time leads to the same problems as before. It is only the front-end software that can meaningfully synchronise input and output, and thus this must supply the relevant information. The simplest solution therefore, which could be provided at little cost, is to have the window manager or tty-driver, upon request, insert an event into the input stream when an update completes. The application would then know that any events received before this were entered while the screen was not up to date.

5.6.3 Peripheral control

Poorly designed interface hardware can destroy attempts at good temporal behaviour, perhaps by having uncontrollable buffering, or insufficiently rich operations; however, in practice this is usually an annoyance which can be overcome. On the other hand, the peripherals rarely contribute anything positively. This need not be the case. Terminals are getting increasingly intelligent and this intelligence can be used. To a large extent, a terminal taking control has the same problems with the underlying system as the operating system has with the application; however, there are cases where it could be worthwhile despite this. For instance, where a PC is acting as a terminal over slow lines or a network, it may well be worth reporting the state of its buffers even when there may be additional buffering in the host. Perhaps, more importantly, the terminal can cooperate either with the operating system to take workload from it, or with the application despite a possibly antagonistic operating system. This latter case is most interesting, as it can overcome the technical inertia of the operating system. For instance, synchronisation, such as suggested above, where output events generate input tokens, can be accomplished trivially (in fact some VDUs have a who-are-you request which can be used for this purpose). Instant feedback is more difficult, as this has to be handled by the terminal if fast scheduling cannot be ensured. However, the solutions proposed for environment control apply here. Either catch-all policies such as no mouse-ahead, or simple descriptive mechanisms such as region-sensitive mouse-ahead, could be provided by appropriate control codes. The use of PCs as terminals, specially designed programmable terminals such as TIN (Macfarlane and Thimbleby 1986). or intelligent workstations such as the BLIT (Pike 1984), allow the possibility of downloading programs of significant complexity which could in particular handle the parts of an interface requiring close synchronisation and rapid response, leaving a bare-bones application in the host.

5.7 Conclusions – a design approach

The above discussion shows that it is possible to talk about and build fully temporal systems, by embedding a steady-state design in a real temporal system. From the above discussion, we have the beginnings of a design approach:

(i) Take a steady-state functionality.

(ii) Decide under what circumstances to allow partial or intermittent update.

(iii) Decide in what circumstances to ignore user commands.

(iv) Choose ways, not subject to partial or intermittent update, to represent the achievement of steady-state, to signify when user input is acceptable and to give feedback when input is ignored.

Stages (ii–iv) are considered separately from the first. In particular, they would be specified and probably documented separately, there should be support for them from packages, the environment, the peripherals or some combination, and further there may be system-wide policies for them, especially for presentation (stage iv).

Following such a design approach should lead to systems with real-time properties that are easier to specify, easier to document and easier to use.

CHAPTER 6

Non-determinism as a paradigm for understanding the user interface

6.1 Introduction

Although this chapter primarily presents conceptual and pragmatic ideas about non-determinism in user interfaces there is another, higher-level, more important point to it. This chapter provides an example of the subtle interweaving of formal and informal reasoning: the way formal analysis can lead to new insights which influence the informal.

We start with an analysis of problems arising from several formal models. In addition to the PIE model we started with, we have now defined two further models, one for windowed systems and one for temporal systems. These models have obvious similarities, and it would be useful to relate them to one another. The first section of this chapter attempts to do so, and quickly we discover that in order to do this we require non-deterministic models to express properties of interest. These models are non-deterministic in a purely formal sense; the rest of the chapter considers the repercussions non-determinism has on our understanding of the user interface.

Non-deterministic models contradict the general feeling that computer systems are deterministic, following a fixed sequence of instructions, so we might wonder whether there is any meaning to this use of non-determinism, or whether it is mererly a useful, but essentially meaningless, formal trick. In the rest of this chapter, we will demonstrate four things about non-determinism in user interfaces:

- It does exist.
- Users deal with it.
- We can help them do this.

- We can use it.

Sections §6.3 and §6.4 deal with the first of these points. The first of these shows that non-determinism is in fact a common experience of users, and we focus on the notion of *behavioural* non-determinism in order to square this with the normal deterministic model of computation. The second section then catalogues various sources of non-determinism in the user interface.

After this we consider how users, abetted by the designers of systems, can deal with this non-determinism. The final point of §6.5 reminds us how non-determinism can be useful in the specification of systems; however, this non-determinism is not intended to be a facet of the actual system, only a tool for development. Section §6.6 goes beyond this, giving an example of how non-determinism might be *deliberately introduced* into the user interface to improve performance. Without being prepared by noting how supposedly deterministic systems behave as if they were non-deterministic, the deliberate introduction of non-determinism would seem preposterous. Instead we are able to assess it impartially and see how it may actually *reduce* apparent non-determinism.

6.2 Unifying formal models using non-determinism

So far in the book, we have dealt with three major interaction models:

- *PIEs* (including *red-PIEs*) – a very general model intended to be applicable to almost all systems.
- *Handle spaces* – for modelling windowed systems.
- *A temporal model* – for considering real-time properties of interactive systems.

In any particular system, we may want to apply properties expressed over several of these models and we could simply map them all onto the system independently; however, it is clearly sensible to consider how they interrelate at the abstract level, and if possible relate them all back to PIEs, which are intended as the general model.

For temporal models we have already seen an example of this, where we abstracted the steady-state functionality and described how it fits into a general design method. However, even there we will find other abstractions we would like to consider.

In the following section we will consider such unification of these models and how it leads us to consider non-determinism. The first two subsections look at temporal models and handle spaces respectively, and consider how these relate to PIEs. In each case we will find that the PIE is insufficient and that a non-deterministic model is necessary. We will have to generalise the PIE model, giving a non-deterministic PIE (ND-PIE). This will come in versions with and without languages, and, in contrast to the deterministic PIE, the version with the language (non-deterministic language!) is the more natural and requires extra limitations to remove the language. We then return to the temporal and windowed models and relate them to the ND-PIE. We find that several properties can be expressed as determinacy requirements on certain derived ND-PIEs. Having considered these two models which started the investigation we look again at the deterministic PIE and see whether any abstractions of this can usefully be described using non-determinism. Again we find that predictability can be seen as a determinacy requirement.

6.2.1 Problem for temporal systems

In the previous chapter we saw how a PIE representing the steady-state functionality of a system could be embedded into a fully temporal description of that system. As we noted, a real user of the system would encounter a more complex functionality depending on the exact timing of input. However, the user will be unaware of such detailed timings (with a clock operating at MHz!), and it is only some abstraction of this temporal behaviour which is apparent. Clearly, from any sequence from C_τ we can abstract a sequence from C by simply ignoring all the τs. This is the user's view of the input (i.e. what was entered but not exactly when). It is not clear what effect to associate with a particular input sequence but for the moment let's use the steady-state effect.

Having removed timing from the temporal system the result is likely to be non-deterministic. For instance, take the bufferless typewriter which requires a single tick after each character in order to reset itself and ignores any characters being entered too fast:

$$C = \textit{characters}$$
$$E = \textit{sequences of characters}$$

$$\begin{aligned}
I_\tau(\,null\,) &= null \\
I_\tau(\,c\,) &= c \\
I_\tau(p\tau) &= I_\tau(p\,) \\
I_\tau(p\tau c\,) &= I_\tau(p\,)\,c \\
I_\tau(pc_1c_2) &= I_\tau(pc_1)
\end{aligned}$$

Thus the input sequence "abc" with different timings may give rise to different effects:

$$\begin{aligned}
I_\tau(\,a\tau b\tau c\,) &= abc \\
I_\tau(\,ab\tau c\,) &= ac
\end{aligned}$$

Clearly, we need a non-deterministic version of the PIE model to model such behaviour.

6.2.2 Problem for windowed systems

A similar problem arises when considering the handle space representation of windows from Chapter 4. Clearly, we can regard the whole system as a PIE with a command set of $C \times \Lambda$ and with a language defined by the valid handles map. More often we would be interested in regarding each window as an interactive system in its own right.

If we "freeze" all other windows we can give the functionality of a window λ in the state e as the result and display interpretations I_r and I_d:

$$\begin{aligned} I_r &= result \circ I_\lambda \\ I_d(p) &= display(I_\lambda(p), \lambda) \end{aligned}$$

where I_λ is the iterate of *doit* relative to λ:

$$\begin{aligned} I_\lambda(null) &= e \\ I_\lambda(pc) &= doit(I_\lambda(p), c, \lambda) \end{aligned}$$

This definition is not very useful: it's not very interesting knowing the functionality of a window system when you only use one window! What we really want to know is the functionality of each window when other windows are being used also. To do this we would consider the possible effect of commands p to a window amidst all possible interleavings from other windows. If all commands are result and display independent in all contexts, then this would yield the same interpretation functions as above. In the general case, however, we would again obtain non-deterministic functionality.

6.2.3 Non-deterministic PIEs

We will now consider non-deterministic generalisations of the PIE model. There are many ways of modelling non-determinism, but one of the simplest is to substitute for some value a set of possible values. However, we have to be careful to choose the right representation of our model, and the right values to substitute. The simplest option is to assign to each input a set of possible effects. That is, we modify the signature of the interpretation to $I_{ND}: P \to \mathbb{P}E$, where $\mathbb{P}E$ is the collection of sets of effects. Unfortunately, this does not distinguish some systems that are clearly distinct. Consider I_{ND} defined as follows:

$$\forall p \in P \quad I_{ND}(p) = \{0, 1\}$$

Does this describe a system that starts off with a value of 0 or 1 and retains this value no matter what the user enters? Does it represent a system that makes an independent choice after each command? Is it something in between? On the basis of the information given, we cannot decide. We therefore need to consider the *trace* of all effects generated by a command.

In Chapter 2, we saw that for any deterministic PIE we can always define a new interpretation $I^*:P \rightarrow E^*$ by:

$$I^*(\,null\,) \;=\; [\,null\,]$$
$$I^*(\,pc\,) \;=\; I^*(\,p\,) :: [\,I(\,c\,)\,]$$

where "::" is sequence concatenation. That is, we define an interpretation giving the entire history of effects for each command history. This is the appropriate function to generalise for non-determinism yielding an interpretation $I^*_{ND}:P \rightarrow \mathbb{P}E^*$. We could then distinguish the single random constant system with interpretation:

$$\forall\, p \in P \quad I^*_{ND}(\,p\,) \;=\; \{\, zeroes,\, ones \,\}$$

where *zeroes* is a sequence of zeroes of length $length(\,p\,)+1$ and *ones* is a sequence of ones of the same length. From the multiple independent-choice system:

$$\forall\, p \in P \quad I^*_{ND}(\,p\,) \;=\; \{\, 0, 1 \,\}^{n+1}$$

where $n \;=\; length(\,p\,)$.

We can see that necessary conditions for a valid interpretation I^*_{ND} are:

effect history is right length for number of inputs:

$$\forall\, e^* \in I^*_{ND}(\,p\,) \quad length(\,e^*\,) = length(\,p\,) + 1$$

history cannot change:

$$\forall\, q \le p \in P,\, e^*_p \in I^*_{ND}(\,p\,) \quad \exists\, e^*_q \in I^*_{ND}(\,q\,) \;\; \text{st} \;\; e^*_q \le e^*_p$$

where $\le$ is the initial subsequence relation. The first condition says that the effect trace must contain one member for each input command plus an initial effect, the second that if some sequence of effects has been the result of a sequence of commands then its first $m+1$ effects must be a possible result of the initial m commands.

From now on we will call a triple $< P, I^*_{ND}, E >$ satisfying these conditions a non-deterministic PIE, abbreviated ND-PIE.

We can say that a ND-PIE is deterministic if for all p there is at most one effect given by $I^*_{ND}(\,p\,)$. That is:

$$\forall\, p \in P \quad \| I^*_{ND}(\,p\,) \| \;\le\; 1$$

If any of the $I^*_{ND}(\,p\,)$ are empty then the resulting PIE has a language.

If we don't want our ND-PIEs to have input languages, then we have to put more restrictions on I^*_{ND}. It is insufficient in the general case simply to ask for I^*_{ND} always to be non-empty. Consider I^*_{ND} where:

$$I^*_{ND}(\text{"}ab\text{"}) = \{\text{"000"}, \text{"111"}\}$$
$$I^*_{ND}(\text{"}abc\text{"}) = \{\text{"0000"}\}$$

If we had typed "ab" and got the series of responses "111", then there is no valid response if we typed a further "c". That is, the ND-PIE accepts a non-deterministic input language. The proper additional rule to prevent this is to ensure that for any input p there are possible extensions to this, no matter what additional input we type:

$$\forall\, p \leq q \in P,\, e^*_p \in I^*_{ND}(p) \quad \exists\, e^*_q \in I^*_{ND}(q) \ \textbf{ st } \ e^*_p \leq e^*_q$$

where $\leq$ is again the initial subsequence relation.

Note the way this is a dual to the "history cannot change" condition. Whether we want such a condition is debatable: it depends on the level of system description we are using and on whether there are any fundamental constraints to user input (like typing at non-existent windows, or trying to use a bank-teller when its cover is down). In fact, each of the ND-PIEs we are going to derive will satisfy this property.

6.2.4 Use for temporal systems

We can now use the ND-PIE to represent the non-deterministic functionality we required in §6.2.1. That is:

$$I^*_{ND}(p) = \{\, I^*_\tau(h) \mid \xi(h) = p \,\}$$

where ξ is the function extracting the user commands from a sequence containing ticks:

$$\begin{aligned} \xi(null) &= null \\ \xi(h\tau) &= \xi(h) \\ \xi(hc) &= \xi(h)\,c \qquad c \neq \tau \end{aligned}$$

With this definition the example of the typewriter which misses characters typed too quickly gives us:

$$\begin{aligned} I^*_{ND}(abc) = \{ & [\,null, a, a, a\,], [\,null, a, a, ac\,], \\ & [\,null, a, ab, ab\,], [\,null, a, ab, abc\,] \} \end{aligned}$$

Not only can we now define this functionality, but it gives us a new way to look at the system. In Chapter 5, perfect buffering was defined. Looking at the definition of I^*_{ND} we see that perfect buffering is precisely the requirement that I^*_{ND} is deterministic.

6.2.5 Use for windowed systems

We consider using a window (with handle λ) whilst ignoring possible interleaved commands to a second window (λ'). This yields a projection from the handle space onto an ND-PIE:

$$I^*_{ND}(null) = \bigcup_{q \in P} \{ [doit^*(e, q, \lambda')] \}$$

$$I^*_{ND}(pc) = \bigcup_{e^* \in I^*_{ND}(p),\, q \in P} \{ e^*::[doit^*(doit(e^*, c, \lambda), q, \lambda')] \}$$

where $doit^*$ is the natural extension of $doit$ to all members of P. Again we can use this to give an alternative definition for a user interface property. Result independence between λ and λ' is precisely the condition that $result \circ I^*_{ND}$ is deterministic.

6.2.6 Non-deterministic properties of PIEs

We have seen that properties over temporal models and handle spaces can be given statements as determinacy properties of ND-PIEs. Are there any interesting ND-PIEs that can be abstracted from simple PIEs?

In Chapter 2 we considered the simple predictability property that a PIE is monotone if:

$$I_p(\, null \,) = I_q(\, null \,) \quad \Rightarrow \quad \forall s \;\; I_p(\, s \,) = I_q(\, s \,)$$

where I_p was the interpretation function "starting" with a command history p. We examined this in the context of the "gone away for a cup of tea" problem, where one has forgotten exactly what command sequence had been entered before the cup of tea. This suggests non-determinism about the value of p, and we can consider the ND-PIE generated by this:

$$I^*_{ND}(\, s \,) \;\hat{=}\; \{\, I^*_p(\, s \,) \,\}_{p \in P}$$

If this ND-PIE is deterministic then the PIE is rather uninteresting, as its functionality is independent of the commands entered. It is a deaf system! The predictability condition can be stated using this ND-PIE as:

$$\begin{array}{l} \forall\, s \in P,\, e_1^*,\, e_2^* \in I^*_{ND}(\, s \,) \\ \quad first(\, e_1^* \,) = first(\, e_2^* \,) \;\; \Rightarrow \;\; e_1^* = e_2^* \end{array}$$

This is a measure we could apply to any ND-PIE whether or not it is derived in this way. It is quite a strong requirement, saying that although the system is non-deterministic, one glance at the first effect resolves all future doubt. We could, of course, go on and give non-deterministic equivalents to the more refined concepts of predictability, for instance defining strategies over ND-PIEs.

One other ND-PIE suggested by the definition of I^*_{ND} above is if we substituted arbitrary PIEs for the collection I_p. For example, given two interpretations I and I' over the same domains P and E, we could define the non-deterministic interpretation:

$$I^*_{ND}(p) \;\hat{=}\; \{ I(p), I'(p) \}$$

If this ND-PIE is deterministic, then I and I' are identical. Later, in §6.4.4, we will see a real situation that could be described using this.

A specific case is where the interpretations are PIEs representing the "same" system with different start data. If we consider the red-PIE, we will often have the case where there is a "bundle" (tray!) of PIEs, each indexed by an element of the result, $\{ I_r \}_{r \in R}$. Each PIE starts with the appropriate result, and is related to the others by:

$$\begin{array}{l} result(I_r(null)) \;=\; r \\ \forall r, r' \in R \quad \exists p_r^{r'} \in P \\ \quad \textbf{st} \;\; \forall p \in P \quad I_{r'}(p) = I_r(p_r^{r'}\, p) \end{array}$$

The ND-PIE generated from these interpretations:

$$I^*_{ND}(s) \;\hat{=}\; \{ I_r^*(s) \}_{r \in R}$$

represents non-determinism about the starting value of the system. Again we will see later how this arises informally.

6.2.7 Summary – formal models and non-determinism

We have seen how a non-deterministic model has been useful in unifying the description of various properties in diverse models. The various examples share one common feature: in each case, the deterministic model is viewed via an abstraction which corresponds to losing some part of the available information. This leads to non-deterministic behaviour. In each case the non-deterministic model used to capture this behaviour was the ND-PIE; however, this is largely because the various models were derivatives and extensions of the basic PIE model. Different flavours of model could be dealt with similarly using different non-deterministic models and perhaps using different methods of expressing the non-determinism.

When viewed in the light of the previous discussion, properties such as predictability and non-interference of windows become efforts to control non-determinism. Either they assert that in certain circumstances the effect is deterministic, or give procedures to resolve it.

6.3 Non-deterministic computer systems?

In the last section, we found that formal models of non-determinism are useful for describing certain abstract properties of interactive systems; however, we were left wondering whether this formal construction held any meaning for the user. We accept that the internal workings of some systems will be non-deterministic, especially where concurrent processes are used, and even that some specialist applications like simulations and certain numerical methods will involve random number generation, but surely most real systems have a deterministic external interface?

6.3.1 The tension for the user

Do users perceive computers as deterministic? In fact, the opposite is the case: most users expect a degree of randomness from the systems they use. Time and again they will shrug their shoulders in bewilderment, "Oh well, it didn't work this time, I'll try the same thing again, it may work now." The apparently random behaviour of such systems conflicts with the alternative model of the computer as a deterministic machine relentlessly pursuing its logical course. Some users are able to cope with this. Expert users may treat it as a challenge, puzzling over the behaviour and experimenting until a logical reason is found. Pragmatists will accept the occasional strangeness and circumvent it, while the awestruck user will regard it as part of the magic and mystery of modern technology. Others may have a more negative reaction. The self-confident may react "this is silly" and lose all confidence in the computer. The self-depreciating may respond "I'm silly", attributing their problems to their own lack of understanding, and possibly retiring from further use. Phrased in these graphic terms the problem seems extreme and demands investigation, but how can it be that thoroughly deterministic programs give rise to this apparent randomness?

6.3.2 Levels of non-determinism

Very few systems are really random: even random number generators are usually based on deterministic algorithms, ERNIE (a random number generator based on quantum effects used in a national lottery in the UK) being a possible exception. Programs that rely on external events could be classified as non-deterministic, for instance the time when a printer signals that it has emptied its buffer, but even then the printer itself will probably behave deterministically. In fact, what is usually termed non-determinism reflects the things that the programmer either doesn't know or doesn't want to know. In other words, it is programmer centred.

We can classify non-determinism according to the level at which it appears:

- *Mechanistic world* – The atomic events measured by ERNIE and the actions of people could certainly be regarded as non-deterministic, and are incapable of prediction even when considering the whole of the computer system. Even here, whether these events are *really* non-deterministic can be argued; however, we are getting into the realms of metaphysics.
- *Computer* – Other events, like printer signals, are deterministic when we take the entire mechanism of computer and printer together, but are not so from the point of view of the computer alone – unless it has a very sophisticated model of the printer, in which case it needn't bother with handshaking at all!
- *Programmer* – The scheduling of programs within a multi-programming system is deterministic from the computer's point of view since it is applying some scheduling algorithm (round-robin, priority-based, etc.). From the programmer's point of view, however, this is non-deterministic and real time may jump suddenly between adjacent program steps. Similarly, file systems may change apparently non-deterministically as other programs operate.
- *User* – Even totally deterministic calculations such as "is $579217^{11}-1$ prime?" are non-deterministic from the user's point of view (Thimbleby 1986a), and are effectively the same as if the computer had tossed a coin to find the answer. Many systems do not use such obviously obscure formulae but manage to produce interfaces that are equally bizarre.

The theme that comes out of the above is that non-determinism is relative: relative both to knowledge and to reasoning abilities.

6.3.3 Behavioural non-determinism

What should a user-centred view of non-determinism be? Imagine two systems which each have two possible prompts, and each day they choose a different prompt for the day. One system bases its choice on whether the number of days since 1900 is prime or not, the other on the decay of a slightly radioactive substance. From the programmer's point of view (and the computer's), we would say that the former is deterministic and the latter not. From the user's point of view the two systems display equally non-deterministic behaviour.

The user sits at the bottom of the hierarchy of knowledge: all the forms of non-determinism are equally random, and should be treated equivalently. That is, we are going to take a *behavioural* view of non-determinism. This recognises that some systems may be "really" deterministic and others may be "really" non-deterministic according to some definition, but if they appear to behave the same, we will regard them as equally non-deterministic. Not only does this mean that we regard some "really" deterministic systems as non-deterministic, but also *vice versa*. For instance, if we had a music system with some random "noise" that led to errors of 1 part per million in the frequency of notes, we would regard the system as being deterministic, as the difference in behaviour would be undetectable.

We could demand tighter views of non-determinism, but for the purposes of this chapter we will adopt the behavioural one. Again, one could use a weaker word to represent behavioural non-determinism; however, the use of such a charged word concentrates the mind wonderfully.

6.4 Sources of non-determinism

In the last section we decided that, taking a behavioural view of non-determinism, it did make sense to describe interface behaviour in these terms. In doing so, we introduced some examples to argue the point. In this section we will catalogue informally some of the sources of non-determinism in the user interface, giving more examples on the way. We will study these sources of non-determinism under six headings:

- Timing
- Sharing
- Data uncertainty
- Procedural uncertainty
- Memory limitations
- Conceptual capture

The first two of these cover the informal equivalents of the two formal problems that started this chapter. The second two consider problems that appear even in the steady-state functionality of single windowed systems, and cover lack of knowledge about *what* you are dealing with and *how* you should do it. The last two are to do with the more complex ways that human limitations can give rise to apparent non-determinism. These two could be thought of as "the user's fault", but the system designer cannot wriggle out of it so easily; good system design should take into account the limitations of its users.

I am sure this list is not complete; however, it gives a broad spectrum of different types of non-determinism to which the reader can add.

6.4.1 Timing

Several of the problems noted in Chapter 5 can be viewed as manifestations of non-determinism. When users type quickly, intermittent and partial update strategies produce different outputs than if they had typed more slowly. If the user is unaware of this exact timing, then the system's behaviour is apparently non-deterministic. This non-determinism is usually deemed acceptable since the final display does not depend on the exact timing, and, further, spells of fast typing may be regarded as single actions anyway. Intermittent update could be said to be less non-deterministic than partial update, since at least all its intermediate displays would have arisen with slow typing. On the other hand, a portion of the screen in partial update is always exactly as it would be if the machine were "infinitely fast", and this portion is therefore totally deterministic.

Similarly, the problems associated with slow machines and buffering can be thought of in terms of non-determinism. A machine that doesn't buffer and loses characters typed too quickly could be regarded as non-deterministic on this score. This is exactly the non-determinism captured by the formal model at the beginning of this chapter. However, a system with buffering can lead to a non-deterministic feedback loop between user and computer, leading for instance to cursor tracking.

Scheduling of multiple processes leads to non-determinism. If for some reason a system makes explicit or implicit use of real time, then, when running on a time-share computer, the run times of its components, and hence possibly its functionality, will be non-deterministic. More often than not this dependency on real time will be in the assumptions made about the relative speeds of certain components, or in the assumption that two statements following one another will be executed immediately, one after the other. The most innocuous case of this is when several concurrent processes print messages to the terminal in random order.

6.4.2 Sharing

Again, the problem of sharing can be regarded as one of non-determinism, reflecting exactly the formal treatment. For instance, as I transact with one process which is my focus of interest, other processes may print error messages on the screen. Because I am preoccupied with the focal process, I may have forgotten that the others were running and thus be temporarily confused. In this case I would not remain confused for long, as I would remember what other processes were running (or ask the system), and infer their behaviour. Thus the non-determinism could be resolved, but not soon enough to stop me acting (potentially disastrously) on a changing system.

This situation is in the middle of the three levels of sharing discussed in Chapter 4. Each of these types of sharing can lead to non-determinism:

- *Single actor – multiple persona* – The user is simultaneously involved (perhaps via windows) with several dialogues, which may share data. When switching from one dialogue to another, he may forget that actions in one dialogue may have repercussions in the other, thus causing apparent randomness. Literally the right hand may not know what the left is doing.
- *Single controller – several actors* – The user sets off several concurrent processes that affect data common to each other and to the user's current view. This is the original case given above, and differs from the single-actor case in that changes may actually occur as the user is actively involved in a dialogue, rather than during interruptions to the dialogue. Because of this it is apparently more non-deterministic, since the system is changing as the user tries to manipulate it.
- *Several independent actors* – This is the case of the multi-user system where not only are several things happening at once, but they are not under a single user's control. This is the most random of all, since the machine processes are at least in principle predictable, but from each user's point of view the other users are fundamentally non-deterministic processes. On the other hand, this is the case where working sets are least likely to overlap, and where protection mechanisms are most likely to exist.

6.4.3 Data uncertainty

A user has a program on a system which was written a long time ago, and the exact contents of it are now forgotten. He wants to make changes to this program and invokes his favourite editor by entering the command "edit prog". This is an example of *data uncertainty*; the user does not know the exact value of the data on which he is going to operate, and a major goal of the interactive session is to reduce the uncertainty, perhaps by scrolling through the file to examine it. This corresponds to the ND-PIE generated from the bundle of PIEs in §6.2.6.

Data uncertainty is, of course, common. The example given is typical of uses over many different types of task. We would not usually class it as a problem: we do not expect to know all the data in a computer system by heart. The importance lies in the user's ability to discover the information: that is, in the *resolution* of the non-determinism, a point to which we shall return in the next section.

6.4.4 Procedural uncertainty

Consider the following two situations:

- A user is within a mail utility and has just read an item of mail which is only 10 lines long and is still completely visible on the screen. She wants her reply to include a few selected lines from the message and does this by using the mail's "e" command in order to edit the text of the message. There are several editors on the system. "I wonder which it will use", she thinks.
- The user who is editing the program (from the example in the previous subsection) decides that the variable names "a,b,c" are not very evocative of their meaning and decides to change them to day-total, week-total and year-total, respectively. He then realises he has forgotten the method of achieving a global search/replace.

In the first situation, the exact nature of the data is known and it is *procedural uncertainty* from which the user suffers. She will look for clues using the knowledge she has of editors on the system. The editor fills the entire screen so it can't be a line editor like "ed", she surmises. Neither can it be mouse-based, like "spy", for it doesn't have menus all over the place. It could be either of the screen editors "ded" or "vi". Tentatively she types in a few characters to see if they're inserted in the text. The editor beeps at her a few times, then deletes half the text... now she knows it's definitely "vi". This form of procedural uncertainty is captured by the ND-PIE which chooses between a set of interpretation functions. It is interesting that the two situations which appear quite different informally, have such a similar informal definition.

The second case is another example of procedural uncertainty of a less extreme and more common form!

6.4.5 Memory limitations

As stated previously, one of the causes of non-determinism is lack of knowledge. This is exacerbated by the fact that people forget. Thus as the user attempts to amass information in order to resolve the non-determinism, her efforts are hampered by her limited memory. A designer must consciously produce features which take account of this. These could include general features, such as an online memo-pad and diary, or more system-specific ones. Further, the user is likely to interpolate the gaps and be unaware of which information is known and which is inferred. This process is distinct from forgetting; we can think of it as *degradation of information*. The designer may need be aware of where such degradation is likely and actually force this information to the user's attention.

6.4.6 Conceptual capture

We are all familiar with the idea of capture, where for example one walks home without noticing when one intended to go to the railway station in the opposite direction. This occurs in computer systems where commonly used sequences of keys can take over when one intends to use another similar sequence. A similar process can also occur at the conceptual level. For example, in a display editor where long lines are displayed over several screen lines, the CURSOR UP key might mean move up one screen line or one logical line. As most logical lines will fit on one screen line the difference may not be noticed. Later when a long line is encountered the action of the system may appear random to a user who has inferred the wrong principle.

6.4.7 Discussion

Of the six headings which we've considered, the first four can be be given formal expression, to some degree or other, using the models developed in this and previous chapters. The last two would require a more sophisticated model of human cognition, with its attendant problems of robustness. There is an obvious parallel between data uncertainty and degradation of information, and between procedural uncertainty and conceptual capture. However, the second of each pair is far more dangerous. The major problem with both these situations when compared to procedural and data uncertainty is that the user may not be aware of the gaps in her knowledge. The situation where a user doesn't know something and *knows* she doesn't know it, is still non-deterministic but the situation where the user thinks she knows what the system is going to do and then it does something completely different, is downright random. We could say that failure in knowledge is far less critical than failure in meta-knowledge.

6.5 Dealing with non-determinism

We have seen that non-determinism is a real problem in the user interface, and that it has many causes. How can we deal with the problems of non-determinism? We will consider four options in this chapter; we can:

- Avoid it.
- Resolve it.
- Control it.
- Use it.

We will consider each of these options in turn.

6.5.1 Avoid it

The most obvious way of dealing with non-determinism is to make sure it never arises. This can be done by designing a system with this in mind, or by adding functionality afterwards. We have already seen examples of this:

- *Timing* – We have said that timing leads to non-determinism. However, the information suggested in Chapter 5 to tell the user when the system is in steady-state and when commands will be ignored, will reduce or remove this non-determinism.
- *Sharing* – Using the definitions of independence we could demand that systems have sharing properties that avoid non-determinism. Alternatively, we could use one of the responses suggested in Chapter 4 to make the sharing apparent, and hence reduce the non-determinism when interference occurs.

In both these solutions, we add information to the interface in order to avoid non-determinism. This is, of course, a general technique restricted only by the display capacity. For instance, if procedural uncertainty is the fault of hidden modes, then we can make the modes visible via a status line.

We can attempt to avoid conceptual capture either by ensuring the models we use are exact matches of the user's model or by making discrepancies very clear. This is, of course, very difficult advice to follow as it is difficult to predict what model the user will infer for the system. However, systems that propose a model (such as the desktop) but fail if the user follows it too far are obvious candidates for improvement. Another situation to be avoided is where two sub-subsystems have apparently similar semantic behaviour, but later diverge. If several possible models have similar behaviour during normal activities we should consider adding features to distinguish the particular model used.

In most systems of any complexity it would be unreasonable to expect complete removal of non-determinism; however, these techniques can reduce the non-determinism or remove some aspect of it.

6.5.2 Resolve it

Assuming the user is in a situation of non-determinism he can try to resolve that non-determinism, attempting by observation or experiment to reach a deterministic situation. Data uncertainty is an obvious candidate for this, and the concept of *strategies* introduced in Chapter 2 can be thought of as an attempt to resolve the non-determinism. When applied to *result predictability* it is precisely the resolution of data uncertainty. The use of the strategy for full predictability can be seen as the resolution of data uncertainty about both the result and internals such as cut/paste buffers. In addition it resolves issues such as mode ambiguity, which is a form of procedural uncertainty.

The designer can help the user in the resolution of data ambiguity by reducing the conceptual and memory costs of strategies. This is aided by the fact that the user will only want to discover some part of the information. Typical techniques include:

- *Improved navigation aids* – By easing the location of information the designer reduces the effort required of the user and makes it more likely that she will be able to remember sufficient information for the task at hand. Further, the job of refreshing memory can be significantly reduced. Mechanisms for this include, depending on the application, search commands in text editors, dependency trees in programming environments, cross-referencing and indices.
- *Place holders* – Because of degradation the user will need continual refreshing of information. Uncommitted navigation aids can be augmented by the ability to lay marks at important positions to refer back to, comparable to bookmarks.
- *Multiple windows* – In a similar vein, a system may allow several simultaneous views, thus making dispersed information simultaneously available. Owing to the limited size of displays all the views will not be present at one moment, and thus we may regard windows as a form of sophisticated place holder.
- *Folding mechanisms* – If the data are structured in a relevant manner, folding mechanisms can significantly aid navigation. Further, if the user has sufficient control over what is and is not folded then unwanted information can be folded away and only the relevant information made visible. Effective folding can thus satisfy some of the requirements for the place holders. It should be noted, however, that in order to achieve these aims, either the user will require control of the structure of the folding, or the fixed structure must be very well chosen (arguably no fixed scheme would satisfy all needs).

Procedural uncertainty is more difficult to resolve. Help systems can be thought of as a way of resolving procedural uncertainty; however, they do of course have to make major assumptions about how much the user knows already. The one sort of procedural uncertainty that the help system definitely cannot resolve, is the method for invoking help. This underlines the need to use consistent and obvious rules for this (permanent icon, dedicated labelled key or the use of "h" or "?").

6.5.3 Control it

Rather than try to remove non-determinism completely, we can try to control it so that the non-determinism experienced is acceptable in some way. This can obtain at the local or the global level. That is, we can ask for complete determinism over a part of the system, or merely that some rules always apply for the system as a whole. We consider the local level first.

It is said that you ought to be able to use a system with your eyes shut. Extending this analogy a bit, we could observe that blind people are able to navigate a familiar room quickly and confidently, whereas they would use a totally different strategy for navigating a busy street. Similarly, when using a computer conferencing facility one might be able to touch-type because the layout of the keys is fixed, and attention is fixed on the screen where unexpected changes will occur. Thus in computer systems, as in real life, we need a solid base of determinacy in order to be able to concentrate on those areas where non-determinacy will occur. We call this requirement *deterministic ground*. Record- and file-locking facilities are an example of how we might for a period enforce determinacy on a shared domain. Similarly, protection mechanisms offer a more permanent way of ensuring some level of determinacy. However, if we look at the three levels of sharing, we see that only the multi-actor case is helped by having private files. So if I have several windows, or have some background jobs running, they are all assigned to the same user, and hence the file system protection would fail to protect me from myself. As an example of this breakdown of security, consider the semantics of the line printer spooler. In some systems, the print command does not make a copy of the file to be printed in a system spool area, but merely makes a note of the request and the file to be printed. The user, however, after issuing the command may reach closure on that operation and then go on to modify or even delete the file to be printed. Perhaps protection mechanisms could be extended to include files private to tasks and windows?

At the global level the effects of sharing are controlled by the accepted procedures that are used for updating them. Some of these are enshrined in the software, and some in organisational and social conventions. For instance, today my bank's teller machine might read £300; however, tomorrow it may read only £20. If I hadn't kept a close tally on my spending, I might not have expected the change and I would regard it as essentially non-deterministic; however, I would have enough faith in the banking system to believe the change to be due to some of my cheques clearing. If this belief were not widely held the non-determinacy of bank balances would become unacceptable and the banking system would collapse. Similarly, early in a day a travel agent may notice 5 seats free on a particular flight, but later in the day, on trying to book one for a customer, the request might be refused. The conclusion drawn is that someone else has booked the seats in the meantime. Thus we rely on the semantics of others' transactions to make the apparent randomness of our view of data acceptable. If the changes we observe in the data are not consistent with our understood semantics we will lose confidence in the data.

I would argue therefore that in order to understand the problems of sharing from a user-oriented view, we should concentrate on defining the semantically acceptable non-determinism of a system. That is, we are prepared to accept *limited non-determinism.* We can apply this to other fields, for instance, if we look at buffering strategies. A perfectly buffered system may well be non-deterministic because the screen does not reflect the current state of affairs; however, we might regard this as acceptable. In contrast, a word processor which ignored all typeahead would be regarded as exhibiting unacceptable non-determinism. The predicate describing perfect buffering is a limit to the non-determinism sufficient to make it acceptable. Similarly, we may not be too worried about the exact strategy a text editor uses when rearranging the display when the cursor hits the screen boundaries. It would be unacceptable if the cursor movement caused part of the document to be altered (I have used a text editor which did exactly this!). Again, the limitation that the cursor movement does not alter the document is sufficient (with others in this case) to make the non-determinism acceptable.

6.5.4 Use it

We have just argued that to a large extent, controlled non-determinism can be acceptable non-determinism. Every user manual uses this fact, as it defines only the external behaviour of the system. So, for instance, different versions of a word processor could use different algorithms, be written in different programming languages and even run on different hardware (in the same box!). Similarly, formal specifications define only the interface behaviour, leaving the internals undetermined. Thus all specification is a use of non-determinism. It is usually assumed that the systems described by formal specifications will be deterministic; it is just which particular consistent system is chosen that is non-deterministic. One could certainly have non-deterministic systems which satisfy a given loosely defined specification, but this is not usually done. There is a paradox here: in order to make accurate statements about a system (and hence reduce non-determinism about its behaviour), specifications are used which are themselves non-deterministic.

There is also a strange duality of non-determinism within the interface. The computer system must, if it is to be useful, be non-deterministic. A completely deterministic system would never tell the user anything that the user didn't know already. Not very useful! On the other hand, from the computer's point of view, the user is non-deterministic, but if the user were not so, the system would always produce the same result. To some extent there is conflict between the two partners' search for determinism, and the programmer usually has the upper hand, forcing the user to answer in specified ways and in specified orders. From the programmer's point of view this could be thought of as producing a more deterministic response from the user. From the user's point of view this is at best restrictive, and possibly may seem non-deterministic because the programmer's arbitrary decisions may have little relevance at the interface. Thimbleby (1980) calls this excessive control by the programmer *over-determination*, and elsewhere I have proposed a technique for returning control of the dialogue sequence back to the user (Dix 1987c). Not surprisingly, this leads to programs with more non-deterministic semantics and which regard the user as a non-deterministic entity.

We can distinguish two aspects of interface non-determinism based on the previous discussion. First are those that are part of the application, and are what makes the application useful. Second are those in the interface, resulting from arbitrary decisions and complexity, which are not wanted. On the other hand, we could use this difference to make the distinction between application and interface, as desired for instance by Cockton (1986), by saying that the application is what we want to be non-deterministic and the interface is what we want to be deterministic.

6.5.5 Summary – informal analysis

We have seen how the user helped by the designer can avoid, resolve and control the non-determinism of interactive systems. We have also seen that non-determinism can be used in formal specification. Earlier we asked whether non-determinism existed at the user interface or whether it was just a formal trick, and decided that it is a real, meaningful phenomenon. We could parallel that now and ask: is the use of non-determinism just a formal trick for interface specification or can it really be used to improve user interfaces? The next section will seek to answer that question.

6.6 Deliberate non-determinism

So far we have dealt with cases where non-determinism has unintentionally arisen in the interface, and we have been interested mainly in removing it and its problems. In the last section, we saw that when developing systems, non-determinism can actually be used to good effect. In this section, we will consider a display update strategy that is deliberately non-deterministic. We will call this strategy *non-deterministic intermittent update*. It has considerable advantages over deterministic strategies from the point of view of efficiency; however, can such a deliberate policy of non-determinism ever be acceptable?

We will begin by describing a basic semiformal framework in which we can consider update strategies. Next we will consider the expression in this framework of total and intermittent update strategies (as described in Chapter 5). We then extrapolate these strategies and define non-deterministic intermittent update. Finally, we consider whether or not it is an acceptable strategy from the user's point of view, and conclude that by deliberately adding non-determinism to the interface we may actually *reduce* the perceived non-determinism.

6.6.1 Static and dynamic consistency

Many interactive systems can be considered in the following manner: there is an object (obj) which is being manipulated, and an associated screen display ($disp$). The way such a system is implemented is often as a state $<obj, map>$ consisting of the object and the additional information (map) required to calculate the current display. For example, if the object were a text then *map* may be the offset of the display frame in the text. As the user issues a sequence of commands (which may be keypresses, menu selections or whole line commands, depending on the system), the state changes in a well-defined and deterministic manner. Thus as a sequence of commands is issued we get a sequence of objects O_i mapping states m_i and displays d_i:

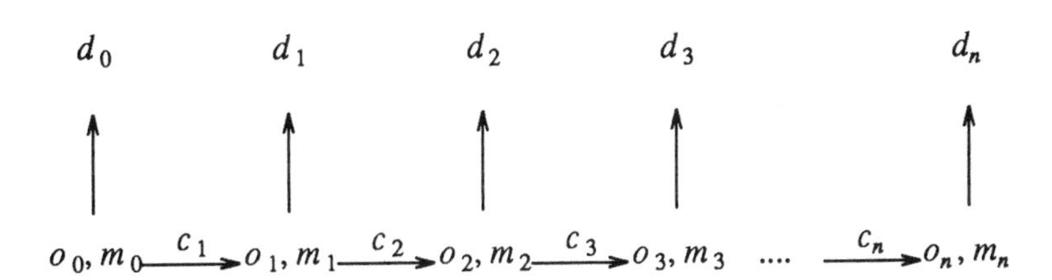

The earlier discussion on specification would lead us to question: what are the *requirements* for this update sequence? Typically they fall into two categories:

- *Static consistency* – At any stage we expect there to be a well-defined relation between the object and the display. For instance, in the case of text with a cursor position and a simple character map display with its cursor, we would expect the characters to match if we overlaid the two aligning the cursors.
- *Dynamic consistency* – Properties should hold between successive displays. Typical of this is the principle of display inertia, which says that successive displays should differ as little as possible.

The designer will (more or less) have an idea of which static and dynamic requirements are important for a particular system; however, these will rarely specify the map completely and thus additional *ad hoc* requirements will be added to define it uniquely. (This is idealised, as often in practice these fundamental design decisions are made as the system is coded, with little reference to global impact.) The additional requirements may themselves be either static or dynamic.

Bernard Sufrin's specification of a display editor (Sufrin 1982) deliberately leaves the specific update strategy only partly defined; instead, he supplies a static consistency requirement and a dynamic "inertia" requirement that if the new cursor fits on the old display frame then the frame doesn't change. Many editors follow this rule of thumb and add additional rules to specify fully the case when the cursor does not fit on the old screen, such as "scroll just enough to fit in the cursor", "scroll a third of a screen in the relevant direction" or "centre the cursor in the new display". All these rules have additional special cases at the top or bottom of the text, and in the case of the first two, when the movement of the cursor is gross.

6.6.2 Intermittent update

As already noted, the time taken to compute the new map and update the display may lead to apparent non-determinism for the user, such as cursor tracking. Clearly we want to avoid this non-determinism.

If, after all other optimisations have been tried, the response is still unacceptable, a possible course is to use intermittent update as described in the last chapter. This corresponds to surpressing some of the displays in the sequence:

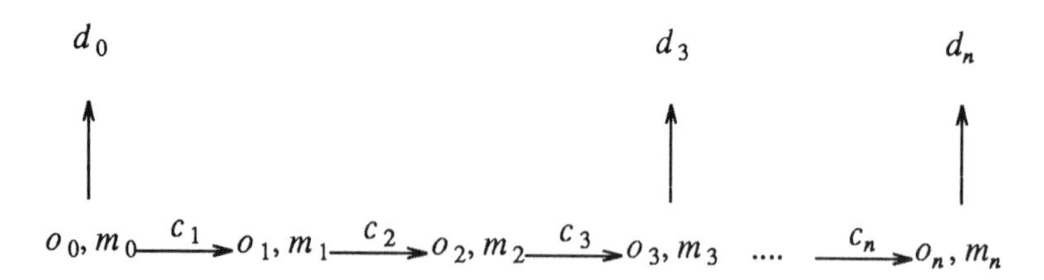

This is mildly non-deterministic, in that the sequence of displays is different depending on exactly how fast the user types. So we already we have a (common) example of deliberate use of non-determinism. However, although this reduces traffic to and from the display device (critical on older low-bandwidth channels), it still leaves the considerable expense of updating the map component. Is there any way we can reduce this overhead?

6.6.3 Declarative interfaces

Some years ago, when I was specifying an experimental program editor a slightly different approach to the standard display inertia was taken. As usual, only basic static requirements were specified, and then a few additional requirements were tried that could loosely be described as "cursor inertia". The first of these was strict cursor inertia, where the cursor position on the screen moved as little as possible: this is a dynamic requirement. This had some very strange consequences: in particular, as the cursor began at the top of the document and hence at the top of the screen, it *always* tried to stay at the top of the screen! A second variation on this was always to position the cursor as near the middle of the screen as possible. In terms of observed behaviour this requirement has its own advantages and disadvantages; however, the important thing about it is that it is a purely static requirement. This means that the display map can always be calculated from the current text alone: it is a purely *declarative interface*.

In principle, one would expect a declarative interface to be very predictable. However, few interfaces adopt this style. Harold Thimbleby's novel calculator (Thimbleby 1986b) does so, but users are often surprised at its features, perhaps because they are unused to the declarative style. The big advantage in performance terms of a declarative interface is that when one only updates the display intermittently, one need only update the *map* intermittently also:

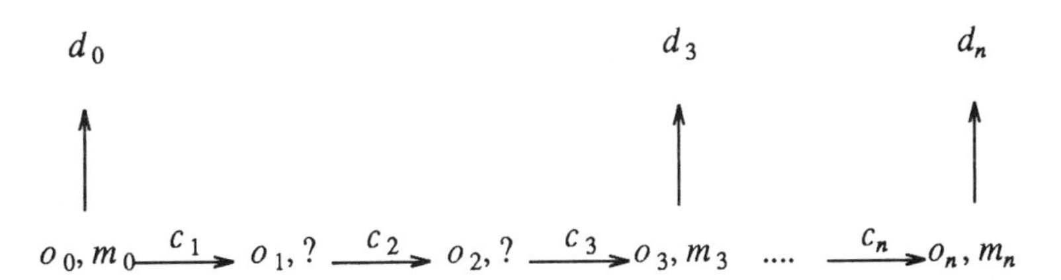

This technique ignores dynamic requirements entirely, so at first glance it appears to be relevant only to totally declarative interfaces. However, given the benefits in terms of performance, we must ask ourselves whether some similar technique could be developed for non-declarative interfaces?

6.6.4 Non-deterministic intermittent update

It is usually the case (and is so in the text editor example) that static consistency is most important semantically. We can imagine then relaxing the demand for dynamic consistency. Normally, it applies between each state transition. However, if we have intermittent update, we could apply it only between displays that actually occur. That is, we effectively bunch a series of commands updating the object and then re-establish the display map based on static consistency, and modify the dynamic requirements to apply to these consistent epochs.

Put in formal terms, previously we would have demanded for each command c_i an object–map pair where each o_i is derived from the previous object via the command, where each $<o_i,m_i>$ is consistent with respect to the static requirements, and where the successive pairs $\{<o_{i-1},m_{i-1}>,<o_i,m_i>\}$ satisfy the dynamic requirements. Now we ask only for an object for each command, a set of epochs e_j and maps at these epochs only, such that at each epoch $<o_{e_j},m_{e_j}>$ is consistent and each successive epoch pair $\{<o_{e_{j-1}},m_{e_{j-1}}>,<o_{e_j},m_{e_j}>\}$ satisfies the dynamic requirements.

6.6.5 Is it a good idea?

What sort of effect would this have in practise? Imagine the text editor with display inertia. The cursor is on the second to last line of the screen, and we slowly enter DOWN, DOWN, UP; at the second DOWN the screen would scroll to accommodate the new cursor position, then on the UP it would stay (by display inertia) in this new position. If, however, we entered the commands quickly and they were bunched for processing, then after the three commands when the static and dynamic invariants are enforced, the new cursor position is within the original display frame and this is retained – the result is non-deterministic! Comparing this to intermittent update, we see that in that case only the intermediate states of the screen differed, whereas in this case the final state differs depending on the exact timing of input.

I would argue that the non-determinism introduced is acceptable for two reasons. Firstly, because it is non-deterministic only within the bounds of the static consistency, it is *limited non-determinism*. Secondly, the exact operations of the dynamic requirements are usually unpredictable anyway, and thus the apparent non-determinism will be no worse, and we may actually reduce it. This is especially true of the principle of display inertia. The reason why it is introduced is to ensure that the location of useful information changes as little as possible in order to aid visual navigation. If intermediate displays are not produced then the deterministic strategy leads to a changing display, whereas the non-deterministic strategy leads to a fixed one; clearly, the latter application of the principle is more in line with its intention, and for the user is probably *more* predictable than its application to imaginary intermediate displays.

6.7 Discussion

We have seen how formal non-deterministic models are useful in describing interactive system behaviour. We asked ourselves whether this had any real meaning. In §6.3 we saw that the appropriate user-centred definition of non-determinism was a behavioural one, and under this definition interfaces did display non-deterministic properties. Further, we have seen many examples of how non-determinism arises in practice. In §6.5 we found that users have many ways of dealing with non-determinism, and that the designer can design systems to aid them in this. Finally, we have seen that it can in fact be beneficial to introduce non-determinism deliberately into the user interface, both to achieve other goals (in the example, efficiency) but also potentially to reduce the total non-determinism of the interface.

What have we gained by this analysis? Firstly, by using the strong word non-determinism rather than weaker ones like unpredictability, we see rather more the urgency of the problems considered. Secondly, we have found that many different well recognised problems can be considered as manifestations of non-determinism; this allows the possibility of cross-fertilisation between the domains. Thirdly, by considering problems in this light, it enables us to see more clearly ways of expressing them, for instance, regarding the problem of sharing as that of specifying acceptable levels of non-determinism. Thus we might describe a mail system, not in terms of multiple users and messages, but rather as a single user with a system displaying limited non-determinism. Finally, the analysis has given us a more pragmatic view of non-determinism and hence the ability to consider the idea of deliberately non-deterministic interfaces.

CHAPTER 7

Opening up the box

7.1 Introduction

When we have considered PIEs and the related models, we have been eager to give behavioural definitions of all properties of interest. We have deliberately made no assumptions about the way a system is constructed, instead concentrating fully on what can be observed from the outside. This is the essence of the black-box view of the interface. A system is described purely in terms of the keystrokes that go into it and the pixels on the screen and ink on the paper that come out. It is a safe route to take, yet it is somewhat extreme, and perhaps over-zealous. Users will not only infer information from the system (captured already in the idea of observable effect), but will also have some idea of the structure of the system. In particular, they may deal with the system at a conceptual level deeper than that of keystrokes. For example, when using many operating systems, the user's notion of command is the entire line, rather than keystrokes. We are reminded of the various interface layers suggested by other authors. Foley and van Dam (1982) refer to three major layers: lexical, syntactic and semantic. The user will be working at or close to the semantic layer: hence it is reasonable to model the deeper layers of the system. This is equally valid for output as well as input, and we should consider the user as using commands on the underlying objects of interest rather than keystrokes on the display. Referring back to the black box, we are saying that it is acceptable, when we are moving towards a valid user model, to "open up" the black box.

Another reason for opening up the black box is in order to refine the model in order to work towards an implementable specification. In many ways the two aims follow each other, as it seems a good idea to follow the intended user model in the detailed specification, even if later on this needs to be transformed somewhat to provide an efficient implementation (see Chapter 11).

As we've said, the system will be layered with the "actual objects" being manipulated at the bottom, accessed via one or more levels of interface. Examples of such layering are:

Input

- Menu selection. Although this involves several user actions (choose menu, open menu, make selection), it is perceived as one "real" action.
- Control-key sequences. For instance, in Wordstar "^KB" is not perceived as two actions but as one, "mark block beginning".
- Command interfaces. As we've mentioned already, entering a command may involve a complex sequence of typing and line editing operations; for a while, the command line itself appears to be the object of interest, but at a different level entering a command is treated as a single atomic action.

Presentation

- Framing. When framing a portion of a text to fit on a display screen, the user is aware that the screen is not the "real" object, but just a view into it.
- Pretty printing and structure editors. At least the idea is that users should see beyond the textual form to the underlying structured object.
- Folding mechanisms. Again the user should see through these to the "real" object.

As users' experience with a system grows, they may understand more of its structure. Let's imagine a simple word processor. Very naive users may even be worried that text scrolled off top or bottom of the screen is lost. Most users will be happy with the idea of a screen window, but will still have a rudimentary idea of what sort of editing and formatting is possible. A little more experience and they will realise that the form that appears on the screen is not necessarily identical to what is printed, and they will become happy with the use of different fonts and complex embedded codes for detailed control of the printed form. Still later they may be aware that both interactive and printed forms are derived from a stream of characters, some printed, some invisible; they will use newlines just like any other character in search/replace operations, perhaps know that centring is caused by an invisible control sequence, and be able to uncentre text by deleting these.

This amount of layering may be thought of as poor design, and many modern systems deliberately try to avoid this sort of scenario, making the system model as close to the interface model as possible. (Hutchins *et al.* 1986) Of course, we can never entirely get rid of layering – at least the finite interface will form a layer – and where systems retain generality and power there is almost always a more complex expert's model. The main improvement is not so much the removal of such layers. It is clear that many systems fail in their layering because the underlying model is implementation driven and the surrounding interface serves unsuccessfully to hide this. A careful, consistent and methodological design of such interface layers based on an intended user model is what is really required.

The rather abstract discussion of relations between PIEs at the end of Chapter 2 forms a basis for considering such layered systems. In the present chapter we take instead the red-PIE as our starting point. We take the result map as being synonymous with the objects of interest, and consider two layered models where the display is a facet of the outer presentation layer, and the result of the inner layer. We use the various abstract relations to model this, but eventually become stuck. We obtain adequate declarative requirements for the relation between the object layer and the display layer, but we find it hard to describe what is in the display layer. We also fail to give more constructive methods that will be of use when moving towards a particular interface design.

This is followed by a section elaborating the model to include separate but linked states for the underlying object manipulation and the display mapping: this uses a *doit* function (state transition) at the two levels, with the outer (display) *doit* defined in terms of the inner (object) *doit* and additional functions. This therefore yields a method of constructing interfaces for existing conceptual-layer object editors.

We then consider the drawbacks and limitations of the line of modelling taken here. In particular, we see that we need models that encompass *display-mediated* interaction. This serves as an introduction to Chapters 8 and 9.

Finally, we look at the use of *oracles* as a way of describing more complicated systems whilst still retaining a linear interface architecture.

7.2 Modelling editors using PIEs

7.2.1 Basic relation between layers

We now attempt to produce simple two-layer models of interactive systems. The intention is to open up the red-PIE model. The inner level describes the functionality of the objects of interest, taken to be the result. It consists of a monotone PIE $< P_{object}, I_{object}, E_{object} >$ together with a result map $result : E_{object} \to R$. Remember that the display will only be a facet of the outer layer. The outer layer consists of the red-PIE as available to the user, $< P, I, E >$, with display map $display : E \to D$ and result map $result' : E \to R$. For any sequence of commands at the outer level, we can identify the sequence of virtual commands to which these correspond at the inner level. This can be represented by a map $parse$. If we want to retain the functionality of the inner PIE, we need to have access to all possible command sequences in P_{object} and hence *parse* must be surjective. Similarly, we can abstract from the total state (E) the state pertaining to the underlying object (E_{object}) using an abstraction mapping *proj*. The result will be derivable completely from E_{object}. This, of course, leads to an *abstraction* relation, as defined in the previous section, between the two PIEs. If the construction is to make any sense at all, the result of the whole system must be the result obtained from the underlying objects. That is:

$$result' \;=\; result \circ proj$$

We can think of this either as a requirement of the system, or more constructively as a definition of *result'*. Either way, from now on we will ignore the result map from E and assume the result is obtained using the underlying result function. We thus have the following picture:

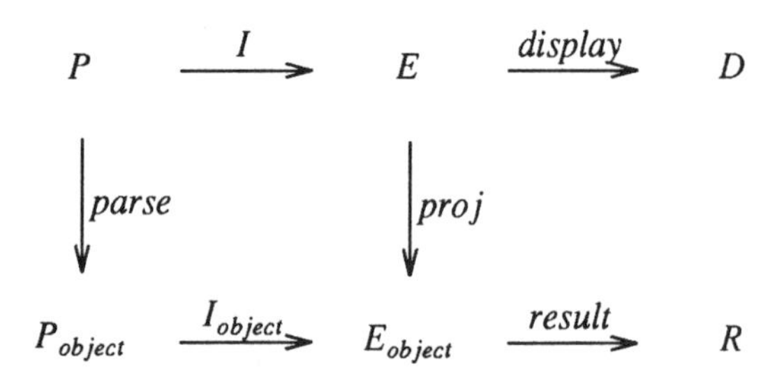

An example of this is the Unix calculator "bc". It is implemented using a simpler stack-based calculator "dc". It works by issuing commands to dc, and for any sequence of commands issued to bc there is a sequence that has been parsed and sent to dc. We can see dc therefore as an abstraction of bc. However, this is not a very good example as few users of bc would see themselves as using dc, and it is more a feature of the implementation than of the interface design. We shall see better examples later. The above diagram is very important: all the rest of the models in this chapter and much of the succeeding two chapters will be just different ways of expanding on it.

Why does it need further expansion? It already describes adequately the relation between the inner and outer functionality. We can therefore demand that our various statements hold at whichever level is appropriate. For instance, we would expect the inner PIE to be reachable, but we may not be so worried about the outer one. Note that unless the *parse* function is trivial E_{object} is different from, but related to, $R^{\dagger}$, and thus this is different from the statements we made about red-PIEs. In particular, in a mouse-based system, the commands at the inner level generated by mouse clicks will be dependent on the present display frame: thus $R^{\dagger}$ would contain this information, but E_{object} need not. Further, we see that it is more likely that we will be able to have strategies strong passive with respect to E_{object} than $R^{\dagger}$, and this then becomes a possible desirable feature of an editor.

What this picture does not tell us much about, though, is the *extra* state needed at the outer level. This is bundled up with the object state in E and we have no separate access to it as we do for E_{object}. Any abstraction sufficient to allow the factoring of the display map will (if the system is at all observable via the display) contain all, or nearly all, of the complete system state. If we are interested only in specifying that a system satisfies certain basic functionality this is fine. If, on the other hand, we are interested in principles relating to how the display map behaves, or alternatively in building the outer PIE, perhaps in a generic way, from the inner one, we will need a model telling us more about the display.

Before we go on to consider these more elaborate constructions, we will consider some of the problems of interpretation that arise when we start to deal with principles stated at multiple levels.

7.2.2 Choice of level, the designer's freedom

While we were taking the commands to be at the outermost, physical level there was not much problem of interpretation. Given any system we could say whether it satisfied a given principle in an automatic fashion. If the principles were to form part of a usability contract then the client could be sure of the product, and the designer would have little leeway in interpreting the contract. As soon as we move on to principles at deeper levels of abstraction, life becomes far more complex.

It may be that we start out with the inner PIE and the designer's job is to envelope it in an interface. It is easy then to draw up the above diagram. However, even then the designer may complain that the inner PIE has too much "interface" in it already. This is connected to the general problem of whether the interface can be or should be "separable" from the underlying application, a premise central to the development of UIMS and discussed for instance by Cockton (1986).

If there is no such fixed starting point, then life becomes far more complex. For instance, there are glaring loopholes that the designer could unwittingly fall into, or deliberately exploit. For instance, the *parse* and *proj* functions could both be trivial, leading to the original red-PIE. Alternatively, the *parse* function may be badly chosen, not picking out a sensible syntax. For instance, in a command-based system the sensible *parse* function would pick out complete lines punctuated with "newline" characters, and then pass these on. However, we could just as easily have defined a *parse* function that punctuated using the character "x". The underlying applications interpretation function I_{object} would have to be fundamentally different. Happily, these scenarios lead to far more complex inner applications and are thus likely to be spotted. Similar, but more subtle, problems are not so easy to deal with. For instance, if we are designing a text editor, should the inner application know about a cursor and have cursor movement commands, or should it be based around context-free commands such as "insert 'a' at the 3rd column of the 10th line". Again, if we are dealing with natural language, should we be parsing at the level of words or sentences?

We have to trust to a large extent the designer's discretion in choosing the appropriate abstractions, and applying the appropriate principles to them. By asking the designer to specify which abstractions are regarded as important and what principles are to be applied, we cannot force a good design, but we are thus approaching a formal methodology which can guide the production of quality systems.

We also should expect the system to have several different layers of abstraction, recalling again the multiple layers advocated in the interface design literature. So for instance, the editor described in Chapter 11 has three such layers of abstraction: the top, full functionality layer; the text layer, where we consider unbounded formatted texts and which also includes the cursor; and a deepest string layer, where the objects are unbounded, unstructured strings of characters, and the commands are context-free (that is, no cursor). The "result" of this deepest level (obtained with a map *string*) is related to the result of the text level (the map *text*) by a pretty print function, a relation which is preserved by the abstraction:

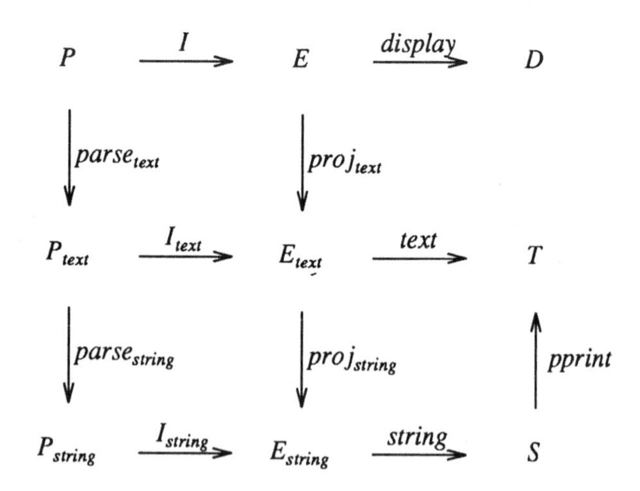

The condition on *string* and *text* is:

$$pprint \circ string \circ proj_{string} \quad = \quad text$$

The notion of pointer spaces introduced in the next chapter will make it particularly easy to design states satisfying this condition. However, that is jumping ahead a little.

7.2.3 PIPEs

We now return to the basic two-layer abstraction in the hope of describing something more of the upper layer:

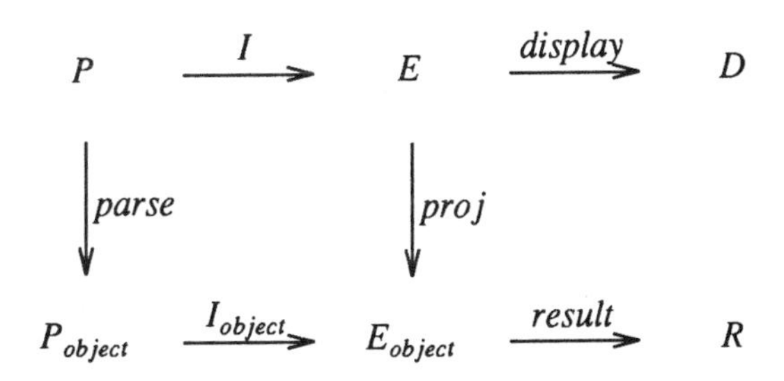

One reason for looking first at relations between PIEs at a formal level, rather than introducing them as needed, was to build discrimination between the various, quite similar diagrams and constructions we will encounter. In particular, almost invariably, when looking at the above diagram for the first time, it is read as a construction for $< P , I , E >$, or in terms of PIE relations, as an *implementation*. That is, the direction of the *proj* map is effectively reversed. The distinction is very important: the abstraction relation given above is very general, and can be applied to almost any system, whereas the implementation relation is far less general. However, as it certainly appears to be useful, we will see how far we can get with it.

Reversing the projection effectively means that the information in E_{object} is sufficient to define the display. Thus rather than proliferating function names, we can assume that the display can be obtained directly from E_{object} via the map *display*. The input layering (*parse*) and the output layering (*display*) are thus distinct, and the whole structure is a lot more linear:

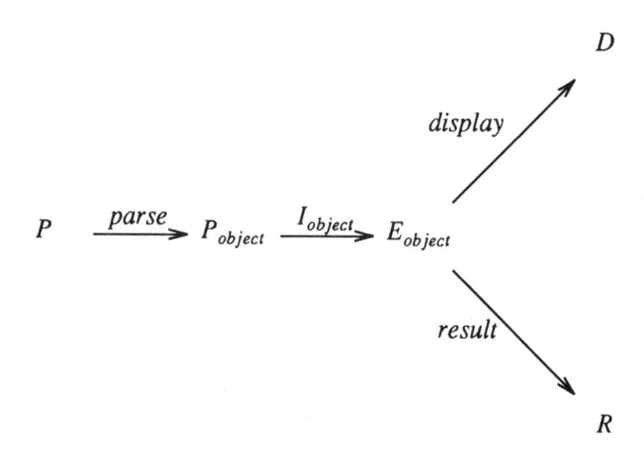

Certain kinds of input layering would just about be described by this structure, for instance, if the relation between keystrokes and object commands were one of simple macro expansion, each keystroke giving rise to one or more commands. It also describes some slightly more complex forms of input layering. For instance, in the editor "vi", keys can be prefixed by a number. This is interpreted as repeating the command that many times. Thus "5j" gets translated into the sequence of commands "DOWN DOWN DOWN DOWN DOWN". In fact, we could have a lot more complex parsing if we wished. However, as in the case of vi, this is at the expense of predictability. In vi, when we have typed the "5" there is no visual indication at all. The only way to discover that the "5" has been entered is to type something else and see what happens. The system is unpredictable. If we choose to implement any system as a PIPE with a parse function of any complexity this is bound to be the result. Unless the output of *parse* contains sufficient information to generate a display for it, there is no way this information can become visible to the user.

Clearly the parse function requires a "short cut" to the display if it is to be both powerful and predictable. We shall try to extend our model to cover this now.

7.2.4 Observable parsing

We need some sort of additional display information from the parsing to be combined with the display from the object in order to give the entire display. If we call the entire display domain D and the individual displays D_{parse} and D_{object}, then D_{object} will be obtained by a simple PIPE, and D_{parse} will have to be obtained using a separate PIE $< P, I_{parse}, E_{parse} >$ with an associated display function, $display_{parse}$:

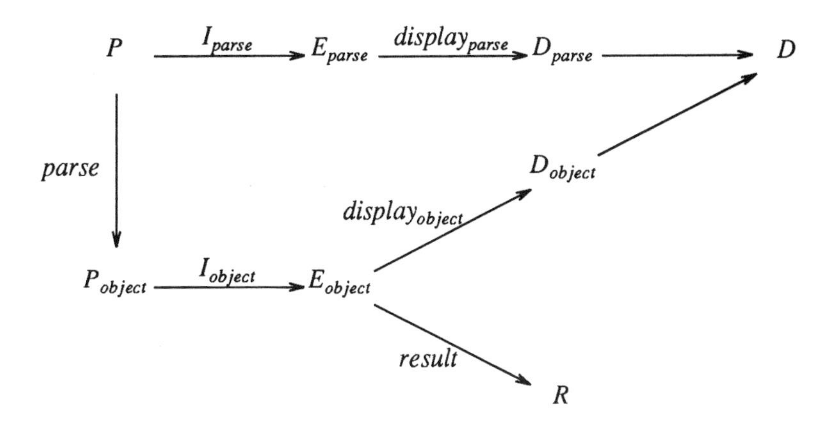

For this to be predictable, we need to be able to observe the state associated with the *parse* function through $display_{parse}$. We can define this state by quotient on P using an equivalence relation:

$$p \equiv_{parse} p' \;\hat{=}$$
$$\forall\, q \in P \quad \exists\, q' \in P'$$
$$\textbf{st} \quad parse(p;q) = parse(p);q'$$
$$\textbf{and}\; parse(p';q) = parse(p');q'$$

So two command histories are equivalent by $\equiv_{parse}$ if equal extensions to them yield equal extensions to their parsed value. We can then phrase predictability in terms of the generated PIE with interpretation $I_{\equiv_{parse}}$. The parsing is predictable if the red-PIE, with display interpretation $display_{parse} \circ I_{parse}$ and result interpretation $I_{\equiv_{parse}}$, is result observable (§3.3). Since this interpretation is monotone the red-PIE is also result predictable.

We also want to preserve the reachability of the inner PIE. This is clearly achieved if given any command history p and its parse p', then given any desired extension to p', we can find an extension to p yielding it. That is:

$$\forall\, p \in P, p' = parse(p), r' \in P_{object}$$
$$\exists\, r \in P \quad \textbf{st} \quad parse(p;q) = p';r'$$

or, in terms of the above red-PIE, $I_{\equiv_{parse}}$ is strong reachable. This reachability condition applies equally to the simple PIPE.

Consider the example of a command line parser. We already have some application defined by PIE_{object}. Its command set C_{object} consists of complete command lines (strings of characters with no newlines in it). We want to define a parse function, and an associated PIE_{parse} to interface this PIE, yet retain predictability. We define *parse*, I_{parse} and $display_{parse}$ thus:

$$
\begin{aligned}
parse(\,null\,) &= null \\
parse(\,p::c\,) &= parse(\,p\,) \quad c \neq newline \\
parse(\,p::newline\,) &= parse(\,p\,) :: I_{parse}(\,p\,)
\end{aligned}
$$

$$
\begin{aligned}
I_{parse}(\,null\,) &= null \\
I_{parse}(\,p::c\,) &= I_{parse}(\,p\,) :: c \quad c \neq newline \\
I_{parse}(\,p::newline\,) &= null
\end{aligned}
$$

$$display_{parse} = identity$$

The full display could then be made by adding the result of $display_{parse}$ to the bottom of the result of $display_{object}$. It is fairly clear that such a construction preserves any predictability and observability properties of the original system. Of course, it is a bit simplified, having ignored command line editing, finite line buffers, etc., but such a full example would follow along exactly the same lines and conserve properties similarly.

In this example the state and display for the parser are synonymous; in other words, the PIE between P and D_{parse} is monotone. Although we'd clearly like it to be always tameable, the monotonicity is not necessary. For instance, the parser could handle a keystroke macro facility, the macro definitions being browsed and edited using the parser PIE. Going further, the typical LOGO interface could be thought of as just such a facility built over a basic functionality of turtle graphics. That is, the commands in C_{object} would be the turtle graphics commands (FORWARD, TURN, PEN DOWN, etc.) and D_{object} would be the actual graphics produced together with the turtle. The result could be the same as D_{object} or perhaps just the graphics without the turtle. All of the LOGO programming language, including all the editing of LOGO procedures, would be part of the *parse* function. The example is a little extreme, but is certainly possible.

In addition to using the "short cut" of I_{parse} for making *parse* predictable we might hope that it could exercise control over the display of D_{object}, for instance screen scrolling. This is certainly the case with the red-PIE window manager, which can be thought of as supplying a parse function with associated display for each of its windows. The combination function is rather more complex than in the previous examples, as it sometimes hides parts of the object's display.

In general, though, we can supply little in the way of display control through this "short cut". This is because display control demands a lot of knowledge about the underlying application, and any display control through the short cut would need to mimic a large part of the functionality of the application. For instance, if D_{object} was taken to be an entire formatted version of the object of interest, we might hope that information in D_{parse} could control the part of D_{object}

displayed in D. However, I_{parse} would not know where the cursor was and I_{object} would not know where the display frame was, and thus the system could not maintain the visibility conditions for local commands described in §3.4.

It would clearly be useful to describe the display and object components in a way that preserves some sort of *independence* between them. In the next section we consider a scheme that allows just such independence.

7.3 Separating the display component

In this section we consider a simple model where the application and display parts of the state are separated. This has been given a section of its own because the model is fundamentally different from those presented before. In all of these we could represent the relationships in terms of function diagrams. In the following we will have to move to a state-transition-oriented approach to capture the effect of the object on the possible display states. This is a more detailed approach to the semiformal model used to describe non-deterministic intermittent update in the previous chapter.

Consider a simple text editor. The display state may be affected both directly, by user-level commands directed at it (e.g. scroll up window), or indirectly because of changes in the object state, (e.g. changing the display frame as the cursor moves out of view). Further, even the direct commands may have their effects modified by the object state (does the scroll hide the cursor?)

7.3.1 Basic model

We capture this behaviour by having two parts to the effect, E_{object} and $E_{display}$, the total effect being the product of these two. (We assume that any parsing has taken place at a different layer.) These have associated state-transition functions, except that the display's state-transition function would depend on the object state as well as the old display state (to account for the limitations on direct display commands):

$$doit_{object} : C \times E_{object} \rightarrow E_{object}$$
$$doit_{display} : C \times E_{display} \times E_{object} \rightarrow E_{display}$$

Ideally we would like these to be totally independent, but the display updates will need some information about the object. Later we will see how this can be minimised.

Further, we require an adjustment function *adjust* that restores any invariants, such as always displaying the current cursor line, broken after object commands:

$$adjust : E_{display} \times E_{object} \rightarrow E_{display}$$

Finally, we have the usual *display* and *result* mappings:

$$\begin{aligned} display &: E_{object} \times E_{display} \rightarrow D \\ result &: E_{object} \rightarrow R \end{aligned}$$

Doit is defined by applying the two state-transition functions, then restoring the invariants using the adjustment function:

$$doit : C \times E_{object} \times E_{display} \rightarrow E_{object} \times E_{display}$$

$$doit(c, e_o, e_d) = e_o', e_d'$$

where

$$\begin{aligned} e_o' &= doit_{object}(c, e_o) \\ e_d' &= adjust(doit_{display}(c, e_d, e_o), e_o') \end{aligned}$$

Typically we would expect commands to affect either the display only or the object only, that is:

$$\begin{aligned} &\forall c \in C \\ &\quad \textbf{either} \ \forall e_o \in E_{object} : \ doit_{object}(c, e_o) = e_o \\ &\quad \textbf{or} \quad \forall e_d \in E_{display} : \ doit_{display}(c, e_d) = e_d \end{aligned}$$

However, in the former case the display state may (and often will) be changed *indirectly* by the adjustment function.

7.3.2 Static and dynamic invariants

Sufrin's Z specification of an editor (Sufrin 1982) fits this model except there are no direct display commands: the display is always changed by side effect and this means there is no $doit_{display}$ function. Further, it does not define the adjustment function explicitly, just the invariants it must supply. It specifies a *static invariance* predicate:

$$invariant_{static} : \ E_{display} \times E_{object} \rightarrow Bool$$

and demands that any editor that follows the specification must have an adjustment function that satisfies:

$$adjust(e_d, e_o) = e_d' \ \Rightarrow \ invariant_{static}(e_d', e_o)$$

In Chapter 6 we discussed the concept of static and dynamic display invariants in fairly loose terms in order to discuss non-deterministic intermittent update. The above is, of course, part of the formal definition of a static display invariant over this sort of display model. We might want to demand additionally that the display state-transition function (if one is needed) also satisfies an appropriate condition:

$$invariant_{static}(e_d, e_o) \ \textbf{and} \ doit_{display}(c, e_d) = e_d' \\ \Rightarrow \ invariant_{static}(e_d', e_o)$$

That is, $doit_{display}$ preserves the invariant.

Alternatively, we could leave *adjust* to handle this. However, if we regarded a direct display command that would violate a static invariant as an exception, we would want it to satisfy an exception principle like "no guesses"; this would be hard to ensure, so it would seem sensible to embed the knowledge in $doit_{display}$ also. Further, we could decide not to apply the adjustment function in the case of direct display commands (assuming they can be identified). Clearly, there is a bit of a trade-off here as the static invariant is spread around.

The Sufrin editor also defines a condition equivalent to display inertia. This is a *dynamic invariant.* We can similarly define a dynamic invariant as a predicate:

$$invariant_{dynamic}: \ E_{display} \times E_{object} \times E_{display} \rightarrow bool$$

such that the adjustment function must obey:

$$adjust(e_d, e_o) = e_d' \ \Rightarrow \ invariant_{dynamic}(e_d, e_o, e_d')$$

The invariants would tend to precede the definition of the display state-transition function and the adjustment function in the design of an interface.

7.3.3 An example

We will now consider a simple example. We assume that $doit_{object}$ has already been defined, and further that E_{object} consists of two parts, a text consisting of a sequence of fixed-length lines, and a cursor position as line and column number, which is always within the boundaries of the text. We don't care what the object commands are or how $doit_{object}$ behaves. We define the display state to consist of a line number at which the display will start. That is:

$$E_{object} = Line^* \times nat \times nat \\ E_{display} = nat$$

In the object state, the first number is taken to be the line number of the cursor and the second the column.

The display will consist of a fixed number (25 say!) of lines and a cursor position similar to the text:

$$D \;=\; Line^{25} \times nat \times nat$$

The display function is designed to obey the obvious fidelity condition:

$$display(\,e_o, e_d\,) \;=\; screen, x, y$$

where

$$\begin{aligned} screen &= text[\,offset, \ldots, offset{+}24\,] \\ x &= line - offset + 1 \\ y &= col \\ text, line, col &= e_o \\ offset &= e_d \end{aligned}$$

However, if the definition of *screen* is to make sense, and if the cursor is to remain within the display bounds, then we must have the relevant static invariant:

$$\begin{aligned} &invariant_{static}(\,e_d, e_o\,) \;\hat{=} \\ &\qquad line - offset + 1 \;\in\; \{1, \ldots, 25\} \;\; \mathbf{and} \;\; offset > 0 \end{aligned}$$

In addition, we might want to demand display inertia, that is:

$$\begin{aligned} &invariant_{dynamic}(\,e_d, e_o, e_d'\,) \;\hat{=} \\ &\qquad invariant_{static}(\,e_d, e_o\,) \;\Rightarrow\; e_d' = e_d \end{aligned}$$

Many adjustment functions would satisfy this: for instance, we could use:

$$adjust(\,e_d, e_o\,) \;\hat{=}\; e_d'$$

where

$$\begin{aligned} \mathbf{if}\;\; &line - offset + 1 \;\in\; \{1, \ldots, 25\} \\ &\qquad \mathbf{then}\;\; offset' = offset \\ &\qquad \mathbf{else}\;\; offset' = max(\,1, line{-}13\,) \end{aligned}$$

which keeps the offset fixed while the cursor is on screen, then scrolls to centre the cursor when it is not.

Similarly, we can extend the command set by the commands CENTRE, SCROLL_DOWN and SCROLL_UP, that centre the screen about the cursor, moves the frame down one line and up one line respectively.

$$doit_{display}(\,\text{CENTRE}, e_d, e_o\,) \;\hat{=}\; e_d'$$

where

$$\begin{aligned} \mathbf{if}\;\; line > 12 \;\; &\mathbf{then}\;\; offset' = offset \\ &\mathbf{else}\;\; offset' = line{-}12 \end{aligned}$$

$$doit_{display}(\text{SCROLL\_DOWN}, e_d, e_o) \;\hat{=}\; e_d'$$

where

$$\textbf{if}\;\; offset > 1 \;\;\textbf{and}\;\; line-(offset-1)+1 \in \{1, \dots, 25\}$$
$$\textbf{then}\;\; offset' = offset - 1$$
$$\textbf{else}\;\; offset' = offset$$

$$doit_{display}(\text{SCROLL\_UP}, e_d, e_o) \;\hat{=}\; e_d'$$

where

$$\textbf{if}\;\; line-(offset-1)+1 \in \{1, \dots, 25\}$$
$$\textbf{then}\;\; offset' = offset + 1$$
$$\textbf{else}\;\; offset' = offset$$

all of which follow the "no guesses" exception principle. If we had left restoration of invariants to the adjustment function, then we would have had SCROLL_DOWN with the functionality "scroll the display frame down one line, unless the cursor is at the top whence scroll it up 13 lines". Not very acceptable (or predictable) exception behaviour!

If we want to use the "no guesses" principle uniformly for all direct display commands, then we could leave static invariant checking out of the individual command cases, leading to a subsidiary function $doit_{normal}$, and define $doit_{display}$ by:

$$doit_{display}(c, e_d, e_o) \;\hat{=}\; e_d''$$

where

$$\textbf{if}\;\; invariant_{static}(e_d', e_o) \;\;\textbf{then}\;\; e_d'' = e_d'$$
$$\textbf{else}\;\; e_d'' = e_d$$

and

$$e_d' = doit_{normal}(c, e_d, e_o)$$

Similarly, the dynamic invariant was of the form: "If possible without contradicting the static invariant...". We could therefore supply a "normal" dynamic invariant $invariant_{normal}$ and generate the full dynamic invariant from it:

$$invariant_{dynamic}(e_d, e_o, e_d') \;\hat{=}$$
$$\textbf{if}\;\; \exists\, e_d'' \;\textbf{st}\; invariant_{normal}(e_d, e_o, e_d'') \;\textbf{and}\; invariant_{static}(e_d'', e_o)$$
$$\textbf{then}\;\; invariant_{normal}(e_d, e_o, e_d')$$
$$\textbf{else}\;\; TRUE$$

Many dynamic invariants will be able to be handled in this manner; however, some might be more complex. For instance, we might have invariants of the form: "If possible make the normal invariant hold. If not make this weaker one hold. If that's not possible make this even weaker one hold...". Of course, we could define similar generators for this form of stratified invariant.

In conclusion, this sort of interface model architecture is fairly flexible and allows the definition of many generic components to make the job of design easier. Further, it maintains a great degree of independence between the display and object at the cost of letting the display "peep" a little into the object's state occasionally. We have been able to factor most of this peeping into the generic components, making it far more acceptable. It does not cater for all interface styles, however. In particular, it is not very suitable for mouse-based systems. We will discuss these shortcomings further at the end of this chapter.

7.4 Display-mediated interaction

So far all the descriptions given have had a strong unidirectional feel: commands change the state of the system, which is reflected in the display. They do not reflect the nature of an interactive system as a feedback loop, with the user completing the loop. Although the system's response *can* be defined solely in terms of user commands, it is more faithful to the spirit of interaction to refer them to the current display. Further, there are some modes of interaction where it is impossible to separate the display cleanly from the object and retain the one-way approach. For example:

(i) Moving the cursor to the top of the display.

(ii) Moving the cursor with the screen: when a display frame moving command would mean the cursor not being on the display, the cursor is moved rather than the command failing.

(iii) Mouse input, where the meaning of location requires knowledge of display contents.

In all these cases, not only do display transitions need knowledge of the object, but the object transitions need to be aware of the display state too.

If we are to encompass such modes of interaction, our models must change to ones where the user's actions on the underlying object are mediated by information from the display context:

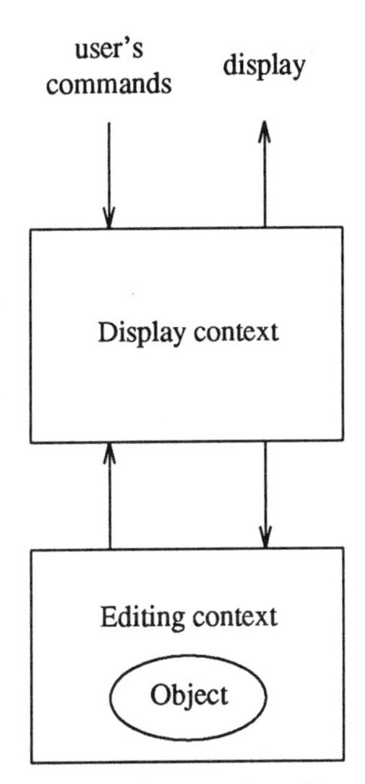

In terms of the model of the previous section, this would make the transitions of E_{object} dependent on $E_{display}$ as well as the other way round, effectively recombining them into a single state. Obviously we need richer models than the simple functional, data-flow-type models we have dealt with so far. These models will inevitably "open up the box" still more, and have architectural implications on the way we specify systems. Before we move on in succeeding chapters to discuss such approaches we look at one more "linear" model.

7.5 Oracles

If display context is important in a system, then a detailed design will have to take account of the feedback nature of the system. However, if we are willing to relax determinism we can use a unidirectional model during the early stages of design, perhaps even during prototyping. We do this using oracles.

Consider first a system that relies on the user clicking over icons on the screen. The icons may appear at different positions, so we cannot simply map the mouse clicks onto the abstract commands denoted by the icons without using knowledge about the mapping between the internal objects and the display. In the next chapter, we will consider controlled ways of describing this mapping, but we may not want the additional complexity at this stage. Instead we add an *input oracle* to the system that the parsing function can ask everytime it wants to know additional information about the system. So, the parsing function receives a double mouse click. It interprets the double click as an OPEN command, but then asks the oracle what lies under the mouse position. The oracle tells it that it is the folder "chapter 7" and so the parse function then translates the mouse click to the abstract command OPEN(*chapter 7*).

A similar situation might arise with a CAD system. The main palette is at a fixed screen position, so the parse function is able to translate mouse clicks into their appropriate abstract commands POINT, CIRCLE, etc. However, when the user clicks over the graphics area it would like to translate this into something like PLOT_AT(*position*). Unfortunately, it does not know about the mapping between screen coordinates and the application coordinate system. Again it asks the oracle.

We can express this architecturally as a modified form of the PIPE architecture:

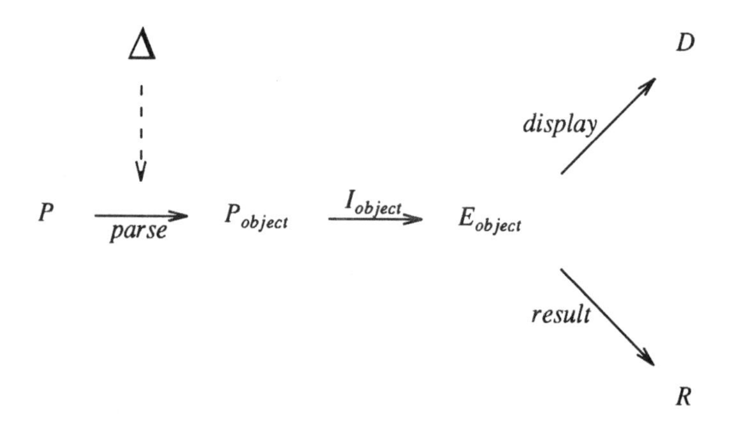

The Δ represents the information in the oracle. But where does the oracle get its information from? One approach is to treat it entirely non-deterministically. The oracle can give any reply. If we want to prove any properties of the system, we can make few assumptions about the oracle. So for instance, if some reply from the oracle would lead to an erroneous state we must assume that this might occur. We must also be pessimistic when assessing reachability or predictability properties. For a typical mouse-based system this would be very restrictive. It is equivalent to playing "pin the tail on the donkey": we are trying to get to a particular state, but our mouse clicks can land anywhere. This can be resolved somewhat if we make some assumptions and restrictions on the oracle. When performing a reachability proof, we note any assumptions that we need to make about the oracle. For instance, in a CAD system we would want to say that the user's mouse clicks can be interpreted as being at any desired application coordinate. When the oracle gets included gradually into the system these assumptions would become assertions to be proved.

If we prefer a deterministic approach, we need to obtain the oracle's value as a function of other parts of the system. Typically it will require information from anything and everything: the complete input history, the current state, display and result, and any internal mappings defined as part of these or the parsing process. If we expressed this functionally it would give us a picture a bit like this:

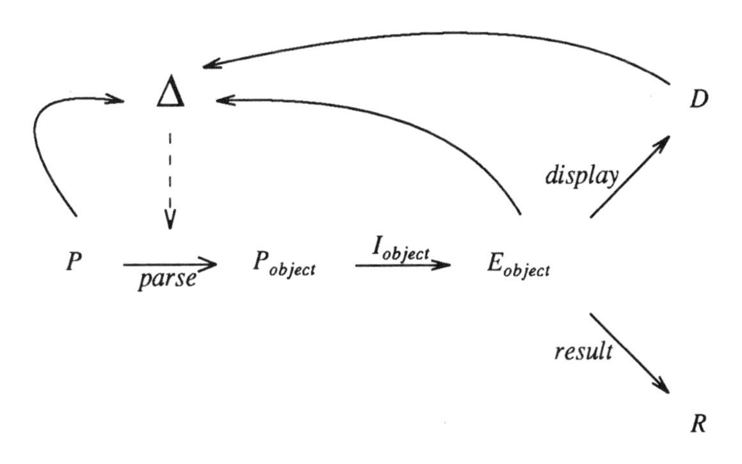

A bit of a mess! However, the whole idea of using an oracle is that the grubby parts of the design are packaged away in one area. Thus for most purposes we would use the former architectural picture, and retain a clean conceptual design during the important early stages. As the design progresses the oracle will be used less and less until eventually it disappears as part of the design. This disappearance will mean that the relevant information is obtained from the other parts of the system, but when we know exactly what information is needed the interfaces of the various parts can be modified to pass it in a clean and well-controlled fashion.

If the parsing function can be expressed in an implementable manner as a function of the user's inputs and the oracle, then the oracle can be used as part of a prototyping process. The oracle's replies could either be simulated by the designer in a "Wizard of Oz" fashion, or we could ask our favourite hacker to produce something quick and dirty to do the job. If the latter path is taken, we can maintain a rigorous formal connection between the specification and the prototype system whilst keeping together all the kludges[†] that have not yet been given a formal treatment.

Oracles can be used in other parts of the system. Assume we are designing a word processor. We have defined an internal text layer with most of the complex functionality, but have not yet decided the exact rules for mapping this onto the screen. We could capture this remaining detail in an oracle, that gives, say, the text coordinate of the top left-hand corner of the display. Again, such an oracle would take information from anywhere it can. It may want to include decisions based on the type of change that has occurred and perhaps explicit user control such as PAGE-UP, PAGE-DOWN keys or an as yet unimplemented scroll bar:

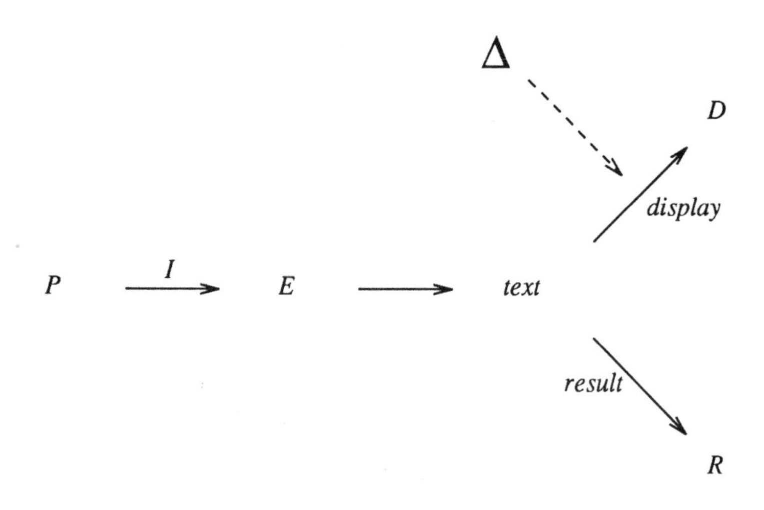

Again, we could imagine using "Wizard of Oz" techniques, or quickly coding up several alternatives during a prototyping exercise. Note how the input oracle was used to cut the feedback loops, whereas this output oracle simplifies the feed-forward.

To some extent the dynamic pointer techniques in the next chapter supply in a clean, formal manner the sort of information delivered by the oracles. They thus make the use of oracles, especially input oracles, less necessary. However, as we said, we may wish to retain the simpler architecture in the early stages of design. The two techniques can work together, however, as dynamic pointer methods

† kludge – a botch or hack; something which works, but inelegantly so.

could be used to implement the oracles, or in generating "default" interfaces for use by the "Wizard" in the simulation of more complex interface styles.

7.6 Going further

We have noted how real interaction (and perhaps we ought to capitalise those two words) is often, and at its best, display-mediated. The currently displayed context is central to interpreting the user's inputs. The next two chapters present different ways of letting the display mediate the interaction. These follow roughly the two ways of viewing a layered system:

- *Editing the object* – Here we want to edit the underlying object and then have the display follow along in a consistent manner. Chapter 8 uses the notion of pointers to express this consistency. To enable the display to mediate interaction there is a well-defined way of translating commands as they pass through the display layers based on *forward* and *back* maps between pointers in the two domains. Further, to ensure fidelity between the basic objects and their display, these are separated out from the rest of the state, which is defined in terms of pointers to them.
- *Editing the display* – Chapter 9 takes the alternative view that it is the display which is dominant, and as it is the *prima facie* interface, the user is considered to be acting on it directly. It uses the notion of views. The display is a view of the underlying objects. The editor generates changes on the objects so that this view mirrors exactly the changes the user makes to the display.

Important though these later chapters are, they should not be seen as invalidating the models in this chapter. The basic abstraction in §7.2.1 is central, and all the models in the succeeding chapters are merely refinements of this. To summarise, the early discussion in this chapter told us that the relation between the functional core and the system as a whole is *not* as a *component*, but as an *abstraction*.

Furthermore, the various other models in this chapter have been shown by example to model significant systems. Even where we want some of the interaction styles that cannot be modelled using them directly, they may still be useful:

- They may be useful to model a *layer* of the full system. For instance, "move the cursor to the top of the screen" can be seen as a shorthand for a sequence of "cursor up" commands. The *parse* function would be complex, but that would not matter if we were interested simply in studying the abstraction.

- They might model a *restriction* of the system. For instance, we might remove the offending commands entirely.
- They allow modelling *non-deterministically* using oracles, which lets us temporarily ignore issues of display context during early specification and prototyping.
- Finally, they give the general *flavour* of the behaviour of more complex models, enabling us to deal better with the complexity.

This last point is important, as we are interested not just in the formal specification of interactive systems, but also in the use of formal methods to foster *understanding*

CHAPTER 8

Dynamic pointers: an abstraction for indicative manipulation

8.1 Introduction

This chapter is about manipulation, by which I mean "doing things to things". This is a rather wide definition, which at its simplest can be thought of as editing, but at its richest encompasses virtually all interaction. Most computer activities involve manipulation at several levels. For instance, you manipulate the characters and paragraphs of a file to make a document, but you also manipulate the entire document when you print it. Another example would be statistical analysis: you manipulate the individual data items to put them in suitable form, and then manipulate the whole data set with the analysis tool.

It is the opposing forefinger which enables us to be manipulative in the physical world, and of course it is our forefinger which allows us to be manipulative in the computer world.

8.1.1 Modes of manipulation

When considering manipulative operations, both in computer systems and in real life, they appear to be of two major forms:

descriptive	–	add £1500 to Alan Dix's salary
indicative	–	delete *this* line

In the former case, the *content* of the desired change is used, and in the second the word "this" would probably be augmented by a pointing finger indicating the *position* of the relevant line. Some operations involve a mixture of these two modes of manipulation, and the imperative part of the operation (the "add" or the "delete") may itself be rendered descriptively (e.g. typing) or indicatively (e.g.

clicking on a button). This leads to a strong hypothesis: *content and position are sufficient for interaction.*

To put it another way: when we want to talk about something it is in terms of *what* it is like (content) or *where* it is (position).

This hypothesis is supported in older systems where position is indicated by line numbers, file names or relative to some "current" position. However, it is even more apparent with modern mouse-based systems where all operations are supposed to be *point* and click.

8.1.2 Mediated interaction

Not only are these systems very indicative, but they have highly display-sensitive input. The meaning of *this* is the thing displayed where the mouse is pointing. We cannot describe such systems in terms of a "pipe-line" model with user's inputs being parsed, then processed and finally outputs being printed. Such a model of interaction is sufficient to describe "what you see is what you get", which relates the different outputs of the system, but it cannot handle these more manipulative systems. In these cases, we expect that any object is manipulable by indicative actions upon its displayed representation, that is: *what you see is what you can grab*. The relevant model for this is one of mediated interaction, where the current display context is central to the interpretation of the user's actions.

8.1.3 Relating levels – translating pointers

These two concepts of mediated interaction and indicative interaction are closely intertwined. In fact, from the example given, the crucial aspect of display sensitivity is knowing *what* the user is indicating. At a cruder level that means translating between the world of screen positions and the world of object pointers. All indicative systems achieve this, and the ways in which they do so are many and varied (and weird but rarely wonderful!). This relationship between levels can be represented by a pair of functions *fwd* and *back* between the pointers to the objects at the levels (*fig.* 8.1). These functions are, of course, connected to the function *proj* which represents the application objects as display objects.

This sort of translation may be repeated several times at different conceptual levels within the system. Typically there seem to be at least three levels. The innermost consists of the application objects themselves. These are then represented in some sort of virtual display, for instance, the idea of a whole formatted document, the entire drawing in a CAD system, or the collection of all windows. Finally, this unbounded virtual display is mapped into the finite physical media.

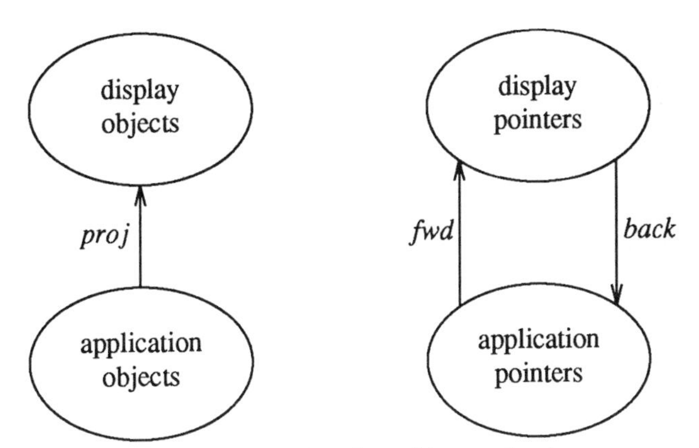

figure 8.1 *Relating levels – object and pointers*

When a user issues an indicative command, the display position can be translated using the *back* maps at each level in turn until the appropriate level for interpreting the command is reached. Similarly, if the system's response is in terms of positional information, this is translated using the *fwd* maps to appear at the screen in a suitable form.

In such a multi-layered organisation, care is needed to ensure that the repeated translations back and forth between these pointer spaces behave sensibly. One expects the *fwd* and *back* maps to be "sort of" inverses to one another. It is surprisingly hard to capture just how "sort of" this is! Further, even having decided how each level should relate back and forth sensibly, it does not follow that the composite translation behaves well. An example of this is *ladder drift*: a phenomenon where the "lexemes" implicitly defined by the pointer relations do not agree at the intermediate level. The result of this is that mapping back and forth through both levels diverges rather than stabilising.

As all indicative systems implicitly have such maps (or even several for different types of pointers), it is clearly important that they be designed, rather than come upon accidentally, in order to avoid the pitfalls described, and to present a systematic user interface.

8.1.4 Pointers in the state

The importance of positional information is clear if we consider the internal organisation of a manipulative system. The components fall into four basic classes of object:

- *Simple types* – For example, mode flags.
- *History* – For undos. This could either be abstracted into a "meta-context" or conversely be thought of as making the domain recursive. In either case it adds no new types into the system.

- *Sub-objects* – Sub-objects of the object type that is being manipulated. For example, copy buffers, find/replace strings, etc. Strictly, these sub-objects may have some context attached to them; however, again this merely makes the domain recursive and does not add more basic types.
- *Positional information* – The cursor position, block pointers, etc.

The confidence that this list is complete stems from the above observation that manipulative actions are either of the content-oriented form "change fred to tom", or of the indicative form "do this here", or a mixture of the two. In particular, the elements that are most distinctively interactive are linked to positional information.

Further, this positional information is the major element in the interpretation of commands in relation to the display. In fact, nearly all such feedback can be regarded as simple conversion between screen position and some positional information in the underlying object.

Pointers are thus seen to be fundamental in understanding manipulative interactive systems, and we will study them in detail. We will study the way pointers change as the objects change, and thus be able to update the editing state which is constructed largely of pointers. We will be able to separate the object from its pointers as required in the previous section. We will study the relations between different objects and their pointers and thus be able to describe the transformations that occur when a command is interpreted in a particular display context, and the fidelity of these transformations.

8.1.5 Dynamic pointers

The pointers that are mentioned are not static: rather, when changes happen to the object the various pointers are changed accordingly, in the sense that they point to the same semantic entities. That is, we are interested in dynamic (semantic) pointers.

We can clarify the concept of dynamic pointers in the context of programming languages, comparing them with normal Pascal or C pointers. These are dynamic in the sense that they can be assigned different values, but are pointers to *structurally static* objects. That is, they may point to different objects and the objects they point to may change value, but the structure of the object is constant.

We want truly dynamic pointers, semantic pointers to *within* a structurally dynamic object, that change their (static) value as the object pointed to changes. A piece of pseudo-program makes this clear:

```
String   A   =  "abcde" ;
Pointer  pt  =  ref(A[4]) ;
<code to insert x at beginning of A>
print ( A, "::", val(pt) ) ;
```

The result would be :

Static pointers: `xabcde::c`

Dynamic pointers: `xabcde::d`

That is, the static pointer still points to the fourth element of `A`, but the dynamic pointer points to the fifth element of `A`: it has changed its value as interpreted statically, but is still pointing to the same semantic entity, "`d`". Examples of such pointers in other contexts include pointers to elements in linked lists (particularly simple, since in most cases the address pointed to stays fixed) and file names considered as pointers into a file system

8.1.6 How are dynamic pointers specified elsewhere?

Suzuki (1982) and Luckham and Suzuki (1979) give formal semantics for normal Pascal-type pointers; however, these are, of course, static pointers. We have to look towards user interfaces to find dynamic pointer specifications.

In Sufrin's editor (Sufrin 1982), the cursor is very neatly handled as an intrinsic part of the object, being the separation point of the text into *before* and *after*. This leads to very clean definitions of operations at the cursor, but is not suitable for generalisation to more than one pointer. In fact, when an additional "mark" pointer is required, this must be added as a special character within the text. This is again very sensible, as on primitive display devices such a mark would have to be displayed in a character cell of its own. This is also the approach taken by Wordstar (Wordstar 1981) when allowing user markers, which are simply embedded control codes. Being an actual character in the text, such markers are guaranteed to change in a way semantically consistent with the underlying text.

The disadvantage of this approach is that the object being edited, and the pointers being used to describe the context of that editing are conflated. This is a problem even in this simple example; however, in more complex domains it becomes insuperable. For instance, the technique would be unmanageable if one allowed several editing windows on the same object, each with its own cursor and marks! In the following discussion it as crucial that the object being edited is separate from its pointers.

Dynamic pointers can be found elsewhere. Any editor with user- or system-defined marks must use them in some sense, as must hypertext systems such as NoteCards (Halasz 1987). They may even be supplied in a programmers' interface: the SunTools graphics package for the Sun workstation has a notion of marks for user-modifiable text regions of the screen that have the properties of dynamic pointers (Sun 1986). However, the techniques used for implementation are *ad hoc* and certainly the importance of dynamic pointers as a datatype does not seem to be recognised.

8.2 Pointer spaces and projections

We now move on to a more formal treatment of dynamic pointers. Instead of launching straight into this, we first produce a model of static pointers and then see how this can be augmented. We then go on to look at the relationships between pointer spaces using *projections*, as is necessary if we want to use pointer spaces for layered designs of interactive systems. The last two subsections describe the way that several projections can be composed, and how this affects the properties of the maps that relate the pointers at the different levels.

8.2.1 Static pointers

In the case of static pointers, clearly we have two sets of interest:

Obj – the set of all objects
Pt – the set of all possible object pointers

For example, in the simple case of character strings and pointers to these we would have:

$$Obj = Char^*$$
$$Pt = \mathbb{N}$$

where the pointer 0 denotes before the first character, and a pointer n denotes the gap between the n th and $n+1$st character.

Of the possible static pointers, only a subset are meaningful for a particular object: for instance, the 5000th character is not very meaningful for "abc". Thus we have a valid pointers map giving the particular subset of Pt meaningful for a particular object:

$$vptrs : Obj \to \mathbb{P}Pt$$

So, in our example, $vptrs(obj)$ would be $\{0,...,length(obj)\}$.

This valid pointers map sounds very similar to the *vhandles* map for handle spaces in Chapter 4 when we considered windowed systems; however, the meaning attached is very different. The handles are merely labels for windows and can thus be renamed at will, whereas the pointers possess a semantic meaning in themselves.

In addition, there will be operations defined on Obj from a set C. Each operation from C will have a signature of the form:

$$c : Params[\, Pt \,] \times Obj \rightarrow Obj$$

$Params[X]$ represents some sort of parameter structure, including (possibly) members of the set X. In the following if f is a function $X \rightarrow Y$ then we will assume there is a canonical extension of f to $Params[X] \rightarrow Params[Y]$, which is obtained by applying f to each component belonging to X.† Similarly, if Z is a subset of X then $Params[Z]$ will represent the subset of $Params[X]$ where each component from X is constrained to lie in Z.

Example

The insert operation on strings may have a parameter structure:

$$Params_{insert}[\, Pt \,] \; = \; Char \times Pt$$

with full signature:

$$insert : Char \times Pt \times String \rightarrow String$$

Clearly, it will only make sense to have parameters containing *valid* pointers for the object being acted on. So we will always have the precondition:

$$params, obj \in dom\ c \quad \Rightarrow \quad params \in Params[\, vptrs(\, obj \,)\,]$$

The reader familiar with type theory will recognise this as the parameter being a dependent type.

For the above example of insertion on strings, we would have for an insertion into the string "*abc*":

$$\begin{aligned} params \; \in \; & Params_{insert}[\, vptrs(\, "abc" \,)\,] \\ & = Params_{insert}[\, \{\, 0,...,length(\, "abc" \,)\,\}\,] \\ & = Char \times \{\, 0,...,length(\, "abc" \,)\,\} \end{aligned}$$

In general, the precondition requiring valid pointers is necessary but not sufficient. An operation may have a stricter precondition. For instance, consider deletion on strings.

$$Params_{delete}[\, Pt \,] \; = \; Pt$$

$$delete : Pt \times String \rightarrow String$$

The precondition obtained from the valid pointers map would be:

† That is, *Params* is a functor on the category of sets.

$$params \in \{ 0,...,length(obj) \}$$

whereas in fact 0 would be an invalid pointer for deletion – there's nothing to delete. A more appropriate condition would be:

$$params \in \{ 1,...,length(obj) \}$$

This model describes the sort of pointer values and operations commonly encountered in programming languages. We will now extend this to allow an element of dynamism for the pointers.

8.2.2 Pointer spaces – static pointers with pull functions

The essential feature of dynamic pointers is that they change with the object being updated. The model of static pointers above is functional, and we cannot change the pointers as such (although later we will consider a model where we can). Instead we supply a function that describes the changes necessary for the pointers to retain their semantic integrity.

This is achieved by augmenting the signature of update operations:

$$c : Params[Pt] \times Obj \rightarrow Obj \times (Pt \rightarrow Pt)$$

The first part of the result is exactly as described for the static case, and we require the same consistency condition on the parameters:

$$params, obj \in dom\ c \Rightarrow params \in Params[vptrs(obj)]$$

The second result from the operations is the *pull function*. This tells us how to change pointers to the original object into pointers to the updated object so that they maintain their semantic meaning. The pull function must always take valid pointers to valid pointers, and this leads to the following condition on all commands c:

$$\begin{aligned} &\forall\ params, obj \in dom\ c \\ &\quad c(params, obj) = obj', pull \quad \Rightarrow \quad pull \in vptrs(obj) \rightarrow vptrs(obj') \end{aligned}$$

Example (revisited)

We can now redefine the insert operation for strings as:

$$insert((c, n), s) = s', pull$$

where

$$\begin{aligned} s' &= s[1,...,n]::c::s[n+1,...,length(s)] \\ pull(pt) &= pt \quad \textbf{if } pt < n \\ &= pt+1 \quad \textbf{if } pt \geq n \end{aligned}$$

Given the assumption argued at the beginning of the chapter that pointers are the principal component of editor state, we can use the pull function as a way of ensuring that this state changes "naturally" with the underlying object. For instance, if we have a set of objects Obj and an editing context $Context[\,Pt\,]$, we can perform the obvious update after a change to the object $obj \in Obj$ due to an operation c:

$$context, obj \;\;\rightarrow\;\; pull(\,context\,), obj'$$

where

$$obj, pull \;=\; c(\,params, obj\,)$$

In particular, we can use $pull$ to provide our complementary function for global commands that we were looking for in Chapter 2. We could, in fact, now give a more generous definition of a global command as any operation where the parameter contains no pointers.

8.2.3 Relationships between pointer spaces – projections

As well as consistency for pointers, we also entered this chapter at an impasse when considering layered editor design. In particular, we wanted to relate the outer display layer to the underlying objects. Further, we wanted the display context to affect the parsing of user inputs, an important instant being mouse commands.

In order to describe such layering we will examine relationships between pointer spaces. In addition, this will enable us to describe one pointer space in terms of another, perhaps a simpler one, or one possessing an efficient implementation.

Any relation will be governed by a structure containing pointers and other data types similar to the parameter for operations, and we will call the set of such structures $Proj\_struct[\,Pt\,]$. The relation, which we will call a *projection*, is a function with three results:

$$proj: Obj \times Proj\_struct[\,Pt\,] \;\rightarrow\; Obj' \times (\,Pt \rightarrow Pt'\,) \times (\,Pt' \rightarrow Pt\,)$$

The first component of the result is the simple relation between objects; the second and third are the *forward* and *back* maps, respectively. These maps relate the pointers of the two objects in a way which is intended to represent the semantic relationship between parts of the objects. Clearly, the back and forward maps are defined only for valid pointers and only valid pointers are allowed in $Proj\_struct[\,Pt\,]$:

$$\begin{array}{l} \forall\ obj, proj\_struct \in dom\ proj \\ \quad proj\_struct \ \in \ Proj\_struct[\, vptrs(\, obj\,)\,] \\ \quad proj(\, obj, proj\_struct\,) = obj', fwd, back \quad \Rightarrow \\ \qquad\quad fwd \ \in \ vptrs(\, obj\,) \rightarrow vptrs(\, obj'\,) \\ \qquad\quad back \in \ vptrs(\, obj'\,) \rightarrow vptrs(\, obj\,) \end{array}$$

If the projection is used to describe a layered system, the target of the projection, Obj', would be "closer" to the interface and Obj would be more an internal object. The system would be built by composing the inner object with the projection. That is, we would be interested in the behaviour of a pair: $Proj\_struct[\, Pt\,] \times Obj$. We can extend any command c on Obj to an operation on $Proj\_struct[\, Pt\,] \times Obj$ with a pull function on Pt':

$$\begin{array}{l} c_{proj}\colon Param[\, Pt\,] \times (\, Proj\_struct[\, Pt\,] \times Obj\,) \\ \qquad\qquad\qquad\qquad \rightarrow \ (\, Proj\_struct[\, Pt\,] \times Obj\,) \times (\, Pt' \rightarrow Pt'\,) \end{array}$$

$$c_{proj}(\, param, (\, proj\_struct, obj\,)\,) = (\, proj\_struct', obj'\,), pull'$$

where

$$\begin{array}{l} proj\_struct' = pull(\, proj\_struct\,) \\ obj', pull \quad\ = c(\, params, obj\,) \\ pull' = fwd' \circ pull \circ back \\ fwd' = proj(\, proj\_struct', obj'\,).fwd \\ back = proj(\, proj\_struct, obj\,).back \end{array}$$

Predictability demands that a user of such a layered system can infer the relevant internal state. If this state is of the form above, we must ask what can be inferred about $Proj\_struct[\, Pt\,] \times Obj$ from Obj', and whether any additional user-level pointers from Pt' are needed to determine it.

If there were a one-to-one correspondence between $Proj\_struct[\, Pt\,] \times Obj$ and Obj', this would give a pointer space structure directly to it. Usually, but not necessarily, this happens when $Proj\_struct$ does not have any pointer components. An example of such a mapping would be certain kinds of pretty printing, when there is a corresponding parsing function. $Proj\_struct$ would then contain information such as line width.

In other cases it will not even be possible to maintain a one-to-one correspondence with $Proj\_struct[\, Pt'\,] \times Obj'$: for instance, in a display frame map where only a portion of the underlying object is visible, in a folding map where the folded information is hidden, or in a pretty printer that displays tabs as spaces. Thus, if we use projections to model editors, observability will depend on a strategy of passive actions, as with the red-PIE. In this case we have a much better definition of passivity, namely an action that alters the $Proj\_struct[\, Pt\,]$ of the state but not the Obj one. In the first two cases, of the display and the folding map, we would be particularly interested in issues of faithfulness between Obj and Obj'. This is made easy by the separation of the projection into object and pointer parts. In the third example, the pretty printing, we would be interested in the existence of a single projection that gave a one-to-one correspondence between Obj and Obj': for instance, a special view showing tabs as a special font character.

8.2.4 Composing projections

One use of projections is to aid in the layering of certain types of specification and modularisation. Bearing in mind that it will almost certainly be necessary to add extra state at each level as well as the projection information, it seams a good idea to study the composition of projections in isolation.

We will consider two projections, $proj$ from Obj to Obj', and $proj'$ from Obj' to Obj''. We consider initially triples of the form:

$$Obj \times Proj\_struct[\,Pt\,] \times Proj\_struct'[\,Pt'\,]$$

We can then derive:

$$\begin{aligned} proj^{\dagger}: Obj \times Proj\_struct[\,Pt\,] \times Proj\_struct'[\,Pt'\,] \\ \rightarrow Obj'' \times (Pt \rightarrow Pt'') \times (Pt'' \rightarrow Pt) \end{aligned}$$

$$proj^{\dagger}(obj, ps, ps') = obj'', fwd'', back''$$

where

$$\begin{aligned} fwd'' &= fwd' \circ fwd \\ back'' &= back \circ back' \\ obj'', fwd', back' &= proj'(ps', obj') \\ obj', fwd, back &= proj(ps, obj) \end{aligned}$$

This is not a satisfactory projection, as it contains pointers from Pt' as well as Pt. However, if the subset of $vptrs'(obj')$ given by $fwd(vptrs(obj))$ is sufficiently rich, we may not need the Pt' pointers at all. Instead, we can project forward to obtain $Proj\_struct'[\,Pt'\,]$ from $Proj\_struct'[\,Pt\,]$. We can then define a proper projection $proj''$ from Obj to Obj'', with projection structure $Proj\_struct''[\,Pt\,]$:

$$Proj\_struct''[\,Pt\,] = Proj\_struct[\,Pt\,] \times Proj\_struct'[\,Pt\,]$$

$$proj'' : Obj \times Proj\_struct''[\,Pt\,] \rightarrow Obj'' \times (\,Pt \rightarrow Pt''\,) \times (\,Pt'' \rightarrow Pt\,)$$

$$proj''(\,obj, (\,ps, ps'\,)\,) = obj'', fwd'', back''$$

where

$$\begin{aligned} fwd'' &= fwd' \circ fwd \\ back'' &= back \circ back' \\ obj'', fwd', back' &= proj'(\,fwd(\,ps'\,), obj'\,) \\ obj', fwd, back &= proj(\,ps, obj\,) \end{aligned}$$

This is identical to $proj^{\dagger}$ except for the term $fwd(\,ps'\,)$, to change the Obj pointers to Obj' pointers in $Proj\_struct'$.

Compositions of projections are particularly useful when we want to consider multi-layer models of interactive systems. If the projection information is being embedded into the state of an editor, then the first form may be sufficient. However, as well as its theoretical interest, the second form has the advantage of only using one set of pointers, so its behaviour is likely to be more predictable and comprehensible to the user. However, any attempt to update the second structure using pointers from the surface Pt, will involve a lot of *back*ing and *fwd*ing. This may lead to odd behaviour unless the two maps have the right sort of inverse relationship.

8.2.5 Relation of back and forward maps and ladder drift

We have said that *fwd* and *back* relate the pointers and we expect them to be "sort of" inverses. There are several possibilities:

(i) Total inverses – the pointers are in one-to-one correspondence, rarely true.

(ii) The *back* map is an inverse of *fwd* – often true of pretty printing, but not of display-type maps where many pointers from the object are mapped to the display boundaries or bottom.

(iii) The *fwd* map is an inverse of *back* – opposite to (ii), ok for display-type maps, but no good for pretty printing where many pointers in the printed version may correspond to one in the object, for instance when padding with spaces.

(iv) There is a subset of "stable" pointers on each side of the projection in one-to-one correspondence, and *fwd* and *back* map the entire set of pointers into these "stable" subsets. This is represented by the two conditions:

(iv_a) $fwd \circ back \circ fwd = fwd$

(iv_b) $back \circ fwd \circ back = back$

Either of (ii) or (iii) implies both these conditions.

Condition (iv) seems a general one that could be demanded of all projections. Amongst other things, it imputes a lexical structure to the pointers, the set of pointers mapped together by *fwd* or *back* representing a lexeme. It does, however, suffer from incomposability.

If some property is useful, and we want to use projection compositions to modularise our design of systems, then it is important that the *fwd* and *back* maps from the different projections compose properly.

Pairs of functions, f and b say, obeying (i), (ii) or (iii) all preserve these properties when composed with similar pairs; property (iv) is unfortunately not preserved. We can summarise these properties of compositionality in a table. We have two pairs of functions $(f{:}Pt \rightarrow Pt',\ b{:}Pt' \rightarrow Pr)$ and $(f'{:}Pt' \rightarrow Pt'',\ b'{:}Pt'' \rightarrow Pt')$. We compose these to give $F = f' \circ f$, $B = b \circ b'$ (*fig.* 8.2).

	(ii) $b \circ f = id_{Pt}$	(iii) $f \circ b = id_{Pt}'$	(iv$_a$) $f \circ b \circ f = f$	(iv$_b$) $b \circ f \circ b = b$
(ii)	$b' \circ f' = id_{Pt'}$	$B \circ F = id_{Pt}$	$FBF = F$ & $BFB = B$ **	$FBF = F$
(iii)	$f' \circ b' = id_{Pt''}$	*	$F \circ B = id_{Pt''}$	–
(iv$_a$)	$f' \circ b' \circ f' = f'$	–	$BFB = B$	–
(iv$_b$)	$b' \circ f' \circ b' = b'$	–	$BFB = B$	–

figure 8.2 *properties of composite projections*

Unfortunately, the condition marked with the single asterisk is the one most commonly encountered in the design of display editors; an object is pretty printed to an intermediate object of infinite extent which is then viewed by a display mapping. *No* general conclusions about the resulting projection can be made, and we have to be careful in the design process to avoid problems.

The difficulty is that the lexemes generated by the two projections do not agree on the intermediate pointers, Pt'. We can see this by way of an example:

$$Pt = Pt' = Pt'' = \mathbb{N}$$

$$f(n) = b(n) \quad = n \quad \textbf{if } n \text{ odd} \quad \textbf{or} \quad n = 0$$
$$\qquad\qquad\qquad = n-1 \quad \textbf{if } n \text{ even} \quad \textbf{and } n \neq 0$$

$$f'(n) = b'(n) \quad = n \quad \textbf{if } n \text{ even}$$
$$\qquad\qquad\qquad = n-1 \quad \textbf{if } n \text{ odd}$$

$$\forall\, n>0 \quad F \circ B(n) = n-1$$
{ the only stable point is 0 }

We can can call this problem *ladder drift*. We can view it pictorially as two ladders, the rungs being the stable points in one-to-one correspondence of the two projections. As we step from ladder to ladder and back we gradually slip downwards: (*fig*. 8.3).

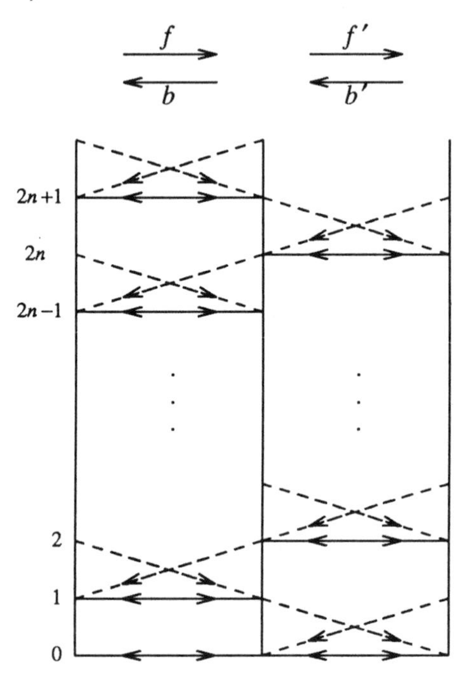

figure 8.3 *ladder drift*

The only way to avoid ladder drift is to have the lexemes partly "agree" on the intermediate pointers. The simplest case is when both lexemes are atomic, which is marked ** in the above table. Another alternative is to have the lexemes generated by one projection always be complete subsets of those generated by the other. Even this lax condition is not met by pretty prints and display mappings. However, they usually satisfy it locally. Pretty prints usually have lexemes contained within lines; the display map has at its extremes two lexemes, one for everything before the screen and one for everything after, each of which consists of whole lines; in the area of the screen its lexemes are atomic. At the boundary of the two, care must be taken to ensure reasonable behaviour. Despite this complexity, it is far easier to be convinced of the reasonableness of these maps when the pointer relation is available separately from the object.

8.3 Block operations

We have already noted that pointers need not point only to positions within the object, but may also point to whole areas. Such block pointers are often used as cursors in syntax-directed editors (Bahike and Hunkel 1987, Parker and Hendley 1987).

Whereas the pull functions for operations concerning atomic pointers are fairly simple, we find that the cases of block movement and copying are far more complex. We will concentrate in this section upon block operations upon a string, partly because it is a well-known structure and the various options are easy to describe. The resulting problems are far from simple to deal with, however, and if anything linear structures have more complex behaviour and more problems associated with their block operations than do operations on tree-like data structures.

8.3.1 Block delete and insert

Before going on to the complicated examples of move and insert we will look at block delete and insert operations, which are simpler:

$$Pt = Block\_ptrs + Atomic\_ptrs$$

$$block\_delete : Block\_ptrs \times Obj \rightarrow Obj \times (Pt \rightarrow Pt)$$

$block\_delete((n, m), obj) = obj', pull$
where

$$\begin{aligned} obj' &= obj[1,...,n] :: obj[m+1,...,length(string)] \\ pull(r) &= r \quad \textbf{if } r \leq n \\ &= n \quad \textbf{if } r \in \{n+1,...,m\} \\ &= r-m+n \quad \textbf{if } r > m \\ pull((r, s)) &= (pull(r), pull(s)) \end{aligned}$$

We might complement this with block insert:

$$block\_insert : Block\_ptrs \times String \times Obj \rightarrow Obj \times (Pt \rightarrow Pt)$$

$block\_insert(n, string, obj) = obj', pull$
where

$$\begin{aligned} obj' &= obj[1,...,n] :: string :: obj[n+1,...,length(string)] \\ pull(r) &= r \quad \textbf{if } r < n \\ &= r+length(string) \quad \textbf{if } r \geq n \\ pull((r, s)) &= (pull(r), pull(s)) \end{aligned}$$

In both these cases the obvious thing to do with block pointers is to modify their endpoints using the pull function for atomic pointers. Note, however, the following behaviour for block insert:

$$obj = \text{"abcdef"}$$
$$b_1 = (1,3)$$
$$b_2 = (3,5)$$
$$sub\_object(b_1) = \text{"bc"}$$
$$sub\_object(b_2) = \text{"de"}$$

$$\textbf{let} \quad obj', pull = block\_insert(3, \text{"xyz"}, obj)$$
$$\textbf{then}$$
$$obj' = \text{"abcxyzdef"}$$
$$pull(b_1) = (1,6)$$
$$pull(b_2) = (6,8)$$
$$sub\_object(pull(b_1)) = \text{"bcxyz"}$$
$$sub\_object(pull(b_2)) = \text{"de"}$$

Is this the correct behaviour for *semantic* pointers?

8.3.2 Block move and copy

In defining these operations, we expect more problems. To help us we consider several different algebraic properties we might want to hold:

(i) $copy(bpt, pt, obj) = block\_insert(pt, sub\_object(bpt, obj), obj)$

(ii_a) $move(bpt, pt, obj) = obj', pull$
where

$$pull = pull_{insert} \circ pull_{delete}$$
$$obj', pull_{insert} = block\_insert(pull_{delete}(pt), s_{bpt}, obj_{delete})$$
$$s_{bpt} = sub\_object(bpt, obj)$$
$$obj_{delete}, pull_{delete} = block\_delete(bpt, obj)$$

(ii_b) same as (ii_a) with delete and insert swopped around.

(iii) $move(bpt, pt, obj) = obj', pull$
where

$$pull = pull_{delete} \circ pull_{copy}$$
$$obj', pull_{delete} = block\_delete(pull_{copy}(bpt), obj_{copy})$$
$$obj_{copy}, pull_{copy} = copy(bpt, pt, obj)$$

(iv) $copy(bpt, pt, obj) = obj', pull$
where

$$
\begin{aligned}
pull &= pull_{insert} \circ pull_{move} \\
obj', pull_{insert} &= block\_insert(\, pull_{move}(\, pt \,),\, s_{bpt},\, obj_{move} \,) \\
s_{bpt} &= sub\_object(\, bpt, obj \,) \\
obj_{move}, pull_{move} &= move(\, bpt, pt, obj \,)
\end{aligned}
$$

If pt is not contained in bpt, (ii_a) and (ii_b) are identical. If, however, pt is contained in bpt then (ii_b) would imply $move(\, bpt, pt, obj \,)$ is the same as $delete(\, bpt, obj \,)$, not quite as intended! Some systems (e.g. ded; Bornat and Thimbleby 1986) deliberately disallow this case, as it is difficult to reason about even informally and is (for similar reasons) often difficult to implement.

In both cases of (ii), the pointers from within the moved block are not moved with the block, but are destroyed – in fact, mapped to the single old block position. In conjunction with them, (iv) says that $copy(\, bpt, pt, obj \,).pull$ maps all pointers within bpt to the old bpt position, even though that area has been unchanged.

Suppose now that (i) is true. In this case (ii_b) and (iii) are identical. Condition (iv) is inconsistent with (ii_a) or (ii_b). This is because (iv) corresponds to a "drag 'em along" policy on pointers, whereas condition (i) together with (ii_a) or (ii_b) corresponds to a "keep 'em still" policy. The latter has the great advantage of monotonicity on pointers.

The "drag 'em along" policy has further strange effects. We consider moving and copying the block (n, m) to the point p. For simplicity, we assume $p \le n$:

$$move(\,(\, n, m \,), p, obj \,) = obj', pull$$

where

$$
\begin{aligned}
obj' = {} & obj[\, 1,\ldots,n \,] :: obj[\, m+1,\ldots,p \,] :: obj[\, n+1,\ldots,m \,] \\
& :: obj[\, p+1,\ldots,length(\, obj \,) \,]
\end{aligned}
$$

$$
\begin{aligned}
pull(\, r \,) &= r && \textbf{if } r \le n \\
&= r+p-m && \textbf{if } r \in \{n+1,\ldots,m\} \\
&= r-m+n && \textbf{if } r \in \{m+1,\ldots,p\} \\
&= r && \textbf{if } r > p
\end{aligned}
$$

$$pull(\,(\, r, s \,)\,) = (\, pull(\, r \,), pull(\, s \,)\,) \quad \textbf{??}$$

$$copy(\,(n, m), p, obj\,) = obj', pull$$
where
$$obj' = obj[\,1,\ldots,p\,] :: obj[\,n+1,\ldots,m\,] :: obj[\,p+1,\ldots,length(\,obj\,)\,]$$

$$\begin{aligned} pull(\,r\,) &= r && \textbf{if } r \le p \textbf{ and } r \in \{n+1,\ldots,m\} \\ &= r+p-n && \textbf{if } r \in \{n+1,\ldots,m\} \\ &= r+m-n && \textbf{if } r > p \end{aligned}$$

$$pull(\,(r, s)\,) = (\,pull(\,r\,), pull(\,s\,)\,) \quad \mathbf{??}$$

The pull for the pointers is obvious given the policy, but the pull for the blocks (marked with **??**) is not so clear. In fact, the non-monotonicity of the pull function has had disastrous consequences on the blocks, as we can see from the following example:

obj = *"abcdefg"*
$b_1 = (\,3, 5\,)$
$b_2 = (\,4, 6\,)$
$sub\_object(\,b_1, obj\,) =$ *"de"*

let $obj', pull = block\_move(\,b_2, 2, obj\,)$
then

obj' = *"aefbcdg"*
$pull(\,b_1\,) = (\,5, 2\,) = (\,2, 5\,) \quad \mathbf{??}$
$sub\_object(\,pull(\,b_1\,), obj'\,) =$ *"fbc"*

As we see, the block pointer gets its ends reversed. If we assume (at the point marked **??**) that this is the same as the properly ordered block pointer, then we get the very odd result for the sub-object of the pulled block. This is clearly not correct behaviour for semantic pointers.

The alternatives for the choice of block pull function are either to treat pointers like $pull(\,b_1\,)$ as invalid and say $(\,5, 2\,) = \bot$, put up with silly behaviour like *"fbc"* above, or have more complex block pull functions. Both those blocks totally contained in *bpt* and those disjoint from it behave as we would expect. Blocks bridging either end of *bpt* could be curtailed in some way, either by removing the bit inside, or the bit outside *bpt*. There are two alternatives for this chopping. We could chop off the inside, that is, a "leave 'em behind" policy for these blocks – within the general "drag 'em along" policy for atomic pointers:

$$pull_{move}((r, s)) = (r', s')$$

where

if $r, s \in \{n+1,...,m\}$ **or** $r, s \notin \{n+1,...,m\}$
then

$$r' = pull(r)$$
$$s' = pull(s)$$

else

$$r' = pull(n) \quad \textbf{if } r \in \{n+1,...,m\}$$
$$= pull(r) \quad \textbf{otherwise}$$
$$s' = pull(n) \quad \textbf{if } s \in \{n+1,...,m\}$$
$$= pull(s) \quad \textbf{otherwise}$$

Alternatively, we could chop off the bit outside, leading to a "drag 'em along" policy for these blocks:

$$pull_{move}((r, s)) = (r', s')$$

where

if $r, s \in \{n+1,...,m\}$ **or** $r, s \notin \{n+1,...,m\}$
then

$$r' = pull(r)$$
$$s' = pull(s)$$

else

$$r' = pull(n) \quad \textbf{if } r \notin \{n+1,...,m\}$$
$$= pull(r) \quad \textbf{otherwise}$$
$$s' = pull(n) \quad \textbf{if } s \notin \{n+1,...,m\}$$
$$= pull(s) \quad \textbf{otherwise}$$

These strategies, although possible, are hardly simple or clear and have very strange behaviour on the block inclusion relationship.

Another way round the problem is to allow complex blocks consisting of lists of intervals.† In the example given above:

$$b_1 = (3, 5)$$
$$sub\_object(b_1, obj) = \text{"de"}$$
$$pull(b_1) = \{(5, 6), (1, 2)\}$$
$$sub\_object(pull(b_1, obj')) = \{\text{"e"}, \text{"d"}\}$$

This solution still has problems, since intervals can become non-intervals; we may need to define block moves and their pulls with non-interval parameters!

---

† I have been told that this is the solution adopted for links in the Xanadu hypertext system (Nelson 1981), although I have not seen any documentation confirming this.

8.3.3 Block move in various editors

It is instructive to look at various editors and see how they handle pointer movement for block operations. We will consider four editors. Vi (Joy 1980) and ded (Bornat and Thimbleby 1986) are two heavily screen-based text editors operating under Unix. Wordstar (Wordstar 1981), in various versions is probably the world's most widely used word processor for microcomputers. Spy (Collis *et al.* 1984) is a mouse-based multi-file editor for bit-map workstations.

- *Atomic pointers* – Vi inherits line pointers from its line editor predecessor; however, it uses cut/paste rather than an atomic block copy or move. It therefore follows a "leave 'em behind" strategy, satisfying conditions (i) and (ii_a). Vi gets round this very neatly, by having only one pointer, the current cursor position, and making all insertions at this point! Ded has only one atomic pointer, the cursor, and it always moves this to the site of block deletion, or movement, irrespective of where it started off. Wordstar has multiple atomic pointers, but these are represented within the text as control sequences, thus they are moved with the rest of the text in block moves, and the user can even choose whether or not to include the pointers at the block boundary. Spy has marks which are moved with the text in block moves, and *duplicated* in block copy.
- *Block pointers* – Vi doesn't have any, however, because it uses cut/paste; if it were to have block pointers they would presumably be handled by the "leave 'em behind" strategy. Ded allows only non-intersecting blocks, and hence avoids the intersecting blocks problem. Wordstar and Spy both have only one block pointer (in Spy's case the selection) and so also avoid this problem.

We see that common editors usually solve these problems in an *ad hoc* manner, sometimes deliberately avoiding them, sometimes not possessing the functionality to show them up. They also tend to incorporate many special cases (especially for the cursor/selection). These inconsistencies are precisely what we want to avoid by use of pointer spaces. Even if some inconsistency is tolerable for monolithic editors, it is not acceptable where generic components are required or where concurrent access to the edited object is required. A good rule of thumb is to adopt the "leave 'em behind" strategy, possibly with special rules for cursor/selection.

8.4 Further properties of pointer spaces

In this section we discuss briefly some more advanced properties of pointer spaces, particularly those concerned with the formal treatment of block pointers. A more rigorous treatment of these properties, many of which are simply generalisations of the properties of intervals, can be found in Dix (1987b).

8.4.1 Types of pointer and relations on pointers

So far, no structure has been assumed on the set of pointers. However, in practice most pointers spaces are highly structured. This structure includes:

- *Special elements* – e.g. beginning and end pointers.
- *Classes of pointers* – e.g. for strings before-character and after-character pointers, and for trees, node and leaf pointers.
- *Ordering and other relations* – e.g. partial or total ordering of Pt or $vptrs(obj)$.

The relation on pointers is most general, classes being unary relations and special elements being classes of a single element. They can thus all be described in the same way, but there is certainly a perceived difference.

The structure of the object may force structure on the pointers: for instance, an object consisting of two major sub-objects may have a set of pointers divided into two classes according to the sub-object to which they point. Conversely, the pointers may be used to impute structure to an otherwise unstructured object: for instance, a string of characters may have pointers corresponding to paragraphs, sentences and words.

We have already considered the case of block pointers onto string; in general, there are likely to be *block pointers* indicating portions of the object. These typically have associated with them a whole set of relations of their own (e.g. inclusion, intersection), and special sub-object projection maps.

8.4.2 Absolute and relative structure

As we have seen, some structure is built into the pointers themselves, and some they inherit from the objects. When the structure is inherited from some non-constant feature of the objects, or when it is defined using the objects, it will be *relative* to a particular object. If, on the other hand, the structure can be defined independently of any particular object, we will say it is *absolute*. If $F(Pt)$ represents the structure operator we are interested in, then we have the following two definitions:

absolute structure:

$$struct \in F(Pt)$$

relative structure:

$$struct \in Obj \rightarrow F(Pt) \quad \textbf{st} \quad struct(obj) \in F(vptrs(obj))$$

Note that an absolute structure is a relative structure with a constant map. Hence we may, if we wish to, deal with relative structures and include absolute structures by default.

An example of a class with relative structure is word pointers on strings: whether a block refers to a word or not depends on the specific string pointed to and not on the pointer. Perhaps a more common form of relative structure is the end pointer for strings, which depends on the length of the string. This does have the useful property that it needs to be parametrised only over $vptrs(obj)$.

There are strong consistency and simplicity arguments for trying to deal with absolute structures as much as possible. However, as the word pointer and end pointer examples illustrate, relative structures are often natural.

8.4.3 Pull functions and structure

For consistency, it would be desirable if pull functions preserved in some sense the structures on the pointers. Taking again strings as examples, most simple manipulative actions, such as *insert* and *delete*, are monotonic on the natural pointer order. Similarly, they preserve block inclusion and end pointers. They do not, however, preserve start pointers when inserting at the string beginning. The start pointer for the original string ends up pointing after the character inserted and is therefore not preserved. We might try to get round this by adding start pointers as special objects to Pt, $Pt = \mathbb{N} \cup \{START\}$, but we would quickly find that adding such elements complicates the description, and hides consistency (*START* behaves like pointer 0 in all respects except update).

On the positive side, note that the start pointer is altered only when it is near to the site of the insert. That is, there is some *locality* of the update outside which the pull is consistent. This idea of locality is very important in the interactive context, as changes to an object can afford to be strange so long as they are strange only in the locality of the point of interaction, and consistent elsewhere. The simple string functions even preserve (up to locality) the relative word pointers.

8.4.4 Block structure

The fact that blocks refer to sections of the underlying object means that they have some inherent structure. We would want to say whether two block pointers *intersected*, (or *interfered* with) one another, or whether one *contained* the other. Similarly, we might want to define an operation that, given two block pointers, gave a pointer as their intersection. These relations would want to satisfy sensible lattice properties, similar to those of sets or intervals. For example, we would want the intersection of two blocks to be contained in both of them.

There are some differences between the operations on block pointers and the natural operations on intervals. We want to be conservative when dealing with blocks, so we want to say that two blocks interfere even if the intersection we return is empty. Similarly, we should not insist that the intersection of two blocks is maximal.

Depending on the structure of the underlying object, we may be able to distinguish blocks that refer to several disparate parts of the object from *intervals*. This sub-class of blocks would have additional properties, especially when there is an ordering on the object and its atomic pointers.

8.4.5 Sub-object projections

Most objects of any richness at all can be viewed in part as well as in whole. Even simple integers can be decomposed into digits or into prime factors. Sometimes we have a relatively unstructured relation between object and sub-object, such as sub-strings of strings. In other cases, the relation is stronger, for instance a sub-tree of a syntax tree corresponding to a *while* loop. If the relationship between object and sub-object is sufficiently natural there will be a relation between their pointers, and we will be able to construct a projection, the defining structure of which will typically be a single block pointer. If the pointers are not rich enough for this already, they can usually be extended to include sub-object projection structures.

In more explicit terms, we are going to associate with a pointer space, $<Obj, P_{Obj}>$, a sub-object pointer space, $<Sobj, P_{Sobj}>$, and a projection:

$$sub\_object: Bpt_{Obj} \times Obj \rightarrow Sobj \times (P_{Obj} \rightarrow P_{Sobj}) \times (P_{Sobj}, P_{Obj})$$

The normal rules for projections apply but, because of the special nature of sub-objects, the *fwd* and *back* maps will have additional properties. In particular, we would expect a one-to-one correspondence between the sub-object pointers and some subset of the object pointers. That is, *fwd* is a left inverse to *back*:

$$\forall\ b \in Bpt, obj \in Obj \quad \textbf{let}\ sobj, fwd, back = sub\_object(b, obj)$$
$$\textbf{then}\ \ fwd \circ back = id_{vptrs(sobj)}$$

Alternatively, we could define a sub-object projection to be any projection satisfying the above condition.

Often there will be a block pointer corresponding to the whole of an object (usually relative to $vptrs(obj)$) that yields the whole object. That is, the *fwd* and *back* for this sub-object are one-to-one, and the resulting projection is one-to-one when considering Obj as whole. If we do not possess such block pointers, we can always construct them.

It will usually be the case that we can consider sub-objects of sub-objects; that is, the sub-object projection is in fact an *auto-projection*:

$$sub\_object: < Sobj, P_{Sobj} > \;\rightarrow\; < Sobj, P_{Sobj} >$$

the original objects, Obj, being a subspace of $Sobj$.

We expect there to be coherence properties for the sub-object projections. If we look at a sub-object B and a sub-object of it, C, then we expect the combined sub-object projections from the object to B and then from B to C to be the same as if we had gone to C directly.

The existence of a sub-object projection automatically generates some relations on the block pointers. There is an obvious containment relation by examining the block pointers of sub-objects, and we can generate an independence relation on blocks by seeing whether all possible values of the sub-objects can occur together.

8.4.6 Intervals and locality information

Blocks, usually interval blocks, have a special role to play in providing locality information. Typically an operation changes only a very small part of an object, and outside this locality the object is unchanged and the pull function takes a particularly simple form. We can express this by having certain operations return an additional block pointer value, its locality:

$$c : Params[\,Pt\,] \times Obj \;\rightarrow\; Obj \times (\,Pt \rightarrow Pt\,) \times Bpt$$

We will refer to this additional component as $c(params, obj).loc$.

This locality information has the property that for any block that doesn't intersect, $c(params, obj).loc$ yields the same object before and after the operation; similarly, the pointers yielded by the sub-object projection are invariant under $c(params, obj).pull$:

let $obj', pull, loc \ = \ c(params, obj)$
if $b \in vptrs(obj)$
and b does not intersect loc
then $sub\_object(b, obj) \ = \ sub\_object(pull(b), obj')$

We could, in fact, define an optimal locality by letting the locality of $c(params, obj)$ be a minimal block satisfying the above. In practice, however, we would return a conservative estimate of the locality, rather than the optimal locality. For instance, for the overwrite operation on strings, overwriting the n th character will normally have a locality of $(n-1, n)$; however, if the character overwritten is the same as the original character the optimal locality is null. In such cases it may often be better to deal with the generic locality rather than the particular one, both for computational simplicity and for ease of abstract analysis.

Locality information can be used to express user interface properties. If the locality of an operation is contained within the display then the user has seen all of the changes. This gives a more precise version of the "mistakes in local commands are obvious" principle stated in Chapter 2. We have confidence not only that we can see when there has been a change, but also that we can see *all* of the change.

8.4.7 Locality and optimisation

Locality information can also be invaluable in performing various optimisations. By knowing where changes have occurred, we can cache sub-objects and know when the caches become invalid. Further, if we know the structure of pointers we can often optimise pull functions outside the locality. For example, if we know that the locality of the string operation, $op(21307, The\_complete\_works\_of\_Shakespeare)$, is $(21306, 21309)$ and that $pull(21310) = 21323$, then we know we can just add 13 to all pointers greater than 21306, leave unchanged those less than 21306 and perform the function call only on pointers in the range 21306 to 21309. Such optimisation can be very important, as a function call is expensive compared with addition.

Frequently such optimisations will be possible where the operation is defined using a projection and the projection has "nice" locality properties. A *self-contained* block is the simplest such locality property. It is a block whose projection is dependent only on the sub-object it contains. An example of a self-contained block, would be a paragraph in text, which can be formatted independently of the surrounding context. If an update operation is performed and the locality of the operation does not intersect with the locality of the block, then the projection of that block is unchanged.

There is a slight complication in this definition. Whether or not a block is self-contained is usually dependent on the object into which it points. That is, self-containedness is usually a relative structure. We must therefore demand that a block is self-contained *both* before and after the operation. For example, if we deleted the line between a paragraph and the preceding one, the original paragraph would be part of a larger paragraph and hence no longer be self-contained itself.

Even where self-contained blocks exist, they may be large. If we change a character within a paragraph, it will not change the formatting of much of the paragraph. Only occasionally will it even mean reformatting to the end of the paragraph. However, not even the beginning of the paragraph will be a self-contained block, as the formatting of its last line will depend on the length of the following word.

We can cater for such cases by considering the notion of the *context* of a block. This is some information that, together with the contents of the block, is sufficient to determine the projection of the block. This context will typically be a surrounding block pointer together with some additional information. Thus if the locality of an operation does not intersect the context of a block, we do not have to update the projection of the block. This generalises the notion of self-contained block sufficiently to be useful. Indeed, block contexts were used in the optimisation of an experimental editor developed at York (Dix 1987b).

8.4.8 Locality information and general change information

The locality information, like the pull function, can be thought of as an example of change information. Similar constructs appear when considering, for example, views of objects. Again a view cannot be defined outside the context of a particular object. Thus pointers and pull functions can be seen as examples of a general phenomenon. On the other hand, I believe that many such constructs can be modelled using pointers and pulls, although sometimes it will be best to import only a subset of their functionality into the new domain.

8.4.9 Generalisations of pointer spaces

The reader will have been struck by the similarity between operations with pull functions, and projections. The former are, of course, a special case of the latter where the two objects are of the same type. The reason for not including an inverse for *pull* in the definition of all operations is that we are after a way to describe the update properties of systems, and hence this is unnecessary. We could imagine cases where it would be of help. For instance, I might change a file and want to trace back a position in the changed document to the original. We might go further and say that both update operations and projections should supply a relation, rather than functions. We could constrain this relation so that it satisfies the pseudo-inverse property of §8.2.5. Again, the reason for not doing so is that we are after a formulation that will enable a system to perform the translations automatically; a relation will yield a choice that the system would not know how to deal with.

We could use relations in two sets of circumstances, though. First, in the early specification of a pointer space or projection, we might know what class of behaviours is acceptable, but not be decided exactly which as yet. This would certainly be the case with the various options for block move and copy, where the range of possibilities is reasonably clear, but not the exact choice. The second place where relations would be useful would be where the pointers are an explicit part of the interface. For instance, it might be sensible for user-defined marks to split when the block containing them is copied, or alternatively for the user to be consulted on what to do with them. This would still be arduous if the user was not given sensible defaults, especially as the number of pointers grows (as in a shared hypertext). Perhaps the best course to take would be to supply both a relation and preferred destinations (in the form of functions consistent with the relation). The designer of a system could then choose whether to use the default supplied, to split pointers over their options, or to offer the end-user choice.

The changes that would need to be made to encompass this generalisation are fairly obvious and would serve only to clutter the exposition. For that reason this chapter (long enough already!) has followed the simple method.

Generalising in a different direction, dynamic pointers are associated with positions. This was the starting point for this chapter. In fact, the axioms given for pointer spaces are very sparse and do not constrain us to this interpretation. There are other structures that are not directly positional in themselves but have similar properties. For instance, we discuss views in the next chapter. Throughout most of that chapter we deal with static views, but many views of objects in interactive applications share many of the properties of pointer spaces. Elsewhere I have shown how dynamic views can be defined in terms of block pointers and sub-object projections (Dix 1987b). This supports the hypothesis that position is central to interactive system design. On the other hand, even if we can describe a phenomenon in terms of positional pointers, it may be better (in view of our behavioural slant) to use just that subset of the properties of pointer spaces that are needed to describe it, and apply them directly to the domain of interest.

8.5 Applications of dynamic pointers

To conclude this chapter we look at the uses of dynamic pointers, both present and future. We first concentrate on the ways that dynamic pointers are used currently in a variety of systems. Some of these have been mentioned before in the chapter, but they are gathered together for reference. After this, we consider the importance of dynamic pointers in the creation of advanced user interfaces. In particular, many of the examples concentrate on programming support environments.

8.5.1 Present use of dynamic pointers

The use of dynamic pointers which we have focussed on particularly has been the design of editors. Here, as we have already seen, are found many examples: cursors, selection regions, marked blocks, display frame boundaries, folded regions. All of these constructs must have some of the properties of dynamic pointers, and would ideally have some sort of consistency between them. Usually their semantics are regarded severally and in an *ad hoc* manner; even an informal concentration on their similarity would rationalise editor design.

Hypertext systems must again use dynamic pointers; these fall into several categories. NoteCards (Halasz *et al.* 1987) is primarily a *point-to-object* system: the text of a card can contain atomic references to other cards in the system. The references to the objects can be regarded as dynamic pointers into the card database, as well as the reference points being dynamic pointers into the card's text. NoteCards also supports what it calls global links. These are links between cards that do not reside at a particular location in the card, and could be described as *object-to-object*. This is the sort of linkage handled by conventional databases. The Brown University Intermedia system (Yankelovich *et al.* 1985) uses a *block-to-block* method where blocks in the text of a hypertext object reference other blocks. However, these references take place through intermediate objects recording information about the link, so this could be regarded as a sugared form of a *block-to-object* link. The awkward problem of how to represent block links is resolved here by indicating only the start position of a referenced block and displaying the whole block on request. This problem of display arises continually when dealing with objects with a large number of links. If the density of links is too high it will be necessary to hide those types of links not of interest, or if one is interested only in the object itself then perhaps all the links. If this is the case then it becomes even more important that the movement of these links is sensible and predictable when the object is edited. In the majority of cases the pointers in the above are implemented either by embedded codes in the objects of interest or as references to representations such as linked lists.

Databases and file systems provide a wealth of dynamic pointer examples. In a way, all file names can be thought of as dynamic pointers. They usually retain the semantic link to the files to which they point, even when those files change their content. They are a particularly simple example, because of the simplicity of typical file system structures. They fail to be true dynamic pointers, in that the file name itself has (or should have) some sort of mnemonic meaning in itself: it both denotes some of the subject matter of the file and often, by means of extension names, may denote type information. Thus the file name is in a sense an attribute of the file rather than a simple handle to it. Those who call their files names such as "jim" and "mary" come closer to the ideal of dynamic pointers! Despite this difference, there are enough similarities to make the analogy useful.

In the underlying implementation of file systems one comes closer to true dynamic pointers. For instance, in the Unix file system (Thompson and Ritchie 1978), each file is denoted by an integer, its *i-node* number. Even when the file name changes, this number remains unchanged and similarly, whereas the location on disc where a file is stored may change as it is written to and rewritten, the *i-node* remains the same.

Again, at two levels databases have dynamic pointers. At the user level these are database key values: these denote a semantic record, even when the record contents change or the database is reorganised. These keys are usually meaningful, like file names: in fact, usually more so. It is often argued that this is a shortcoming in database design: for instance, a record keyed by surname and forename might change its key value when the subject married. Some more advanced relational database designs codify this meaninglessness by supplying system-defined *surrogate* keys for each entity. (Earl *et al.* 1986) This gets close to the sort of referencing between records and their sub-records in network databases, again a place where the integrity of the semantic link must be maintained despite changes in the database values.

Finally, we could look at the example (anathema to computer scientists) of the BASIC programming language (Kurtz 1978). In this, all statements are numbered. When, as often happens, the programmer finds that the original choice of numbering left too little space for the statements required, most systems supply a command that changes the numbering to space out the program. In the process all statements that refer to line numbers have their operands changed accordingly. That is, the number "100" in "GOTO 100" is, in fact, a dynamic pointer to the statement numbered "100"!

8.5.2 Future use of dynamic pointers in advanced applications

Once one recognises dynamic pointers as a central data type, many different applications become obvious, some being self-contained, like the editor, others requiring that the pointers be an integrated part of the environment in which the application operates.

In the previous sub-section, we noted that file names are really an attribute of file objects, rather than just a handle to them. In direct manipulation systems there is less need for all objects to be denoted by file names. Naming is still important, as the work on window managers and the problem of aliasing emphasises (§1.5.2). However, names could become more of an extended annotation of the file contents, rather than the principal method of manipulation, a job taken over by the mouse and icon. In addition, temporary objects, which have no identity of their own but have importance merely for their contents, can remain anonymous, being denoted only by their spatial arrangement on the screen. In such cases the distinction between objects and the windows and icons representing them can become slightly blurred, and it is not surprising that the model for representing windows and that for dynamic pointers are similar. It is likely that the distinction between such systems and hypertexts may also become blurred: for instance, the directory or folder would become a simple list of dynamic pointers to other objects, similar to the file-box in NoteCards (Halasz *et al.* 1987). It would be quite natural that files describing the file system (perhaps containing dependency information, such as in the Unix utility "make" (Thompson and Ritchie 1978), could be created using pointers rather than names. It is unclear whether names in such a system would be an attribute of the object or of the link; perhaps both would be allowed.

Pointer spaces and projections will be indispensable for representing concurrent editing on multiple views of the same object: for instance, when manipulating a graphical object directly and at the same time editing its representation in a graphics representation language. Similarly, one might want to edit an unformatted text and its formatted form concurrently. Such dual-representation editing overcomes some of the doubts raised about the limitations of direct manipulation for representing the full functionality desired, freeing the user to choose among the different representations as is most natural and useful.

As we've already noted, an environment where dynamic pointers are normal would lead to the development of facilities such as find/replace as general tools. Contrast this for instance with Unix, which has general-purpose find tools (grep, fgrep, egrep) and global context editors (sed, awk) at the command interface level (Thompson and Ritchie 1978), but because they are not easily interfaced at the program level their facilities are repeated throughout the different editors of the system. Obviously, for some task-specific reason a designer may want to incorporate slight variants of existing tools within new applications, but she should not be forced to do so.

Thinking of such simple text tools at the command interface level, leads nicely to considering the more complex tools of a programming environment. Multi-stage compilers already have to retain some of the notion of a *back* function to the original source in order to give sensible error messages. For instance, the "C" language pre-processor (Kernighan and Ritchie 1978) embeds control lines into the output stream so that the later stages can know from which line of which file errors originated. Other similar tools in the Unix environment do not have such facilities, so for instance using the various pre-processors to the typesetter "troff" (Ossanna 1976) means that error messages are recorded at line numbers that bear no relation to the original text. A consistent framework at the level of the environment would make all these tools easier to define and use.

When error messages are reported (assuming they refer to the correct locations) dynamic pointers are again useful. For instance, the "ded" display editor (Bornat and Thimbleby 1986) has a facility for taking the error messages from a compiler and stepping through the source to the line of each error message. Unfortunately, as the file is corrected, the line numbers make less and less sense in relation to the new file. In essence, the *pull* function has been ignored. If the underlying implementation had been in terms of dynamic pointers this failing would have been avoided: the error messages would have been parsed into dynamic pointers to the original source file. (Assuming they were not already supplied in this form by the compiler.) Then as the file changed when the errors were corrected, the dynamic pointers would have kept pace with the semantic position in the file to which they refer. The Microsoft QuickC environment (QuickC 1988), where a compiler, editor and debugger are all designed in a consistent framework, does behave in the appropriate manner, and positional elements such as breakpoints and error lines retain their semantic position through changes. Of course, the dynamic pointers are part of the QuickC environment, rather than of the operating system or file system. Therefore, if you choose to use your own editor rather than the QuickC editor the semantic positions are lost.

Where compilation is integrated into the editing environment, dynamic pointers can aid in incremental compilation and error reporting. Alternatively, they may actually form the basis of integration of editors and compilers, regarded as separate tools within a dynamic pointer-based environment. In the next sub-section we see a detailed example of how locality information can give textual editing advantages similar to syntax-directed editing in terms of incremental parsing, and superior in some important cases. It also gives an example of how the locality information can be used to give improved error reports based on assuming the changed region is at fault, and warnings where unintentional syntax slips are suspected.

Similarly, dynamic pointers can be used for proof reuse. The correctness proof between various stages in the refinement of a formal specification could be very large. It would be tedious to have to redo the whole proof after every change even when one knew that the proofs would be nearly identical. High-level proof heuristics and plans (Bundy 1983) will, of course, save this work, but in many cases it will be possible to use the proof exactly except that the pivot points for the application of the proof steps will have changed marginally. If the pivotal expressions are represented by dynamic pointers, then the positions after the change will probably lead to a completely faultless reapplication of the proof. In the cases where this fails, it will at least remove some of the tedious work from the shoulders of either the programmer or the more intelligent theorem prover.

8.5.3 Incremental parsing and error reporting using dynamic pointers

We have considered many possible uses of dynamic pointers and we will now look in detail at their use for incremental parsing and error reporting. We could expand in a similar fashion on several of the points raised above, but the present discussion can serve as an illustration.

Syntax-directed editors can achieve a large degree of incremental semantic checking and compilation because of their knowledge about the syntactic objects changed and the extent of such changes. Often, as users are not totally happy with the syntax-directed approach for all editing, they are allowed the selection of regions for standard text editing. (Bahike and Hunkel 1987) Unfortunately, such regions have to be completely reparsed and analysed when the text editing is complete. Using the *locality* information that can be supplied by dynamic pointer operations, together with the *projection* information between the existing parse tree and original text and the *pull* function relating that to the new text, one can reduce the work needed dramatically. This information can be used with both small-scale and large-scale changes:

- *Small-scale changes* – The advantages here are similar to those of syntax-directed editing. It is possible to infer from the locality of a change to the text what corresponding section of syntax tree should be changed, and just reparse this.
- *Large-scale changes* – One of the reasons for switching out of syntax editing mode may be that the syntactic primatives make it difficult to change the flow of large-scale control structures, where the blocks that these control are unchanged. As the controlled blocks can be arbitrarily large this may be very costly in terms of reanalysis. However, the locality

information can indicate which pieces of text corresponding to well-formed syntactic entities are unchanged, and thus only the surrounding context will require reanalysis.

The case of large-scale change is perhaps less obvious and is more important, so we consider an example program fragment:

```
while ( x > 0 ) {
        y = f(x);
        /* lots more statements */
            ......
        /* finished all the hard work now */
        x = x-1;
       } /* end while */
```

The programmer then realises that the variable x can only have the values 0 or 1 when the while loop is encountered, and amends the program by changing the `while` statement to an `if` and deleting the line `x = x-1`:

```
if ( x > 0 ) {
        y = f(x);
        /* lots more statements */
            ......
        /* finished all the hard work now */
       } /* end if */
```

By examining the locality of change the parser/compiler can realise that the sub-expression `x > 0` and the main group of statements have been unchanged, and thus retains the old syntax tree for them, just reparsing the token `if` and the statement brackets "{}".

Where the grammar is ambiguous this approach may yield the non-preferred interpretation and thus the syntax tree may need slight jiggering. Alternatively, even though the sub-expression is still the same, it may be the case that there is no proper parsing based on the assumption of the sub-tree being the correct parsing for its text fragment; this would force one to reanalyse more (unchanged) text. Again, it depends on the grammar whether this is likely or even possible. In the second case, there may be no correct parsing because of a syntax error. Working on the assumption that the previously parsable section is correct, one may be able to make far better error reports (and perhaps intelligent suggestions) than normal. In either case, even where there is no error, the change in meaning of the sub-expression may be intentional, but there is a good chance that it is accidental and warnings could be given. For example, consider the "dangling else" problem, a classic example where the syntax tree might need jiggering. The programmer starts off with the following fragment:

```
if ( x != 0 )
        y = f(x);
else
        y = 1;
```

Then, realising that `f(x)` causes an overflow if the global variable `z` is zero, the programmer amends it to leave `y` unchanged in this case:

```
if ( x != 0 )
    if ( z != 0 )
        y = f(x);
else
        y = 1;
```

The original parse tree would have the following form:

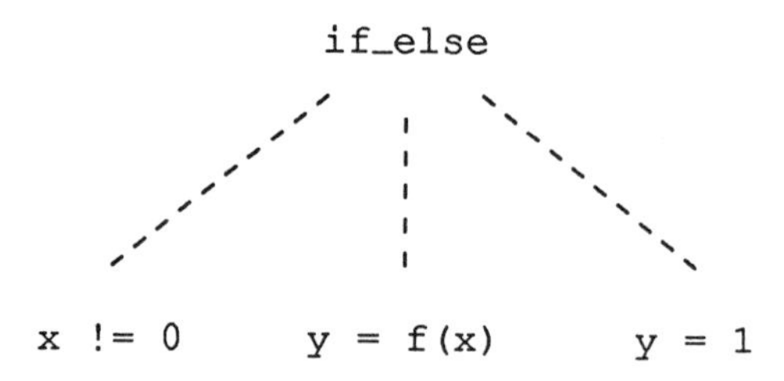

Only the text corresponding to the `then` arm would have been altered, so the amended tree would be:

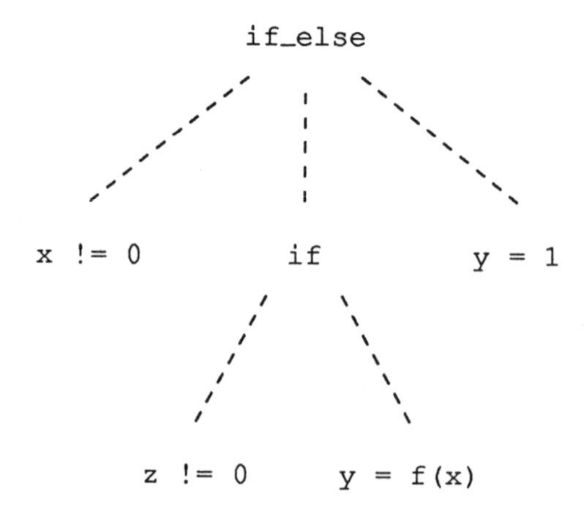

However, the analyser would realise that this was taking the wrong rule for disambiguating the dangling `else` and it should be bound to the innermost `if`. The tree would finally read:

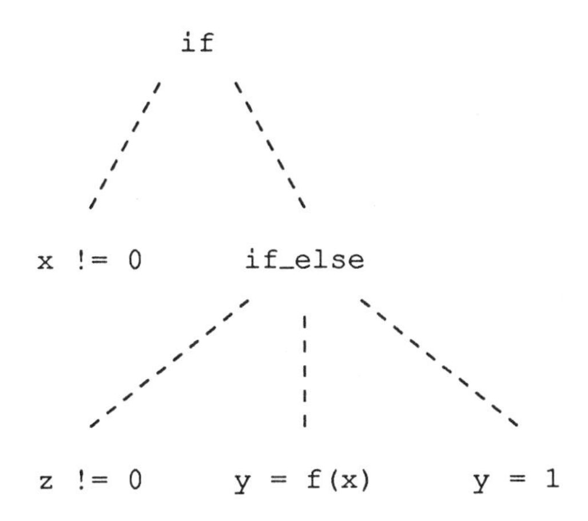

At this stage it would be worth giving a warning, as although this change in the form of the syntax tree might have been just what was intended (and a good reason for using text editing rather than syntactic operations), it could easily have been a mistake. In this example, the reparsing was of course wrong with respect to the programmer's intentions and the warning would have been appreciated.

8.6 Discussion

We have seen in this chapter how dynamic pointers can be given a clean semantic definition in terms of pointer spaces. The pull function that is associated with any operation provides a natural and consistent way to update the state of an editor or application. In order to understand the mapping between display and application and layered design in general, we considered relations between pointer spaces called projections, and the composition of these.

The translation of block pointers was found to be quite complex and this is evidenced by the various inconsistent ways atomic and block pointers are updated in existing systems. Using dynamic pointers to discuss these issues outside of a particular system helps us to form an understanding of the issues, even if no single answer can be found.

We have discussed how some of the more detailed analysis of dynamic pointers using *sub-object projections* and *locality* information about projections and operations can help us to express interface properties and can be used for system optimisations.

Finally, we considered many examples of how dynamic pointers are and could be used. Clearly, one could go on thinking of such uses. Some of those suggested can be implemented now under existing environments, but many rely on the pointers being an integral part of the environment. I believe that advanced interactive environments will contain entities similar to dynamic pointers, and if they are not recognised as a single unifying concept they will instead be implemented in diverse incompatible ways, yielding a clumsy and unpredictable interface.

CHAPTER 9

Complementary functions and complementary views

9.1 Introduction

At the end of Chapter 7, we took two branches. One was based on the "editing the object" approach and was followed up in Chapter 8. This essentially took the view that all our update commands are addressed to the underlying object and the display acts merely as a way of viewing the effects of those updates. The other branch was the "editing the view approach" and assumes that it is the user's view of the object that is primary. Updates are then addressed to the view, and the underlying object must keep pace with these changes. This is arguably a more user-centred approach, and is investigated in this chapter.

I say arguably because if we are to have a good user interface the two approaches will converge. On the side of editing the object, the presentation attempts to uncover the object's structure on the display. On the side of editing the view, if the global effects of updates to the display are to be predictable, then again the map between object and display and the effects of updates must be transparent. Good design will probably consist of a blend of the two approaches, and it is hardly surprising that at the end of this chapter we end up using the pointer space designed out of the former strategy to model structures resulting out of the latter.

In the rest of this section, we introduce the concept of a *complementary function*, which acts as a unifying concept for several problems in addition to editing the display. The problem addressed by the rest of this chapter can then be seen as one of finding suitable translations from update functions to their complements. We then look at some of the principles that we will wish to apply to translation strategies. In particular, predictability will play a pivotal role in driving the chapter.

Researchers into database theory have already investigated this problem and we use their terminology. However, we will see that the databases normal to this field differ in several respects from those common in general interface design. That is, our structures are both *dynamic,* and subject to *structural change.* Despite this, we look principally at the case of static views, as these are far more tractable and at least give us the general flavour required for more complex cases. The databases with which we deal differ markedly from typical DP databases in their structural complexity and scale. For instance, we may be interested in program syntax trees, or a small set of numbers, rather than large homogeneous relations. This is particularly important since many of the views defined over databases are slices put together out of these large relations, whereas those in which we are interested may be far more complex (even when static).

The first major section gives a short algebra of views, so as not to clutter up the succeeding sections with too many definitions. Of particular importance are the definitions of *complementary* and *independent* views. The reader could choose to skip this and refer to it as necessary.

Section §9.3 reviews possible translation strategies, starting with a rough taxonomy of strategies based on the level of generality at which they operate. It then examines the advantages and disadvantages of *ad hoc* translation schemes and schemes based around product databases. Finding these wanting, we go on to discuss the use of a complementary view which stays constant as the view of interest is updated. This is seen to have many advantages and bridges the gap between totally *ad hoc* methods and the restraints of product database formulations.

Although the use of complementary views allows a level of predictability over the updates to the underlying data, it has unpredictable failure semantics, and §9.4 introduces the new concept of a *covering view* that supplies sufficient information to predict failure. The security implications of this are discussed.

9.1.1 Complementary functions

In addition to the problem of editing an object through its display, there are several other problems which, although differing markedly in some respects, all share a common structure:

- We are viewing a representation of an object on a display and then change it (perhaps insert a character into a text file). We wish to update the display image by making small changes rather than completely recreating the picture.
- We are editing some view of a large database; we wish the underlying database to change in a way consistent with the changes in the view. We will see that this is usually far simpler than editing the display.

- We want to say that the changes to an object are confined to a locality.
- We are editing several objects, all of which must remain consistent with one another. If one is changed, the rest must change to remain consistent.

Each of the above problems reduces to the following: we have two sets (A, B) representing the objects/views and a relation (r) between them; we have a transformation function $f : A \rightarrow A$ and wish to find a *complementary* function, $cf : B \rightarrow B$, which is consistent in that:

$$a \leftarrow r \rightarrow b \quad \Leftrightarrow \quad f(a) \leftarrow r \rightarrow cf(b)$$

That is, the following commutes:

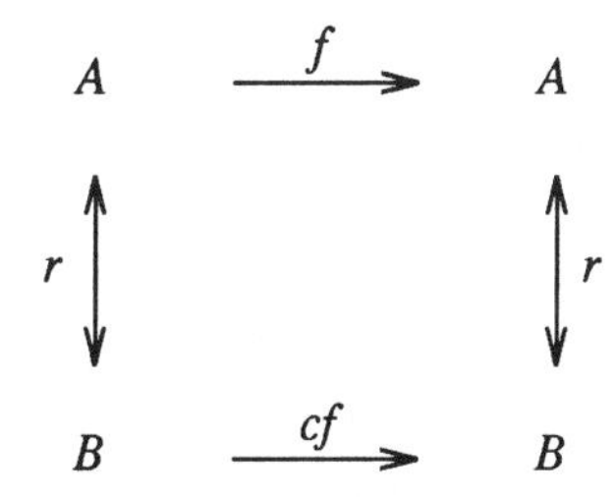

In cases (i) and (ii) the relation will often be directed: in case (i) A is the object, B the view and r is the viewing function $A \rightarrow B$; in case (ii) the function is the other way round.

We could call this the "spider's legs" problem. The legs represent the various views and they are joined by the body (the relation): as we pull one leg the others must follow. Note that this is equivalent to the problem of polymodality in logic programming. The finding of a complementary function is sometimes called translation (Bancilhon and Spyratos 1981).

9.1.2 Principles

Several problems may occur in defining such a translation scheme:

- *Uniqueness* – Unless the relation is functional $A \rightarrow B$ there will be many choices of cf for a given f.
- *Failure* – Again, unless r is a function $A \rightarrow B$ there may be *no* complementary function possible that satisfies the constraints.

Similarly, when we have chosen a particular translation scheme, we still have the same problems in new guises:

- *Uniqueness* – If there are several updates that achieve the same effect on A, do their complementary functions achieve the same effect on B?

That is, for any pair of updates f,g and objects a, b such that $a \leftarrow r \rightarrow b$, is it true that:

$$f(a) = g(a) \quad \Rightarrow \quad cf(b) = cg(b)$$

- *Failure* – Even where there is a consistent choice of a complementary function, it may not be defined for a particular translation scheme. For example, we may disallow the editing of key fields in a record, perhaps forcing the user to delete and re-enter such records. Further, for a particular update the complementary function may be partial. We may allow the editing of the department field of an employee record, but insist that the new department should already exist, even though in principle it would be possible to create a new department spontaneously.

We can relate these issues to the usability principles of predictability and reachability (Chapter 3) and to sharing properties (Chapter 4):

- *Predictability* – Does the user know what translation scheme has been chosen? Does the scheme admit a model by which the user can infer its behaviour? Can the user predict whether a given update will *fail*? If not, what additional information is required?
- *Reachability* – The choice of translation scheme will affect the updates that are possible to B via the complementary functions, and also, because of the choice of failed updates, those that are possible to A.
- *Sharing* – In the case of several views being edited, the choice of the translation scheme will affect sharing properties such as the commutativity of updates on different views.
- *Aliasing* – Do the users know what they are seeing? That is, do they know what the viewing function is? If they do not, there is no way they can relate what is seen and manipulated in the view to the underlying state of the system.

9.1.3 Complementary views – what you can't see

Aliasing reminds us that it is important that the designer and user are agreed as to what is in a view. For example, my bank statement has a running account balance. My interpretation of this view is how much money I have got, so if the amount is always positive I think I am in the black. However, the bank regards monies from cheque deposits as being available only when the cheque has cleared, and thus although the amounts shown on my statement are always positive I get charged interest and transaction charges.

The problem is partly one of observability: the statement does not tell me my available balance, and thus does not tell be whether I am "really" in credit. To obtain this information from the statement means I have to remember what proportion of my deposits are cash and which are from cheques. It is also an aliasing problem, as a bank statement based on the actual date of deposits and one based on available balance would both appear the same. That is, the *content* of the view does not tell us about the *identity*.

Aliasing is a thorny problem and may only become apparent when a breakdown occurs (as when I was charged on my account). However, the designer and the user do at least have the common view before them and this can serve as a focus for thrashing out a common interpretation.

We can characterise the problem of aliasing as *knowing what you can see.* A new issue that will arise as the chapter progresses is the importance of *knowing what you can* **not** *see.* Whereas for observability the important issue for designer and user is agreeing on what they do see, the crucial issue for update semantics is agreeing on what they do not see.

This seems a rather surprising statement, but the thing that makes it possible to understand change is knowing that certain other things do not change. When we edit one file, we expect other files to remain unchanged. When we pull the plug out of the bath, we expect the water to stay in the sink. So, for each view we will look for a *complementary view* which remains constant through updates. It is hard enough to get over the problem of aliasing. The designer and the user have trouble agreeing on what they *can* see. It is clearly even more important that explicit effort is made in the design and documentation of systems so that the user knows about this often implicit but crucially important view. Knowing what stays constant, i.e. knowing what you do not see, is fundamental.

9.1.4 Extra complexity – the text editor

The data handled by typical interactive systems turn out to be far more complex than a typical DP database. This is the case even for apparently simple data structures such as the text of a document. We'll take a look at a text editor and the choice of complementary function for the two modalities.

The text consists of an unbounded number of lines all of the same fixed length (for simplicity) and a cursor position within the text. The display consists of a fixed number of lines of the same length as the text and a cursor also. The invariant relation will be that the display should be identical to the portion of text it would cover if the cursors were lined up.

- *Editing the text, display following* – Take, for instance, inserting a new blank line after the cursor. We are then faced with problem of non-uniqueness: should we lose the top line of the display, or the bottom? Usually the choice is different depending on the position of the cursor in the display.

- *Editing the display, text following* – This case is more difficult. If we again consider inserting a line on the screen, and say the bottom line of the screen is scrolled away, would we expect it also to disappear from the text? If not, why not? It would certainly be consistent with the invariant: obviously, some choices of complementary function are better than others! Even more bothersome is the case of deleting a line. Usually a new line "appears" from the hidden text to fill the gap. We are hardly then editing the view. In fact, to describe this adequately (and line insertion), we need a slightly more complex idea of complementary function. When we insert into our (say) 80x25 screen we obtain a 80x26 screen, and deletion similarly gives an 80x24 screen; we then expect these to bear a relation to the actual new display. Thus our screen update is from the real display (D) to some sort of extended display (X). We get the following diagram:

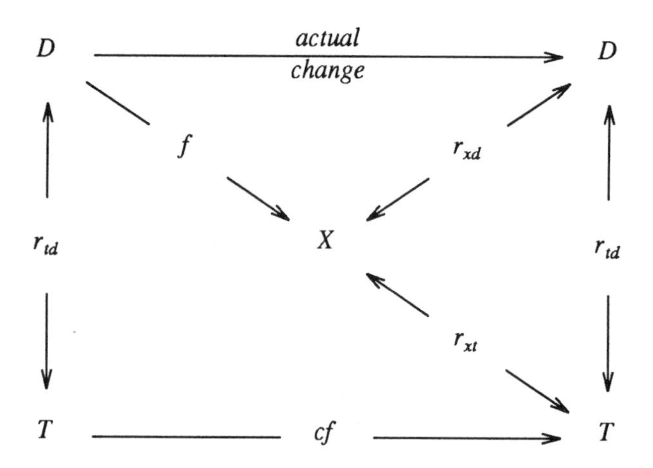

Given such complexity, is editing the view, rather than the object worth thinking about? For simple cases (like text!), probably not, as it is reasonable to expect the users to perceive the underlying object as the item with which they are interacting. However, as the underlying object and the map between it and the view get more complex, it becomes more and more likely that the user will manipulate a surface model. For instance, when editing a file one rarely thinks in terms of the reads and writes to the disk and the allocation algorithms for disk space that are going on underneath; the file system is just such a complex view of the disk.

One other thing that is worth noting from the example is that we tend to think of the relation between text and display as functional: "this display is the text between the 53rd line and the 77th". This is a valid perspective and the one we will use for much of this chapter. However, in the case of the display of a text, the view moves. We want dynamic or semantic views, not static ones. Fortunately, in many situations the forms of view do not require this complexity.

9.1.5 Differences from standard databases

The example above shows that we should not expect all the properties of DP-style databases to hold for the more general case. In most DP situations, the structure of the database is fixed for long periods of time, and changing that structure is seen as a major, expert task. If we consider general systems for which we may want to design an interface, this may not hold. For instance, we may want to design an interface for manipulating a file system. We clearly want to create and delete files as well as edit them.

In addition to this structural change, we also may have far greater structural complexity. So for instance, a portion of a program syntax tree is far more complicated than a list of personnel records.

9.1.6 Summary

The concept of a complementary function over a relation captures many diverse problems, but we here confine ourselves to the case where the relation r is a function and usually refer to it as v, the principal view.

We must be prepared to examine the cases of dynamic, structurally changing and structurally complex views. However, we initially look at the simple case of possibly complex but static views. Further, we assume until §9.5 that the database is structurally static. This means that all views are valid for all database states.

9.2 Algebra of views

The rest of this chapter considers complementary functions on databases resulting from updates to views. This section gives a short exposition of those aspects of the algebra of database views that will be needed later. The definitions are not new; they can be found elsewhere (Bancilhon and Spyratos 1981).

A view is regarded simply as a function on a database, $v: Db \rightarrow range(v)$. If the view is being used to produce some sort of output for the user, then the form of the map will be very important. However, as in the case of the definitions over PIEs, we do not consider fidelity conditions here. Other views will be wanted for their conceptual or definitional value, and are not intended to represent a user image. In these cases fidelity will be irrelevant and only the information content of the view is important. Hence two views would be equivalent if they had the same information content.

9.2.1 Information lattice

Where we are interested only in the information content of views, we can define an information ordering on them. One view is considered a subview of another if the former can be derived from the latter:

{ *subview* }:
u is a subview of v, $u \leq v$ **iff**:

$$\exists\, g : range(v) \rightarrow range(u) \quad \textbf{st} \quad \forall\, db \quad g \circ v(db) = u(db)$$

As when we considered PIEs, definitions concerning information can be made using functions (as above) or more implicitly:

$$u \leq v \quad \hat{=} \quad \forall\, db, db' \quad v(db) = v(db') \quad \Rightarrow \quad u(db) = u(db')$$

That is, we can always work out what is in view u if we know what is in view v.

This clearly gives a partial order to the views. This partial order admits a least upper bound or supremum by using the product view:

{ *product* }:

$$(u \times v) : Db \rightarrow range(u) \times range(v)$$

$$(u \times v)(db) \quad = \quad (u(db), v(db))$$

This is clearly the minimal view (with respect to $\leq$) of which both u and v are subviews.

If the views are for the user, then this corresponds to a new view where both the originals are visible. So it would still be the least upper bound even if we added quite strict fidelity rules on the information order.

There is also a maximal element, amely the identity function on the database, id_{Db}. In both the case of the supremum and the maximal element, whether they are acceptable or not as a "real" visible view depends on the set of views regarded as acceptable. In the case of normal databases the former is probably acceptable, whereas the latter, implying that the user can see all the database at once, is arguably not reasonable. For the case of display views of texts, even the supremum represents a display twice as large as is possible. However, we only need to use the supremum to make definitions, so it can be used even when unacceptable as a normal view.

9.2.2 Independent views

It will be important later to know whether two views are interrelated or not, that is, whether the views can vary independently of one another. We can define independence of views:

{ *independence* }:
u, v are independent **iff**:
$\forall\ a \in range(u), b \in range(v)$
$\exists\ db$ **st** $a = u(db)$ **and** $b = v(db)$

That is, for any given pair of possible values for u and v, we can find a database simultaneously yielding those views.

As well as using this concept later when we consider translations of views, it is clearly important when we consider sharing. For example, consider two views being presented in two windows. The database is being manipulated by editing the views and we assume all updates to the views are possible. Clearly, the two windows cannot be display independent unless the views are. If the views were not independent then there would be circumstances when a change in one view would yield a situation where there would be no consistent database state yielding both views.

As in the case of display independence in Chapter 3 and independence of blocks in the previous chapter, pairwise independence does not imply independence of more complex configurations. For example, consider a database consisting of three numbers, a stock level, an amount on order, and a minimum stocking level:

$$db = stock\_level, amt\_ordered, min\_stock$$

There is the sensible constraint that the actual stock level plus the amount on order must always exceed the minimum stock level:

$$stock\_level + amt\_ordered \geq min\_stock$$

If we consider the three views giving the three numbers separately, then each view is independent of the other two;‘ however, they are clearly not independent as a group. Thus in this example (and in general):

u independent of v, **and** u independent of w
does **not** imply u independent of $v \times w$

Similarly, we could create arbitrarily complex situations with n-way independence but not $n+1$-way independence.

In this example, we could have "normalised" the views, to make them independent, by considering the view set:

$$\{\ stock\_level,\ amt\_ordered,\ stock\_level + amt\_ordered - min\_stock\ \}$$

These views are independent in all configurations, but the last view is, of course, by no means sensible. For example, updating the stock level when an item was sold would implicitly reduce the minimum stock level! This example gives us some inkling of the complexity of view editing.

9.2.3 Complementary views

In terms of the information lattice, the opposite of two views being independent is their being *complementary*. That is, together they contain all the information in the database:

{ *complementary views* }:
u and v are complementary **iff**:
$$\forall\ db, db' \quad u(db) = u(db') \textbf{ and } v(db) = v(db') \quad \Rightarrow \quad db = db'$$

or, in terms of the supremum:

$$(\,u \times v\,) \quad \equiv \quad id_{Db}$$

A third, functional formulation is:

$$\exists\, f : range(u) \times range(v)\ Db \quad \textbf{st} \quad \forall\ db \quad f(u(db), v(db)) = db$$

The function f is, of course, unique and is the inverse to the supremum, so we could say:

u and v are complementary **iff** $(\,u \times v\,)^{-1}$ exists

In general there may be many views complementary to a given view. Consider a database consisting of two integers, x and y, constrained so that $x > y$. If we have one view giving x (call it X), then the view giving the value of y is complementary to X, but then so is the view giving y cubed. These two views contain exactly the same information so could be said to be equivalent. But even if we consider information content, the view yielding $x+y$ is also a complement to X but is different from Y. Further, we may add information to either of the views, and they remain complements.

In fact, view independence and complementariness satisfy the obvious conservation properties with respect to subviews:

if $u \leq u'$ **and** $v \leq v'$
then
u' and v' independent $\Rightarrow$ u and v independent
and
u and v complementary $\Rightarrow$ u' and v' complementary

9.2.4 Summary

In summary, we have defined four basic concepts:

- *Information or subview ordering* – This says when one view contains enough information to determine another.
- *Supremum over this ordering* – This yields a view ($u \times v$) containing as much information as u and v put together.
- *Independence* – This tells us when two views can take any pair of values.
- *Complementary views* – This is when two views are sufficient to determine the entire database and, in particular, implies that there is an inverse to the supremum.

9.3 Translation techniques – complementary views

This section reviews different ways of translating updates to views into updates on the underlying database. It starts off by recalling the translation problem, and putting it in terms of the notation that is used in the section. This is followed by a subsection which classifies translation strategies into different levels.

Sections §9.3.3, §9.3.4 and §9.3.5 deal in turn with *ad hoc* translation, translation when the underlying database can be regarded as a Cartesian product of views, and finally, translation using a complementary view.

9.3.1 The view update problem – requirements

To recap the situation, we are concentrating on a view v on a database Db. We wish to find a complementary function for any updating function f on the view. In the terms used by the database community, this process of finding the complementary function is called *translation.* Thus the problem which we address is:

{ *translation* }:
given a view v and a function $f : range(v) \rightarrow range(v)$,
find a translation $Tf : Db \rightarrow Db$ such that:

$$f \circ v = v \circ Tf$$

If we think of the view update as giving us an explicit change for a part of Db, the translation problem can be seen as determining what happens to the things that are not mentioned explicitly in the view, or are specified only partly by the view update. It is thus a form of the *framing* problem.

9.3.2 Translation taxonomy

Two major determinants of the update strategy are:

(i) What update operations are available as translations?

(ii) How much information about the update is required to calculate the translation?

Dealing with point (i) first of all, it is clear that there may be limitations on the sorts of updates that are possible. For instance, in the case of updating a display in line with a text, there are the available terminal codes, and in the case of a view on a database, there may only be certain operations allowed on the data by its underlying implementation. It will often be assumed that any transition between consistent states is possible. This assumption may or may not be reasonable. On the positive side, we must assume the set of updates available are complete, in that any consistent state can be obtained via some series of updates; if this is not so then we must assume there is something seriously wrong with our implementation. It is fundamentally unreachable. On the negative side, even though in principle such a sequence exists, it may be arbitrarily complex and therefore hard to generate. Such complexity is potentially disastrous in a shared database, where a large part of the data may be locked during a very long transaction. An important practical consideration for any translation scheme, is that the complexity of the update should be of the same order as the complexity of the view function.

Considering (ii), we need to clarify the level of instantiation at which we wish translate. Typically, updates are parameterised functions, for example:

"insert the character _?_ into the text"
"change date to __?__"
"delete the record with key __?__"

Thus the general update is of the form:-

level 1 – "add record": $param \times A \rightarrow A$

We can then instantiate this for a given parameter:

level 2 – "add record *Fred born on 29/2/60* ": $A \to A$

and finally, for a given view instance, $a \in A$:

level 3 –

birthdays	
NAME	DATE
Jill	3/7/53
Tom	27/10/61
Glenda	15/4/47

→

birthdays'	
NAME	DATE
Jill	3/7/53
Fred	29/2/60
Tom	27/10/61
Glenda	15/4/47

In the expression of the translation problem, we have dealt implicitly with level 2; however, if the translation function *Tf* is not a basic update operation on *Db* but instead does different updates depending on the the exact value of the view, then it is operating at level 3. On the other hand, if it is generic over a parameterised class of update functions then it operates at level 1.

When a translation scheme operates at level 1 or 2 , giving for each update operation a fixed sequence of database updates irrespective of the database state, we can say it is *context independent*. For example, when keeping a text in line with a screen display, many operations yield context-independent updates; so that "insert _?_" at the cursor point on the display translates to "insert _?_" at the text cursor, irrespective of the current text and display states.

On the other hand, some translators at level 3 may need the value of the view to decide on the update, but may not need to know the particular update function (e.g. *add_record*) used. That is, they have the property that they are not dependent on the *type* of the update, only on the previous and updated states. We could call such translations *process independent*. An example of this would be files as views of a file system. When an edit is finished, the update to the file system is dependent only on the final value of the edited file, not on the particular edits which were done to the file.

Other translations will be hybrid: for instance, when keeping a screen display in line with a changing text, the update "delete the line below the cursor" may become something like:

> "delete the line below the cursor,
> then :
> if the cursor is near the bottom of the screen add the relevant line of the text to the bottom of the screen,
> otherwise add the relevant line of the text at the top."

It should be noted that a translation may be either context or process independent at the behavioural level, yet have an implementation that does not posses this property. This is particularly the case if the update strategy is defined using operations that are not available on the database directly, but have to be simulated: in particular if one assumes free update of the database.

9.3.3 Ad hoc translation

We can approach each task of update translation in a one-off way. In the short-run, this is the simplest translation strategy. For instance, we might supply an index of variables used in a program:

```
amount      occurs 2 times
    amount  is static variable in main
            declared line 3
            used lines 7 and 8
    amount  is parameter to do_work
            declared line 73
            not used
total       occurs 1 time
    total   is static variable in do_work
            declared line 77
            used lines 86, 87, 93 and 101
```

We begin to ask what updating various fields might mean. We decide that updating the variable name `amount` in the "`occurs`" line would be regarded as a global renaming of all occurrences of the variable name. However, it would have to fail if this resulted in any name clashes. Similarly, we might want to allow editing of the variable name in the "`is static in main`" line. This would rename the variable just in that context; again it might have name clash problems, but would also mean moving this to a new subheading. Editing the procedure name within which a variable is used could be interpreted as a renaming of the procedure; however, editing the line numbers where a variable is referenced makes no sense at all.

Even where updates are allowed, we would need to be careful. For instance, if we wished to edit the variable `amount` to read `total` we would have an intermediate state which would lead to name clashes. The edits would therefore have to be committed in non-atomic units.

Clearly it is a complex job deciding what is updatable, and what those updates mean. Features of translation that are apparent here are:

- Only a subset of possible update operations are allowed. Further, this example has assumed that updates are the natural ones for the view. In fact, the set of updates will usually be defined more rigidly by the underlying data objects and are not necessarily sensible ones for the view.

- Those updates that are allowed may fail sometimes. (This is a feature of all translation strategies, and of normal updates.)
- The semantics of updates may have to be changed radically: for instance, editing a variable name in an individual use would relocate that use to a different variable heading. (Again, this is not solely a problem of *ad hoc* strategies.)

Ad hoc strategies have several disadvantages:

- *Predictability* – It may not be clear to the user what the effect of an update will be. In particular, it may be that different ways of achieving the same change in the view may have different effects in the underlying database.
- *Reachability* – It may be very unclear what changes can be achieved through the view.
- *Dishonesty* – It is easy to fool oneself and others that it is the view that is being edited, whereas this is really a subterfuge for editing operations on the underlying object. This is particularly a problem if the concentration is not on "what would editing this value mean", as in the example above, but instead on "what can we do with database updates".

Perhaps the most important of the criticisms above is the predictability of the consequent changes in the unviewed parts of the database. For instance, in our stock control example, it would be sensible for any update of stock level to generate an order and hence change the number on order, but not reduce the minimum stock level. Making this sort of inference in an *ad hoc* system can be difficult.

On the other hand, an *ad hoc* approach has important advantages:

- *Generality* – There are no fixed limitations to the sort of view we can tackle. But it may be difficult!
- *Task orientation* – It is designed for a particular task and can be fitted to it. For instance, it may be deemed inappropriate to edit procedure names via a variable names index. This form of task-specific limitation may be difficult to achieve using a more unified strategy, and even where it is possible, it is less likely to be noticed.

Because of these disadvantages, and in spite of these advantages, we now look at more unified and generic approaches to translation.

9.3.4 Product databases

The simplest case of view update is when Db can be expressed as a Cartesian product and the views as "slices" of this. That is, $Db = \prod_{\mu \in M} A_\mu$ and:

$$\forall v \in V \quad \exists \mu_1,\ldots,\mu_n \ \textbf{st} \quad range(v) = \prod A_{\mu_i}$$
$$\textbf{and} \ v(<a_\mu>_{\mu \in M}) = <a_{\mu_i}>_{i \in \{1,\ldots,n\}}$$

In this case, for any given v we decompose Db into a product $A \times B$ where A is the product of the A_μ "sliced" by v and B is the rest. Then we have simply $v(<a,b>) = a$ and for any update $f : a \to a'$ we have a translation to $Tf : <a,b> \to <a',b>$. A simple (flat) file system is an example of a product database where the individual files are the views.

If the situation is more complex and there are invariants to be preserved of the tuple, then it is tempting simply to amend this so that for any consistent pair $<a,b>$ the update $f : a \to a'$ has a translation $Tf : <a,b> \to <a',b>$ if the latter is consistent, otherwise it fails. This strategy works quite well if the consistency relation is simple.

An example of this is a relational database with the views being the relations. If we can normalise the relation so that the only consistency requirements are those of non-null key fields and existent or null alien keys, then the former is within a single relation and the latter is the only cross-relation (and hence cross-view) constraint on update. The reason why this particular relation is acceptable is that the reachability is preserved; in order to delete a tuple, we can simply go round the views nulling all references to it, and then remove it. Similarly, to add a set of mutually referencing tuples we can add them with nulled alien keys and then go round filling them in when all the tuples are in place. Not all cases are so simple, and it is easy to think of consistency requirements that totally disallow certain updates. One of the aims of normalisation is precisely to make these consistency requirements simple, and to push as much semantics as possible into the relational structure.

In principle, any database with views could be regarded as the Cartesian product of those views with suitable consistency requirements, but unless these requirements are very simple (as above) there is not necessarily much to be gained in terms of update specification. The approach taken by Khosia *et al.* (1986) is considerably more complex, being based on modal logic. However, their view that a database should be specified first as a collection of views, has some of the characteristics of a constrained Cartesian product. Their update rules between views are *ad hoc*.

9.3.5 Complementary views

The general idea that, for a *particular* view, we can decompose the database into a product is more useful. In fact, the attempts to deal with the update problem are precisely along these lines. Essentially we look for another view (u say), such that between them v and u determine the database uniquely. This is precisely the condition that u is a complement to v. We can then say that for any db with $a = v(db)$ and update $f : a \rightarrow a'$ the translation is such that $u(db)$ remains constant. If no such update is possible, then the original update to the view fails:

{ *translation* }:
let $a = v(db)$, $b = u(db)$ **and** $f : a \rightarrow a'$ be an update
if $\exists\, db'$ **st** $v(db') = a'$ $u(db') = b$
define $Tf : db \rightarrow db'$
otherwise *fail*

That is, the following diagram commutes or the update fails:

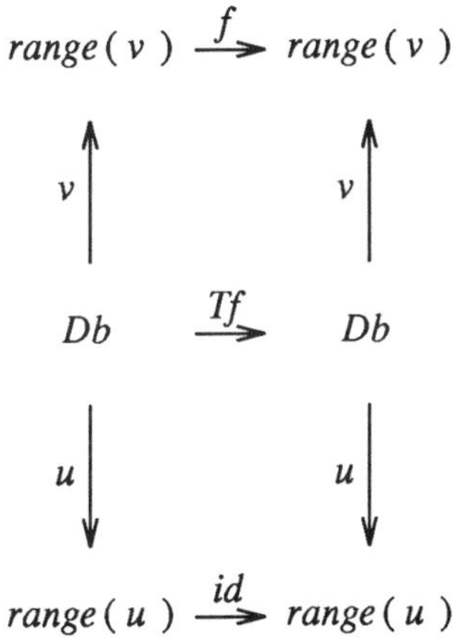

$$
\begin{array}{ccc}
range(v) & \xrightarrow{f} & range(v) \\
\uparrow v & & \uparrow v \\
Db & \xrightarrow{Tf} & Db \\
\downarrow u & & \downarrow u \\
range(u) & \xrightarrow{id} & range(u)
\end{array}
$$

Note that this is precisely the same situation as the simple product, as we can regard the valid database states as equivalent to the set of pairs $<v(db), u(db)>$. Alternatively, we could say the product database translation is a special case of this, where we choose for any view v with $range(v) = \prod A_{\mu_i}$, the complement u where $range(u) = \prod_{\mu \neq \mu_i} A_{\mu}$.

Using complements is not only a nice way of obtaining a translation, it is essential (either implicit or explicit) to achieve reasonable properties. Bancilhon and Spyratos (1981) show that any translation rule T that satisfies:

{ *morphism* }:
$\forall f, f' : A \rightarrow A \quad Tf \circ Tf' = T(f \circ f')$
and
$\forall f : A \rightarrow A \quad f(a) = a \Rightarrow Tf(db) = db$

where $v(db) = a$, must be given by such a complement. Further translations generated by complements are *process independent*, which enhances their predictability.

There are, of course, many such complement views for a given view. The bigger the view, the more updates it fails. It is therefore sensible to look for a "minimal" view. We can measure minimality using the information ordering, and then ask for a complement that is minimal with respect to this. However, even such a minimal complement cannot be guaranteed to be unique.

The special case of the unconstrained product is obtained if the views are also independent. If a complement is independent, then it is automatically minimal although it is still not unique. However, Cosmadikis and Papadimitriou (1984) show that where the database is relational, there is a unique independent complement so long as we only look at monotonic views. (Here monotonic means that as one adds tuples to the database, tuples are added to the view.)

This special case is not likely to be of general use for complex views, however, so we must be content with having a non-unique complementary view and telling the user what it is. Thus the procedural uncertainty of the user is reduced from the question of "What will this update do?" to "What will stay constant when I do this update?". The latter is static information and more easily conveyed and memorised than the former. It does depend somewhat on the users' preferred learning strategy but at least they have the choice.

9.3.6 Discussion

Clearly, translations generated by complementary views have a lot of advantages in terms of predictability and the consistency of update semantics. The unique complementary independent view derived by Cosmadikis and Papadimitriou is based on assumptions unlikely to obtain for the majority of complex views found in general interface design. The choice of a complementary space becomes a major decision. In particular, it is where task knowledge can be brought into the design process. For instance, if we were trying to form a complimentary view for the variables' index view in the example earlier, we would probably put the procedure names in the complementary view. In fact, this would be an especially interesting choice, as it deliberately makes the complement non-minimal, and is an indication that even where minimal or independent complements can be found, this is not necessarily what is wanted.

Of course, this is hardly a novel point, as most operating systems distinguish between the permissions to view and modify data.

Product databases have a lot of advantages:

- All views have an independent complement, implying no failure for internally consistent view updates.
- Reachability, as we can change components independently.
- Procedural uncertainty is minimised, since the complement is obvious.

However, they also have some disadvantages:

- No complex, cross-database views, implying no views like indices, lists of cross-references, etc.
- No hiding, formatting, etc., so all views give complete information on some part of the database.
- No structural change: the product and its views are fixed.

We can alleviate some of these disadvantages. The lack of complex views can be alleviated by having additional *read-only* views that are not necessarily projections. These could be used for navigation but not update. Hiding and formatting can be achieved by allowing several views of each primary view, whose update relative to the primary view is defined either *ad hoc* or using a complementary subview. The semantics of the subview updates can be cross-checked against the explicit view if they are unclear. The issue of structural change is difficult for complementary view translation as well as for product databases, and we consider it in a separate section (§9.5.3).

The real problem with using product databases, is that it is not always easy, or meaningful, to express a design in that way. Having said that, the normalisation process in relational databases attempts to approximate a product database, so it is not an unreasonable goal. Still, for general interfaces we are likely only to approximate the product design in some places, and have to search for more complex complements elsewhere.

For some views it is very difficult even to find a complement. For example, the proper complement for the variables index is not very obvious. If we find ourselves forced to use an *ad hoc* translation strategy, we can still impose a loose complementary view methodology. That is, we can concentrate on what is to be left unchanged in the database when a view is updated, and design our translation around that general idea. It is likely to be helpful for the user to know what is constant, even if this does not uniquely define the update that has occurred. This is an example of searching for a deterministic ground, as discussed in Chapter 6.

In the next section, we consider further the properties of complementary views, in particular their sharing properties, and the predictability of update failure.

9.4 User interface properties and complementary views

We have already noted how the use of a complementary view, where it is possible, increases the predictability of an update strategy, as the user need only know what the complement is and can then infer the result of view updates. In this section we look further at the interface properties of complementary views.

We begin by looking briefly at aliasing problems with complementary views. After this we consider sharing properties and how these relate to sharing of windows as defined in Chapter 4. We then turn our attention to the predictability of update failure, and introduce the concept of a *covering view*. Finally, we assess the security implications of imposing a covering view as an interface requirement.

9.4.1 Agreeing upon the views: aliasing and worse

The designer of a system presents a view to the user. The user sees the contents of the view as the system runs, and infers some idea of what the view is. Of course, as we have seen before, aliasing means that the user and the designer need not agree. Content does not determine identity. Two views may look the same and yet be different. Note that this form of aliasing is more complex than, say, knowing one's position in a document. There we assumed that the user knew the *kind* of view, and the problem was knowing where in the document the view was. Here there may even be confusion about the nature of the view itself.

Imagine we have a programming system that, when the user clicks upon an identifier, gives an index of other references in the program through which she can further navigate. One day, the programmer is looking at a paper listing and wants to get to the line containing the identifier `sum`. Coincidentally the variable `sum` is on the screen, so she clicks this. Unfortunately, it does not reference the line which she wants. Why? Well, the programmer had assumed that the indexing was lexical, to all variables with the given name. In fact, the designer's model was that only the incidents of the same semantic variable were given. The form of the indices for the two views is the same, a list of references, but the views are very different.

Note that this form of aliasing is closely linked to *procedural uncertainty* and *conceptual capture*, as discussed in Chapter 6. This can be attacked only partly through the immediate user interface; the full solution must involve both study of the natural models of users and the production of suitable documentation to explain the nature of the view. A much broader idea of the "interface" is required to address problems of aliasing that arise through procedural uncertainty than those that arise through data uncertainty.

Now, if it is hard for the designer and the user to agree about what they *do* see, how much more difficult it is for them to agree about what they *do not* see. The complementary view is, by its nature, not presented in the interface, and thus misconceptions about what does not change may take a long time to emerge, and the possibility of confusion is enormous. Note that this is not just a problem when the designer explicitly uses a complementary view approach. Whatever view update strategy is chosen, the user is not and *cannot* be aware of all the hidden ramifications of their actions. In the simple case of a product database (if the user is aware of it) the user's actions can be made *local* to the particular attribute. In a more complex setting, the notion of locality depends crucially on viewpoint and the problems are precisely akin to those of determining the complementary view.

There is no easy answer to this issue of agreement between user and designer. The best hope lies in the designer being aware of the problem.

9.4.2 Sharing properties of complementary views

We consider again the sharing properties of windowed systems, where each window is a view of an underlying database. In the section on the algebra of views, we said that if all updates succeed, then view independence guarantees display independence (§9.2.2). Of course, when we consider complementary-view-based translation, some view updates will fail. This means we can be more generous about the views that do not share, as some view updates that would have resulted in display interaction will now fail. Display independence says that the display of one window stays the same as commands are executed in another window. If all the commands are view updates and the complement to the view is constant, then if the first window is a subview of the complement we can guarantee display independence. That is, assuming the complement to the view v is u:

the window of v' is display independent of the window of v **iff**
$$v' \leq u$$

Note that the condition is not only sufficient but necessary.

PROOF:
If v' is not a subview of u, then we can find two database states db and db' such that:

$$u(db) = u(db') \quad \textbf{and} \quad v'(db) \neq v'(db')$$

However, because v and u between them determine the database we must have:

$$u(db) = u(db') \quad \Rightarrow \quad v(db) \neq v(db')$$

Thus the update:

$$f: v(db) \rightarrow v(db')$$

would succeed and would change the value of v'.

Result independence is far more difficult. Consider the following example of two pairs of independent complementary views. We have a database consisting of three, unconstrained integers, (x, y and z). We will consider the two pairs of views:

x with complement (y, z)
y with complement $(x, y \times (x + z))$

These views are independent of one another and satisfy the conditions for display independence. We consider the two updates:

$$f_x : x = 1 \rightarrow x = 2$$
$$f_y : y = 1 \rightarrow y = 2$$

If we start in the database state (x=1, y=1, z=1), we consider the effects of the translations of these two updates performed in different orders:

$\{ f_x$ first then $f_y \}$:
$Tf_x : (1, 1, 1) \rightarrow (2, 1, 1)$
$Tf_y : (2, 1, 1) \rightarrow (2, 2, -\frac{1}{2})$ $\qquad 1 \times (2+1) = 2 \times (2-\frac{1}{2})$

$\{ f_y$ first then $f_x \}$:
$Tf_y : (1, 1, 1) \rightarrow (1, 2, 0)$ $\qquad 1 \times (1+1) = 2 \times (1+0)$
$Tf_x : (1, 2, 0) \rightarrow (2, 2, 0)$

That is, the two updates do *not* commute, and hence windows based on these would not be result independent. This is all the more surprising, as our complements were independent and hence "good" ones with no failure. In this example, as the database was a simple product with no constraints, we could have defined the complements so that the updates to x and y did commute (though we might have good task-specific reasons for our definitions). In more complex databases such a choice may not only be inexpedient, but be impossible.

We can give a sufficient (but not necessary) condition for result independence of two independent views v and v' with complements u and u', respectively:

$$\exists\ u_{glb} \ \ \textbf{st}\ \ u_{glb} \le u \ \ \textbf{and}\ \ u_{glb} \le u' \\ \textbf{and}\ \ u \le (v \times u_{glb}) \ \ \textbf{and}\ \ u' \le (v' \times u_{glb})$$

Informally, u_{glb} is complementary to $v \times v'$ and thus all updates are observable from v and v'. However, again, unless one is using a simple policy for finding complements (such as the product database), such a view may not be possible or desirable.

9.4.3 Predicting failure – covering views

The big advantage of an independent complement is in terms of predictability. If the views are not independent then there are updates that are sometimes legal and sometimes not, depending on the value of the (invisible) complementary view. If we wanted to see all this complementary view in addition to the principal view, then by its definition we would be viewing the *entire* database; exactly what the views of the database are there to avoid. Clearly, we must look for a *covering view* (w) that is small enough to present to the user along with the principal view, and yet has enough information to predict whether a particular update will succeed or fail. We can formulate the property of the covering view as follows:

$$\{\ covering\ \}: \\ \forall\ db_1, db_2 \ \ \textbf{st}\ \ (v \times w)(db_1) = (v \times w)(db_2) \\ \exists\ db_1' \ \ \textbf{st}\ \ u(db_1') = u(db_1) \\ \Rightarrow \ \ \exists\ db_2' \ \ \textbf{st}\ \ u(db_2') = u(db_2) \\ \textbf{and}\ \ v(db_1') = v(db_2')$$

That is, if for any two database states both the principal view and the covering view are the same, then any consistent update on the first database is consistent for the second. Note again that for any given view and complement, even a minimal covering view is not necessarily unique.

Example

Consider the case of a relational database with the following relations:

employee	
EMP_NAME	EMP_DEPT
Diane	Sales
Fred	Accounts
Tom	Sales
Jane	Clerical
Helen	Sales

department		
DEPT_NAME	BUDGET	other ...
Sales	500	
Clerical	250	
Accounts	3000	
Stores	20	
?Personnel??		

The update view is employee, and every EMP_DEPT must be a valid DEPT_NAME. We try to enter a record for Gillian in the personnel department (just formed); this succeeds or fails depending on whether there is an entry in the department relation for Personnel or not. A suitable covering view for this is clearly the set of department names.

Another example would be if u were a statement in a Pascal program (and Db the set of valid Pascal programs); then a possible covering view for u might contain a list of all the variable names in the scope of the statement and their types.

In fact, we can ask for a covering view even when the translation is not generated from a complement. In this case we have to amend the definition to say that w is a covering view for an update f to a view v if:

$$\begin{array}{l} \forall\, db_1, db_2 \;\; \textbf{st} \;\; (v{\times}w)(db_1) = (v{\times}w)(db_2) \\ \quad (\, \exists\, db_1' \,\textbf{st}\, Tf : db_1 \rightarrow db_1' \,) \;\; \Rightarrow \;\; (\, \exists\, db_2' \,\textbf{st}\, Tf : db_2 \rightarrow db_2' \,) \end{array}$$

That is information from u and w together is sufficient to tell whether a particular update *to* u will succeed.

NOTE: the covering view deals only with the data uncertainty about the complement's effect on update failure. If users are unaware of what view they are manipulating, or what the constant complement is, then the covering view can only help but may not be sufficient to make this clear. In general, procedural uncertainty is very difficult to handle (viz. 200-page manuals!).

9.4.4 Security

Of course, one of the reasons for restricting access to a view of the database is security. The covering view isn't updatable, but might require information that is not intended to be available to the user.

This is not such a drawback as it seems: the very fact that the updates on the view behave differently depending on the hidden information implies that, in principle, it is possible to infer some of the hidden information. It is precisely this information that is required in the covering view. Thus if one finds that there is some item of information in the covering view that is regarded as confidential, one should think hard about whether that information was indirectly available anyway. This is similar to the case of statistical queries to databases, where one is allowed to ask for summary statistics of the data but is not supposed to have access to the data itself. Depending on the queries allowed it may be possible to infer the raw data from the summary statistics, breaking confidentiality (Jonge 1983). This argument should not be taken to its extreme: for example, if one enters a random user name and password to a system, then it will behave differently when they are valid and when they are not; however, it would not be reasonable to demand that all the user names and passwords be public! Clearly, one would trade off predictability against security in this case. Even so, predictability analysis and the creation of suitable covering views would actually form a powerful test of system security.

9.5 Dynamic views and structural change

In most of this chapter, we have assumed implicitly that it is meaningful to talk of *a* view, which can be applied to any database state. Typically, views may be more dynamic than that. We can consider three types of dynamism:

- *Content* – The view is defined by some feature of content: for example, the set of all records of people in the Personnel department.
- *Identity* – The view is defined indicatively: for example, *this* collection of records which I want to deal with no matter how their content changes. (The collection may initially be defined by content but is not constrained by it.) Again, in editing text we want to display *this* region of text, whatever changes happen elsewhere.
- *Structure* – Much of the discussion of database views assumes a constant database structure. In general systems this may change, and the set of views available at any time can change accordingly. Many database packages deal very badly with schematic changes to live databases.

The first of these categories is included because it has been described elsewhere as a dynamic view (Garlan 1986). However, that definition is in fact consistent with the definitions of static view given earlier in this chapter. The complication of it, compared with more syntactic views, is that update semantics are less clear. However, it can be dealt with using complementary views quite adequately. The second category is clearly related to the issue of dynamic pointers. We will reserve the title *dynamic view* for this category of view. The third category of structural change is very hard to deal with and we will see by example how easy it is to make a mess of this.

9.5.1 Dynamic views

We look at general views where we want the semantic identity of the view to be preserved. As with pointer spaces we have two distinguishing factors, the set of views available changes with the object viewed, and the notion of "sameness" between views depends on the operations that caused the database change.

The first of these can be captured by using a set of *valid views* for each database state, in a similar fashion to the valid pointers to an object we considered in the previous chapter. For the "sameness" property we can use the idea of a *pull* function again. For each update F, we associate a pull function $F.pull$ which takes views to their "natural" equivalents:

$$\begin{aligned}&\forall\ db \in Db,\ v \in valid\_views(db)\\&\quad \textbf{let}\quad db', pull = F(db)\\&\quad \textbf{then}\ pull(v) \in valid\_views(db')\end{aligned}$$

When we discussed various update strategies, we regarded *process independence* as being a useful trait. That is, the form of the update was unimportant, only the initial and final states mattered. This will almost always *not* be the case for dynamic views. We may easily be able to perform two different series of updates which would yield the same final database state, but affect the dynamic views in a totally different manner. We must not only know *what* an update does, but *how* it does it.

As an example of the importance of this, consider two records in a database:

$$\begin{aligned}&< TOM, 1973 >\\&< FRED, 1971 >\end{aligned}$$

We start with a view looking at Tom's record. We then edit the database exchanging first the names, and then the dates. A view defined by content would still look at $< TOM, 1973 >$, but a view based on identity would be of the record that now says $< FRED, 1971 >$.

The problem of view update translation is now correspondingly more complex: given $a = u(db)$ and $f: a \rightarrow a'$ we want a translation $Tf: db \rightarrow db'$ such that $(Tf.pull(u))(db') = a'$.

It is similarly more complex to talk about complements for such views. We cannot talk about *a* complement of a specific view, but instead of a complement *function* from views to views. This complement function must of course be preserved by the *pull* functions. If we let the set of complement views be $V_c : Db \rightarrow B$, a possibly different set of views than V, we get the following condition:

$$\begin{aligned}&comp:\ V \rightarrow V_c\\&\forall\, v \in V, F \in U, db \in Db\\&\quad comp(F.pull(db, v)) = F.pull(db, comp(v))\end{aligned}$$

For the map to define a complementariness relation, we need the view and its complement to determine the database state:

$$\begin{aligned}&\forall\, db, db' \in Db, v \in valid\_views(db), v' \in valid\_views(db')\\&\quad \textbf{let}\ \ u = comp(v),\quad u' = comp(v')\\&\quad \textbf{then}\\&\quad v(db) = v'(db')\ \textbf{and}\ u(db) = u'(db') \quad \Rightarrow \quad db = db'\end{aligned}$$

That is, if we have a value for the view, and a value for the complement, there is only one possible set of database state and view with these as their values. We can then use this to define a translation scheme for update, and the new view to use. It does not give us the general pull function for other views. This may not matter if there is only one view of interest, but if this is not so we need more structure again to obtain the relevant function. Alternatively, we can use the existing knowledge of pointer spaces and model dynamic views using these.

9.5.2 Using pointer spaces to model dynamic views

The properties one wants of dynamic views are clearly similar to those of dynamic pointers, so it is natural to consider using pointer spaces and subobject projections to give dynamic views. The notions of determinacy and independence generated by subobject projections are consistent with the definition of information ordering and independence for views. The only difference is that the subobject projection, as a function of a block b, is defined only for objects with a given set of valid pointers. This is precisely what we expect for dynamic views. Again, even when a sequence of operations maps back to an object with the same valid pointers, there is no guarantee that statically identical blocks are semantically the same (i.e. pulled to each other). So, for instance, we might operate on the string "abcde" by *delete*(3,_) followed by *insert*(2, "c" ,_):

$$abcde \xrightarrow{delete(3)} abde \xrightarrow{insert(2,\,c)} abcde$$

However, $pull((2,3)) = (3,3) \neq (2,3)$. As block pointers behave so much like the intended behaviour of dynamic views, it is natural to identify the object component of their subobject projection with the view.

This approach is taken further in Dix (1987b) and demonstrates the usefulness of the dynamic pointer approach. However, the use of dynamic pointers cannot hide the fundamental complexity of dynamic views. When we considered static views, we saw how difficult it can be even for the designer and the user to retain a common understanding both of what they do see and of what they do not see. This is even more difficult and requires more care on the part of the designer when the views are dynamic and essentially "move about" the database.

9.5.3 Structural change

Some years ago a colleague was demonstrating a powerful (and expensive) commercial relational database that he was evaluating. After seeing some of the basic operations, I asked how well it coped with changes to the database schema. Within seconds he added a new field to an existing relation. I was duly impressed. However, I noticed that the database manager had filled the empty fields with a special NULL entry. One of the options for a field was that NULLs were not allowed. What would happen if you asked it to add a new non-NULL field. He tried it and, sensibly enough, the database refused. Presumably you obtained a non-NULL field by first creating a normal field, filling it with non-NULL entries and then changing its attributes. What would happen if you made a mistake? One of the fields in the relation we were looking at had some NULL and some filled entries. Try changing that field to be non-NULL, I suggested. He did and the database complied. All the records with NULLs in them were removed!

Structural change is hard. The designers of the above database had obviously thought about it but had not considered all of the consequences. Further, error recovery mechanisms may not be as good when updates involve structural change.

A computer installation I once worked in used a large mainframe with the manufacturer's standard operating system. One of the attributes of a disk file was the number of tape backups that would be made automatically when the file was updated. Apparently this afforded a high degree of data security. However, one day an old version of a large file was required, the tape was found and an attempt made to restore it. Unfortunately it failed. In order to restore a file it had to return it to the same device with the same attributes, such as block size and file type, as when the backup was made. These attributes had changed and the values when the backup was made had been forgotten. The number of possible attribute combinations was large, and the information was not explicitly available from the backup tape itself. The relevant data were recreated from paper records!

The recovery mechanism above was designed to work only in the structurally static case. If the operating system's designers had thought about the possibility of structural change, they could easily have added some standard tape header giving the necessary attribute and device information.

If we have a structural view of some underlying data, the update semantics are clearly quite complex. The notion of a complementary view does not help us that much: the things we do not see may not change, they may simply cease to exist! A small part of the view, say a file icon, represents a vast and valuable piece of data. Small changes to the view may mean enormous changes to the data. In simple file manipulation systems this is recognised and often special dialogues are initiated if the effect of an action is gross. Most computer filing systems are configured as, at most, a simple hierarchy. With these, it is not too unreasonable to expect users to understand the impact of their actions and to expect systems to detect potential troublespots. If the organisation of data becomes more complex, say with an object-oriented database or a hypermedia system, it becomes harder for the user to appreciate the potential scope of their actions and harder for the system to tell which actions it should warn about and which it should not. For instance, in the Unix file system a single file may have several "links", that is, it may have several access paths through the otherwise hierarchical file system structure. Removing any link but the last merely removes an access path: the file is still there and (if you can find it) can be recovered. Removing the last link destroys the file, yet there is no readily apparent difference between these two operations. The consistency is very useful at a systems programming level, but is potentially disastrous at the level of the user interface. As with simpler systems, if the designer is too zealous in producing dialogue boxes and confirmation messages the user will, of course, reply habitually, and the benefit is lost. Finding the appropriate balance is not easy, and there appear to be no easy formulae.

The trouble with updating structural views is that they are powerful. On the one hand, this power can be of great *value*, allowing the appropriate structuring of information and adding to its usefulness thereby. For instance, the database described above allowed (albeit disastrously in some cases) structural change. If this were not supported, a costly copying process between the old and new formats would be needed. Of course, with that power comes also *risk*. Inevitably, powerful actions will occasionally go wrong, and this emphasises the need for equally powerful recovery systems. Recovery systems that work only when the data structure is static and assume that structural change will be performed only by faultless "wizards" are not worth a lot. The Macintosh wastebin is an example of a recovery mechanism for a structural view, so it is not impossible (however, even this tends to "empty itself" in certain circumstances).

We come back to the fact that structural change is hard. This means that we have to think more about it. So many potential designs are considered only in structurally static scenarios; this may well represent a large proportion of the use, but side-steps some of the most difficult design issues, relegating them to a phase late in the production process when they may be fudged.

9.6 Conclusions

We have investigated the problem of updating an object through views to it. Although *ad hoc* methods have some advantages, it is necessary to use some more predictable method to reduce procedural uncertainty. If the database can be seen as a product then the problems of update predictability and reachability are easy. Further, we can easily test for sharing and interference between updates. Traditional database theory tried as far as possible to simulate this position by the use of normalisation rules, and in a sense follows the route (strongly favoured in programming language design too) of pushing semantics into syntax. This often means trading power for predictability. This might be a good goal to aim for, but is unlikely to hold for many more complex views.

We have seen that researchers into database views have used the concept of a complementary view, which stays the same whilst the principal view is updated, as a way of defining the update for the database. This is more likely to succeed as a strategy than searching for a product formulation, and is less restrictive, yet it retains some of the predictability properties. In particular, it is only necessary to know what the complementary view is, to know the effect of updates.

Unfortunately, the failure semantics of updates are not in general predictable for complementary view updates and thus it is necessary to introduce the idea of a *covering view*. This contains sufficient information from the complement to predict which updates to the principal view will fail. The covering view would not be updatable in itself, and perhaps would have weaker sharing restrictions.

We saw that both the primary view and its complement have *aliasing* problems. Agreeing on the view presented to the user has some difficulties, but at least there is a common reference and differences in interpretation are likely to become apparent. The complementary view, by its very nature, is unseen and thus agreement is far more difficult. In both cases the aliasing is a form of *procedural uncertainty* and thus difficult to address purely within the on-line system. Solutions are bound to encompass the whole analysis of user's models and the production of documentation and training.

Although DP databases tend to have fixed structures, and the views on them tend to be gross and static, the same is not true of more general databases (e.g. graphics, program syntax trees) and thus we need to look at the case of dynamic views and structural change. Both these problems are considerably more difficult than the static case, and even that is not straightforward.

In conclusion, it is wise to search as far as possible for a product formulation of one's problem space and, failing that, look for complementary views. Even where these fail and an *ad hoc* approach is necessary, it will probably be advantageous to take a complementary-view-flavoured approach, and concentrate on what is to be left unchanged by an update. There is little reason for not including covering views, unless the noise these cause far outweighs the annoyance of unpredictable failure, or if it involves an extreme and pedantic case such as that of passwords and user names. For two rules to sum up the chapter:

- Make sure the users know *what* they see.
- Make sure the users know what they do *not* see.

CHAPTER 10

Events and status – mice and multiple users

10.1 Introduction

If we look back over the models presented in the previous chapters, we find an imbalance between the interactions from the viewpoints of the computer and the user. In the PIE model, the user entered a sequence of commands, and the computer responded with a sequence of displays. At first glance, the formal models look fairly symmetric between input and output. This impression changes if we examine the domains of these sequences and their interpretation.

One obvious difference between these two sequences is the small number of possible user commands compared with the vast number of possible displays. That is, there is a mismatch in bandwidth between the user's inputs and the computer's responses. Although this feature will not be central to this chapter, it is an important issue in its own right (Thimbleby 1990). The mismatch is reasonable, since it is quite possible for computer systems to produce displays at a rate of several million pixels per second whereas typing at such a rate would be hard. Of course, we will not be able to take in information at that huge rate, but even so, our ability to gather information from a display is still far greater than our rate of generation. If we want to allow the user to control the system (rather than *vice versa*) we need to design the dialogue carefully so that the computer's high output bandwidth can help users increase their effective input bandwidth.

The second difference is one of persistence. Considering the red-PIE model, although it said nothing about the precise temporal behaviour of the displays and commands, there were implicit assumptions in the definition of properties. For instance, the simplest form of predictability was the ability to tell where you are from the current display. The scenario was: I interact with an application, go away for a cup of tea (a not infrequent occurrence), then upon returning want to know where I've got to. If I use the current display to obtain this information, it can only be the persistent part. Any beeps or whistles, although very important to signal errors whilst two-finger or copy typing, would be useless as status indicators upon my return. Other forms of predictabilty involve exploratory activity by the user and thus assume the user's presence. For these forms of predictability, persistence is not so important. Paradoxically, such exploratory behaviour does not typically encounter non-persistent parts of an interface.

To distinguish these levels of persistence, we shall talk about *status* and *events*. Status refers to things which always have a value, such as the current display or the temperature. Events are things that happen at a particular time, for instance user's key strokes or the computer's beeps. In fact, the distinction is nothing like as clear-cut: events are rarely instantaneous, and may often be regarded as rapidly changing status. We briefly examine this interplay between the two in the next section.

Succeeding sections look at how to include status inputs such as mouse positioning within formal models. We can use these models to classify the way systems respond to status inputs. In particular, we find there is a close correspondence between the complexity of the models required to describe a particular system and the hand–eye coordination required.

Finally, we consider how the distinctions between status and events can help in the understanding of multi-user systems such as mail or conferencing.

10.2 Events and status – informal analysis

As we have already noted there is an interplay between status and events which clouds the immediate distinction. The way we classify a particular phenomenon will be partly subjective, and in particular depends on the temporal granularity with which we view it.

Consider for example a mechanical alarm clock. If we ignore its inputs (i.e winding it up, and setting times) and concentrate on its outputs, we have two distinct channels: its face, which displays the current time, and its bells, which wake us up in the morning. This seems like a fairly clear example of status and event. The time displayed varies continuously and is always available, while the alarm rings at a particular hour of the morning.

Let us look in more detail at the alarm's bell. If I happen to be rather sleepy, the clock may ring for a while. That is, it becomes a status output. The important events then take place when it starts ringing and when I eventually turn it off. So, if we look at the clock with the granularity of hours in a day, the ringing of the bell is an event, but if we look at it with a granularity of seconds (or minutes!), whilst I am actually waking it is a status.

The clock face is even more complex. If, say, the alarm is not set, but I want to do something at five o'clock. I keep looking at the clock, and when the hands pass the hour I react. I have interpreted a change in status (through the five o'clock mark) to be an event. This turns out to be quite a general phenomenon: any change in status can be regarded as an event. Notice also that in order for the change in status to become an event, I had to watch the clock. Probably I would do this periodically, so there would be a slight gap between the objective event of the clock passing through five o'clock and the subjective event of my noticing it. Again, these are general phenomena. Frequently, some procedure is carried out to monitor a status indicator without which changes in status could not become events. Further, such procedures will typically induce lags between the events noted by different observers.

Now, if we look closely at the clock face, we notice that the second hand moves rather jerkily. Inside the clock, several times a second, the spring vibrates and moves a ratchet which allows the clock gears to move on one notch. This discrete movement is visible on the second hand, and present but rather hard to see on the minute and hour hands. In fact, if we look closely at a large grandfather clock the jerky movement of the minute hand becomes visible, and the hour hand probably jerks noticeably on Big Ben. That is, at a fine grain, the change in status of the clock face is due to events – the ticks. Of course, we could drop another level again and describe the position of the cogs and springs to yield a status description of the entire workings of the clock.

So, even on something as everyday as an alarm clock, we see quite a subtle interplay between status and event descriptions. As most digital computers operate in discrete time steps it would be possible to describe them completely in terms of the events specifying the changes at each instant. This is very obvious if we consider a terminal attached to a time-shared system. The changes to the terminal's screen are sent character by character down a communications line. In principle, if users had access to this event information they could reconstruct what their screens should look like.

From a formal point of view, we could model the display in this way. User input events give rise to system output events which specify the way the screen is to change. In fact, this is precisely the view programmers are usually given of the interface. Even apparently continuous graphics displays are usually the result of a stream of change requests from the program. This does not describe the way the user sees the system. Most of the changes happen at timescales well below the user's limits of perception, and even when they do not they are often intended to (shades of the infinitely fast display). So, most of the time it is right to view the display as status, and any model that treats it as a sequence of change events will be inadequate for understanding user behaviour.

On the other hand, just as the movement of the clock hands through five o'clock was interpreted as an event, the user may interpret certain changes in the status of the display as events. So, the user may not regard the individual characters which are drawn on the screen as events (although the programmer might). However, if all these changes together mean that a box has appeared in the middle of the screen with the words "fatal disk error", this is a significant event. Note that even if the change in status has happened in a very "un-eventy" way for the program, perhaps by changes to a memory-mapped screen, the user is still free to interpret the change in status as an event.

The granularity with which we choose to classify events and status depends in part upon the tasks which we are considering. So, for the alarm clock, if the task were getting to work on time then the ringing of the alarm clock would be an event. If, on the other hand, the task were the act of waking and getting out of bed, a status interpretation would be suitable. Thus if we wish to inform the user of an event, we must think about the pace of the task to which it contributes. For highly interactive rapidly paced tasks we would need events that would be recognisable at fine granularity: a buzzer, for example. If the task is long lived and low paced, such as the building of a tower block (or cottage), then an event, say the completion of a phase, would not warrant such an intrusion; a memo would probably suffice.

So, the distinction between event and status depends on granularity and interpretation. For the user, it also depends on salience. The appearance of a message at the top of my screen may be viewed as a status change event objectively, but it has little value unless I notice it. An event for *me* is something that I notice and act upon.

10.3 Examining existing models

We can now examine the models we have already considered in the light of the status/event distinction. This distinction can also shed light on other dialogue specification formalisms.

We started the chapter by noting that the PIE model was basically an event-in status-out model. Most of the other models follow in a fairly similar vein and to a large extent share this property. The exception is the complementary view model of Chapter 9, so let us start there.

10.3.1 Updating views – status-in status-out

In Chapter 9 we considered the paradigm whereby the user manipulates a view of the system. Normally we expect that, as the underlying state of the system changes, then the view changes along with it. The view-update paradigm suggests that, in addition, we allow the user to alter the view and have the system maintain a state consistent with that view. The precise manner by which the user interacts with the view is not part of this model, and may involve event- or status-like inputs and outputs. However, at the level at which it describes a system, this model comes pretty close to a pure status-in status-out model. What is more, both the user's status input and the user's status output are the same thing, i.e. the view.

Obviously, in this paradigm we can think of the changes of view as events, but the way that the model emphasises the static relationship between view and state encourages a status-like interpretation. This is further emphasised if we also recall the condition of *process independence*. This condition said that the changes in the underlying state brought about by a change in the view were independent of the way in which we performed those changes, and were a function only of the initial state and the current view. This form of history independence is not essential for a status-based system. There is no reason why the current state should not be based on the entire history of user status input. However, it does very strongly encourage such a viewpoint.

10.3.2 Status outputs and event outputs

Returning now to the PIE model and its derivatives, we first consider the output domains. As we have noted, several of the properties suggest that the display component is status-like. It is possible to add event-like components to the existing model. For instance, we could define the display of a word processor to be $Bell \times Screen$, where *Screen* is the screen that is displayed and *Bell* is a flag indicating whether the bell rings as the screen is updated. This display domain could be used to describe the behaviour of the system after an error, with the *Bell* component alerting the user to the problem. In fact, the possible error behaviour rules we considered both in Chapter 2 and when considering temporal behaviour assumed that some such component exists. If we have such a display definition, we have to be very careful when applying predictability rules. Some would require that we limit the display considered in the rule to the status part only. In particular, if we want the user, upon returning to the system, to be able to infer anything from the current display, then the display considered is without any event part.

If we consider more complex event behaviour, the system may produce events at instants not directly in response to user events. This could, of course, be captured in the temporal model of Chapter 5. However, the rules developed in that chapter were aimed at status output systems and would require modification to encompass event outputs. In particular, the main emphasis was on the steady-state behaviour, that is, the display after the system has settled down. This assumes that transient behaviour, including event responses, is relatively unimportant. For example, consider a spoken interface to a directory enquiries system. If the user prods the system with a speech input, the system then responds with speech. The steady state would consider the spoken response as part of the transient, unimportant behaviour. Thus the steady-state functionality, which is supposed to capture the essence of the system, would be... silence. This sounds rather like Beckett!

So, with a little stretching, we can begin to apply our models to event-based outputs. However, the above discussion does warn us that properties that are appropriate for status-like outputs are not necessarily appropriate for events and *vice versa*. It seems wise therefore to distinguish the event and status outputs of a system within a formal model, even if they appear otherwise identical within the model. This is rather like the way we distinguished the display and the result components in the red-PIE model and its derivatives. Looking at the model alone, the two differed only in their name. However, they differed greatly in the interpretation we laid on them and in the way they were used in framing properties.

10.3.3 Status inputs

As one would expect, many of the same warnings apply when we consider status inputs. The most common form of status input is a mouse position, or other form of pointer, but if we took as an example a computer-assisted car, we would also need to consider pedals, steering wheel, etc. Bill Buxton suggests that computer interfaces could make use of many more different input devices, many of which would be continuous status devices. We will consider properties for status inputs in the next section, so now we will just look at the way they can be included within our existing frameworks.

The PIE model is very heavily entrenched in an event input world. Even within its restrictions, we can describe status input to some extent, but not very convincingly. In the last section we considered the way the screen was updated by many events. In a similar way we can convert the user's status input into a series of events representing changes. For instance, if we were building a red-PIE model of a mouse-based system, we might have the command set as $Keys + MouseButtons + Moves$. *Keys* and *MouseButtons* would represent the events of hitting the keyboard or clicking a mouse button, respectively. *Moves* would capture the mouse movement with commands like *MouseUp*,

MouseDown, etc. This is precisely the way many window managers and graphics toolkits deal with the mouse.

These mouse movement commands "work" in the sense that we could write down interpretation functions and display and result maps which would mirror the behaviour of a system. They do not, however, capture the essence of the system for the user. When I move a mouse I do not feel that I am entering a sequence of up, down, left and right commands. What I am doing is moving a pointer on the screen. Arguably, this is true even for text cursors with cursor keys. Much of the routine cursor movement is proceduralised: the discrete commands become unconscious and the effect for a skilled user is a smooth movement of the cursor about the screen. Again, the level of analysis is all-important.

One of the intentions in the study of dynamic pointers was to help deal with mouse-based systems. Do they help us here? By talking about the relationship between positions on the display and locations in the underlying objects they give us a vocabulary for relating certain types of status inputs to status outputs. However, they do not address the specific distinctions between status inputs and event inputs. Where we considered the form of manipulation in detail we assumed that the mouse position was interpreted in conjunction with some event-based command. In fact, we will see that this represents an important subclass of status input systems.

A more generalised view of status input can be derived by considering the temporal model, the τ-PIE. If we add for each time interval a value for the user's status input, we allow system descriptions which include many types of behaviour, dependent on the precise timings of the user's status. Whether the full generality of such responses is a good thing, will be considered in the next section. For now, let us note that by adding status to each time step, there are no pure τ or tick time intervals. Considering event outputs required a radical re-evaluation of desirable properties. Adding status inputs requires a similar re-evaluation and, in addition, leaves some of our constructions in need of a complete overhaul.

10.4 Other models

We now take a quick look at some other dialogue formalisms, and some of the issues they highlight. Several formalisms make use of process algebras, such as CSP. Alexander's SPI (Alexander 1987b), which has already been mentioned, is an example of this. By the nature of the underlying formalism such descriptions are event-in event-out. Status information is typically conveyed by change events in such descriptions. Most of the formalisms which I classified as of "psychological" origin tend also to regard the user's interaction in an event-driven way. Typically, there is little if any description of the system's response either as status or event, the emphasis being on the user's tasks or goals and their relation to the user's event inputs.

General computing specification formalisms, presumably because of the discrete nature of most computing, tend to deal only with event inputs. For example, specifications of interactive systems in Z (Sufrin 1982, Took 1986a), and in algebraic formalisms (Ehrig and Mahr 1985), effectively map user events to individual schema or functions which then act as state transformers. The outputs tend to be functions of the state, a constantly available status. This situation is not inherent in these formalisms, which are quite capable of describing status inputs and event outputs; it is just that the event-in status-out description is often easiest.

Sufrin and He's model of interactive processes, (Sufrin and He 1989) which is defined in Z, blurs somewhat the event/status position of its outputs. It inherits the display/result distinction (but calls them view and result) and, like the red-PIE model, describes these as functions on the state. However, unlike the PIE model, these functions are not continuously available, but are defined only for certain states. Now, as the display is clearly not meant to black out during the intermediate states, I take the unavailable states to represent the constancy of the display, and the available states to represent the display *changes*. In fact, they describe the availability of the display and result as "just after" the state change event. That is, they are very close in nature to status change events. On the other hand, Sufrin and He define properties similar to predictability properties, which suggests that at least the display is of a status nature. Also, all the examples used by Sufrin and He have a display for each user command; the non-display states seem to be there to allow for internal events and state changes.

Both the Sufrin and He model and the process algebras lead us to consider whether it is better to deal with the display and the result as event or status.

10.4.1 Display – status or event?

For the display, we have already dealt with something very similar to the Sufrin and He model when we considered alternative definitions of the PIE using functions from command history to effect history. The case where the effect history does not "grow" with the command history is precisely the same as the unavailable states here.

As we have noted, the distinction between status and event is partly subjective. A visual display is "always there" and so my preference is to regard it as status. I would prefer to reserve event outputs for more obvious things such as auditory feedback. There are in-between cases, such as a flash of the screen (sometimes used as a silent bell) or the *appearance* of a dialogue box which marks an exceptional behaviour.

A preference for a status interpretation of the display does not stop us from modelling it using status change events, but does influence our perception of those events. A slightly more profound worry about intermittent display events is whether we should consider using a model that appears to sanction user commands with no feedback. Any event-in event-out model should include an *interactivity condition*. This would require that each user input event is followed by at least one system output event.

10.4.2 Result – status or event?

We now turn to the result component. The result has a much more immediate interpretation as an event. Certainly in the classic example we have used, the word processor, the result, i.e. the printed document, occurs at some (infrequent) point in the interaction. We can, of course, ignore here the time it takes to do the printing. It makes sense therefore to have the result as an event that occurs when the appropriate print subdialogue is complete. Similar arguments would apply to a text editor which works on a copy–rewrite basis. The alterations the user has performed are written to the file system only at the end of the interaction, or at specific periods when the user asks for the file to be written.

On the other hand, there is frequently an idea of the potential final result. So, when we use the word processor, we know that there is some document that *would* be printed if we issued the appropriate commands. This has a validity at two levels. On the one hand, there is the internal state of the system, which will certainly have some component corresponding to this potential result. On the other hand, from the users' point of view, at any point in time they will have a fairly strong idea of what the potential result will be. Thus both from the system's and the user's perspective the result is status-like.

We can link these two concepts of the result if we have some notion of a "normal" mechanism for obtaining it. For instance, a word processor may have a PRINT command, or a text editor a SAVE. We would regard the potential result of these systems to be the physical result when the appropriate command was issued. Breakdowns can occur in several ways. There may be more than one "normal" mechanism for obtaining a result. For instance, the text editor may have an EXIT which saves the file, but also a QUIT command with no save. In such cases we have to consider that one method of obtaining a physical result is more "normal" than the rest. Furthermore, the form of the final result may be determined intrinsically by the exact dialogue by which it is obtained. The print subdialogue of many word processors includes the choice of many important layout parameters such as page length, or multi-columning. In these cases, any notion of the potential result as status can be only an abstraction of the full event result.

These last two points are to do with an ambiguity over the notion of the potential result. If the user and the designer differ in their understanding of this notion, breakdowns in use are likely to occur. Within the limits of the user's perception, we can assume that the user and system agree about the current state of the display. There is no such guarantee for the potential result. We rely on the display to give appropriate cues for them to synchronise. We recall that a large part of Chapter 3 concentrated on this issue of what we can tell of the result from the display. Of course, users do not notice everything on the screen, so that even if the display has sufficient information on it to perform this synchronisation it may not be salient. In empirical studies of a reference database system at York, it was found that a significant class of error could be attributed to a disagreement between the user's and system's idea of the current result.

So where have we got to? Physically, results tend to be events but conceptually they are often best described as status. An important issue for any design is ensuring that the two views are in agreement.

10.5 Status inputs

In this section, we consider different models of status inputs. These models are related to and express different complexities of interactive behaviour of status inputs. As the most common status device is a mouse or other pointer device we concentrate on these, but much of the discussion is valid for any status input device.

We have already noted that we could include status inputs in the τ-PIE framework simply by adjoining a status to each time period. Let us now expand on this. At any instant we will assume that there is a current value for the status device and at most one event input. The event inputs we will take from a set of commands C and add a tick τ for the instants when there is no event. This is exactly as the τ-PIE. We then add the status inputs from a set Pos (for position). Thus the user input at each instant is a pair:

$$C_\tau \times Pos$$

where

$$C_\tau = C \cup \{ \tau \}$$

The user's input history is then a sequence of these:

$$H_{Pos} = (C_\tau \times Pos)^*$$

Just as in the τ-PIE we can define the behaviour either using an interpretation function over histories, or as a state transition function *doit*:

$$I_{st}: H_{Pos} \rightarrow E_{st}$$

$$doit_{st}: (C_\tau \times Pos) \times E_{st} \rightarrow E_{st}$$

We obtain the display and result component from the minimum state E_{st} in the normal way. By interpreting some abstractions of E_{st} as event outputs and others as status we could obtain the complete possibilities of event/status input/output; however, for the purposes of this section we are interested only in the input side, and will ignore any distinctions in output.

An additional distinction we need to make on the input side is between those event commands which are physically connected to the status device and the rest. We will call these sets of commands *Button* and *Key* respectively, because when we consider the typical mouse and keyboard setup the mouse button events are clearly more closely connected to the mouse position than the keyboard. Such a distinction has a degree of subjectivity about it, but is not too difficult to make. There are awkward cases, as for example when a shift key on the keyboard is used in combination with a mouse button click. I personally find such combined keyboard–mouse chords rather abhorrent, but they are certainly common. If a system *must* have such events, they would belong in the *Button* class. The important thing is that this is a *physical* connection between the commands and the status device; there may or may not be a corresponding *logical* connection. However, it is precisely the mismatch between physical and logical proximity that I dislike so much about keyboard–mouse chords and which will be discussed later.

This model is now capable of expressing virtually any input behaviour. The properties which we describe in the rest of this section can be framed over this model. Sometimes, rather than doing this directly, we can develop a more specific model for a class of behaviours which we can relate back to this general model of status input. We start with those systems, or parts of systems, which are least dependent on status information, moving on to more and more complex status dependence.

10.5.1 Position independence

The simplest thing to do with status information is to ignore it! If the status is ignored throughout the whole system, we can just use the PIE or τ-PIE model. This is the situation for applications boasting "no mouse support", but is not very interesting.

Typically, some parts of an interface are position independent, in particular the keyboard inputs. We can make position independence a property of commands, by simply demanding that the interpretation of the command does not depend on the current value of the status input. In the case of mouse-based systems a command is position independent if it has the same effect no matter where the mouse is.

We can express this property using the model described above:

position independence:

$$\forall\, p, p' \in Pos,\, e \in E_{st}:\; doit_{st}(c, p, e) \;=\; doit_{st}(c, p', e)$$

Within a single application, the same functionality may be obtained using position-independent or position-dependent commands. For instance, in Microsoft Word4, the PRINT option can be selected by a position-independent keyboard accelerator or by a position-dependent sequence of mouse clicks over menu options. The latter are position dependent because the interpretation of the mouse clicks is determined by the menu selection over which the mouse pointer lies.

As we have noted, if all the commands were position independent the system would be rather boring (with a few caveats) from a mouse's perspective. However, for some commands, and in particular the keyboard commands, position independence seems positively beneficial. Position-dependent commands may obviously require quite precise hand–eye coordination. This coordination of the mouse position with input which is physically remote from it seems rather questionable. One requirement we may therefore want to make of status inputs, is that all commands in *Key* are position independent.

Some windowed systems apply a "click to type" paradigm. In order to select a window for keyboard input some definite action must be taken, either clicking a mouse button over a specific part of the window's border, or just anywhere on the window. Other systems use instead a "focus under the mouse" paradigm, whereby the window which is under the current mouse pointer at any moment is the active one to which all events, including keyboard events, are addressed. In such systems, every event is effectively position dependent. If we look at the applications themselves, they frequently apply position independence for the keyboard, but this property is "spoilt" by the environment. The danger of such systems is that while the user's attention is on the keyboard, an inadvertent nudge against a pile of papers may unintentionally move the mouse and redirect the input. On the other hand, the movements required are often gross and it appears a relatively benign form of position dependence. We will see in a while that by reinterpreting the position status, the model can be adjusted to leave the keyboard commands themselves position independent. This adjustment will still preserve an element of warning about the behaviour, but will centre this on the mouse movement rather than on the keyboard. This seems to correspond to one's intuition about the situation, that it is not the application or the keyboard's "fault".

10.5.2 Trajectory independence

Now we shift our attention to events that are dependent on the position. An important subclass of these comprises events which depend only on the *most recent* position.

Let us consider a mouse-based word processor. At the top of the screen it has a line of buttons. Clicking the mouse over these uncovers further menus. As we move the mouse pointer over menu selections they are highlighted and we make a choice using a second click. Text is highlighted by depressing a mouse button over one end of the required portion, dragging the mouse to the other end, and then releasing the button. The highlighted text is operated on by various of the menu and keyboard commands. The text entry itself is position independent.

Notice two things from this description:

- As the mouse moves over the menu selections and as the mouse drags out the text selection, the display varies continuously with the mouse movement.
- When a menu selection is made, or when the text is selected, it is only the position of the mouse when the event takes place that is important. The intermediate mouse positions do not affect the final state.

The second of these says that the final state of the system can be determined by looking at snapshots of the position at the moments when events occur. If we define the *trajectory* of the mouse, as the history of exact movements of the mouse between events, then the second condition says that the commands in the word processor are *trajectory independent*.

Although the trajectory is not important in determining the final state, the first of the two observations reminds us that it *is* important for the user. It is the detailed feedback from the intermediate states which makes the mouse both usable and enjoyable. Indeed, constant feedback is the very heart of interactivity.

A simple statement of trajectory independence might look something like this:

$$\forall\, p \in Pos,\, e \in E_{st}: \quad doit_{st}(\tau, p, e) \;=\; e$$

This will not do, however, since the display is determined from E_{st}. We need to distinguish the part of the state that is purely to do with feedback from the functional part. We have already made similar distinctions in earlier chapters:

$$E_{st} \;=\; E_{feedback} \times E_{functionality}$$

The trajectory independence condition would then be applied to $E_{functionality}$ only. We could, of course, "cheat" and choose the $E_{feedback}$ to be *everything* and the $E_{functionality}$ to be empty. This would make the system totally, and vacuously, trajectory independent. We can protect ourselves from such inadvertent cheating by demanding that $E_{functionality}$ has at least sufficient information to determine the result mapping.

This split is similar to the layered design we described in Chapter 7. Here though, the functionality we wish to capture in the inner state is more than in E_{obj}. That was supposed to be the state pertaining to the actual objects of interest, ignoring their representation. Here we would wish to include features such as the position of the menu on the screen in order to interpret the mouse position when a button was pressed. In other words, this is really a further layer outside the models we were considering there.

Rather than looking at predicates over the general model, we can look at a simpler model which implicitly captures trajectory independence. This will also bring us back to something which can be directly related to the models used in Chapter 7. This model will apply only to systems which are *totally* trajectory independent.

10.5.3 A model for trajectory independence

The crucial property of trajectory independence is that the long-term behaviour of the system is determined only by the status at the moment when events occur. So we first look only at those instants. We have an internal state E_{ti} that is updated at each event. The update function makes use of the event (from C) and the status (from Pos):

$$doit_{ti}: \quad C \times Pos \times E_{ti} \rightarrow E_{ti}$$

The result is a obtained as a mapping from E_{ti}:

$$result_{ti}: \quad E_{ti} \rightarrow R$$

This differs from the general status input model in that we have dropped the τ events which signified non-events. This is because we are looking only at the instants when events occur. In fact, so far it is exactly like a red-PIE, with command set $C_{ti} = C \times Pos$. However, the display component will differ somewhat from that of the red-PIE.

The first of the two observations we made earlier was that the value of the status input was constantly reflected in the display. In order to capture this we make the display a function not only of the current state (from E_{ti}) but also of the current position (from Pos):

$$display_{ti}: \quad E_{ti} \times Pos \rightarrow D$$

So, in the case of the word processor above, in the state when the pull-down menu had appeared, the display would be of the menu (dependent on state) with the entry under the mouse highlighted (dependent on status).

By separating out the event updates from the status feedback, we implicitly capture the position independence of the model. It can be related back to the full status input model by setting:

$$E_{\mathbf{st}} \;=\; E_{ti} \times Pos$$

$$doit_{\mathbf{st}}(\tau, p, (e, p')) \;=\; (e, p)$$
$$doit_{\mathbf{st}}(c, p, (e, p')) \;=\; (doit_{ti}(c, p, e), p)$$

$$display_{\mathbf{st}} \;=\; display_{ti}$$

$$result_{\mathbf{st}}(e, p) \;=\; result_{ti}(e)$$

This simply adds the current position as part of the state. However, both the result and update parts of the model ignore this part of the state. In other words, it is just a simple way of achieving the split described in the previous subsection. This relationship between the trajectory-independent model and the full model is rather like the embedding of steady-state behaviour in the τ-PIE model.

We have already begun to tie this model in to the red-PIE. We can deal with the display component in two ways. One is to set the display from the PIE point of view equal to the *function* from Pos to D:

$$D_{PIE} \;=\; (\, Pos \rightarrow D \,)$$

$$display_{PIE}(e) \;=\; \lambda p \;\; display_{ti}(e, p)$$

This is not a totally indefensible position, but talking of the display as a function may appear a trifle odd. Perhaps a more natural approach is to consider an abstraction of the display. For example, in the word processor, we can think of the display with the contents of the menu items displayed, but ignoring the highlighting. In order not to lose important information, such as the fact that *something* is highlighted, we may need to add a little extra to this abstract display, but on the whole this would probably correspond to how a user might describe the display after an event. Again this reminds us of the way we can abstract away "awkward" real-time parts of the display (such as a clock) in order to obtain a system with steady-state behaviour.

So, we have not only produced a simplified model which describes trajectory-independent systems, but we also have the means to fit this class of status input systems into the layered models of design introduced earlier in the book.

10.5.4 Spatial granularity and regions

We have just considered *when* positional information might be used. We will now move on to *what* part of that information is used. To begin with, let us think about a simple CAD system. Like the word processor it has pull-down menus, and perhaps a fixed menu of shapes it can draw as well. Instead of a text window it has a graphical area with points, lines, circles, etc. These shapes are drawn by selecting the appropriate tool icon and then clicking at the appropriate anchor points in turn: so to draw a line we would click first at one end and then the other. We could allow movement and reshaping of existing shapes by depressing a mouse button over appropriate anchor points and then dragging before releasing the button.

This could easily be cast into the trajectory-independent framework above. What we need to concentrate on here, though, are the differences between the icon selection and the point selection:

- *Size* – There is an obvious difference in size. The icon is just a bigger target and therefore requires less fine hand–eye coordination. The problems with achieving such fine control are often addressed by including grids for initial plotting and by having zones of attraction around the anchor points for later selection.
- *Semantics* – The use of positional input to choose the icon is just a way of indicating which icon is of interest. The same object could have been achieved with a keyboard accelerator or by speaking the name of the shape. With the anchor points, however, their very nature is positional, so the semantics of the anchor points have a close correspondence to the form of input.

These two differences are somewhat interlinked. Intrinsically positional parts of an interface tend to be fine grained, although there are grey areas. Text selection is just such an area, which is why we switched to CAD as an example! If we consider only granularity, text selection would lie somewhere between icon selection and point selection, and we might just regard characters as small icons. However, the location of characters within a document is intrinsically positional, and hence we should class it with point selection when we consider semantics.

There are obvious reasons for using the mouse for intrinsically positional features, but why is it used for less semantically necessary ones? In a CAD system where the mouse is used extensively for object-level manipulation, it makes ergonomic sense not to shift the user between input devices. This does not apply to a word processor, where typing is a major activity. In fact, in many mouse-based word processors keyboard accelerators are available in order to short-cut the use of the mouse and menus. In such systems the major reason given for the use of menus and icons is cognitive: it is easier to recognise than to remember, thus clicking over an icon that says PRINT (and perhaps has a suitable picture on it) is easier for the user than remembering whether to type PRINT or LIST. A further related reason is to increase the effective input bandwidth. We recall the mismatch in bandwidth between input and output. Iconic interfaces and menu-based systems are one way of using the output bandwidth effectively to increase the input bandwidth. This is especially obvious in a file-system browser, where clicking on a file is an apparently more effortless task than typing the appropriate filename. This stretching of bandwidth is connected to the use of context in determining possible user inputs. The selection of a particular file icon, for example, is made easier because the directories or folders of interest are displayed. In terms of information flow, this is a form of clever coding which makes use of the redundancy of the user's input language. Non-graphical systems use similar contextual devices such as the idea of a current directory, current drive or current active file, as well as the use of "wildcards" in filename specification. Graphical systems extend this by allowing multiple simultaneous contexts, and by making the "coding" more easily discernible.

Whatever the reasons for the use of position dependence, if the granularity of the target is quite large then the hand–eye coordination required is far easier. For parts of the system, such as icons and menus, where the mouse can be anywhere in an area of the screen, we can factor the dependence of the system on the position by defining a *region* mapping:

$$region: \; Pos \; \rightarrow \; Reg$$

For a particular programming environment, the elements of region might include menu buttons "File", "Edit", "Options" but also all-embracing things like "Program window". Having defined such a mapping we can distinguish those commands which are merely *region dependent*, like menu selection, from those which require fine grain positioning, like anchor point plotting:

region dependence:

$$\forall\, p, p' \in Pos : \; region(p) = region(p') \;\Rightarrow\; doit_{ti}(c, p, e) = doit_{ti}(c, p', e)$$

A comparison of the numbers of region-dependent and fine-grained commands could form a good measure of the motor control required for an application. One could not use such a measure blindly, however, as one should look at the semantics of the commands and whether fine-grain control is appropriate because the feature is intrinsically positional. Again, one can cheat. One could say that every pixel in a graphics window was an individual region. This would fail the semantics test, as the pixels, although discrete, represent an intrinsically positional world. For the same reason, although it might be appropriate to factor text positions (possibly using *back* maps from the pointer space projection), they are an intrinsically positional attribute and should not be regarded as lots of screen regions.

In the examples given above, the intrinsically positional elements have all been at the application object level whereas the region elements have been interface objects. This is not always the case. Application objects frequently contain discrete subobjects which may be referred to by screen region. Also there are positional interface objects such as a scroll bar.

10.5.5 Trajectory dependence

We now move on to the final category of status input behaviour, *trajectory dependence*. This is where the outcome of an interaction depends on the precise path of the status device between events. The most obvious example of trajectory dependence is in freehand drawing or spray cans in graphics packages. In some ways this is more than an example and is archetypical of trajectory dependence in the same way that point plotting is archetypical of fine-grain position dependence. Freehand drawing is trajectory dependent not because a designer thought that it was a good idea, but because its semantics demand a trajectory approach. Non-trajectory-dependent forms of freehand drawing would be obscure to say the least.

It is frequently the case that drawing or painting is activated only whilst a mouse button is depressed. Thus the period for which fine hand–eye coordination is required is marked for the user by the conscious act of holding down a button. Furthermore, the button is on the device which requires the control: that is, the physical and logical proximity coincide. This "pen-down" paradigm is often used elsewhere, both in other trajectory-dependent situations, such as gesture based input, and also in trajectory-independent contexts such as window movement. To describe such situations, it is easiest to regard the holding down of the mouse button as another status input with a simple on/off value. We can then ask for trajectory independence when the button is up.

Freehand drawing and the like are not the only examples of trajectory dependence. We have already mentioned gesture input but there is a further class of trajectory-dependent situations based around revealing screen objects. A plethora of menus, scroll bars, icons, etc. may soon clutter up a screen, leaving little room for "real" information. Many designers choose to conserve screen space by hiding various features which appear only after some user intervention. Sometimes this intervention is of a trajectory-independent nature, perhaps clicking on a menu bar which makes a pull-down menu appear. However, sometimes it is trajectory dependent. For example, some window managers uncover any window over which the mouse moves.

Line drawing intrinsically requires trajectory dependence. These other features do not. The intention is presumably to make the user's job "easier" by not requiring an explicit mouse event. This is a very dubious practice and requires careful consideration. It is a problem firstly because of the extreme level of hand–eye coordination required. Position-dependent commands require fine control for only the fraction of a second that it takes to depress or click a button. Trajectory dependence requires consistent control over some period. Secondly, the problem is further exacerbated by the lack of *proportionality* which is typical of trajectory dependence.

There are several forms of proportionality, and here I am thinking of the ability to undo erroneous inputs. We have dealt with various types of undo properties, and the ability to undo easily is regarded as an important feature of direct manipulation systems (Shneiderman 1982). Not only should users' actions be undoable, but as a general rule there should be some proportionality between the complexity of an action and the complexity required to undo that action. Any true measure of complexity requires psychological insight, but a rather crude interpretation is all that is needed here.

Consider a trajectory-independent system. Imagine the user is moving the mouse and no other events occur. The user wishes to move to a position p but gets to p' instead. All the user needs to do to correct the situation is to move the mouse back to p. Because the intermediate positions are not important for the long-term development of the system, the mistakes in positioning have no long-term effects. Note that mouse movement is sufficient to undo mouse movement, and that small positional errors (from p to $p+\delta p$) can be recovered using small movements (δp).

10.5.6 Some trajectory-dependent systems

With a little understanding of the problems of trajectory dependence, we can go on to look at a few examples. Several windowed systems have "walking menus". Some options in a menu have submenus associated with them. Instead of having to select the relevant menu item to make the submenu appear, the user has simply to slide the mouse off the side of the item. The system is position dependent, as the same screen position may be associated with several submenus depending on which main menu item the user moved off. However, if the user slides into a submenu unintentionally then the mouse can be slid back again to regain the main menu. Thus, even though the system is trajectory dependent it still retains the proportionality properties that mouse movement can undo movement errors, and that small errors have small corrections. Further, the corrections are the opposites of the movements that caused the problem, so there is a further naturalness about the situation.

The Xerox InterLisp windows make use of appearing scroll bars. This is a deliberate policy of reducing screen space usage. With menus there is little alternative. It would not be possible (let alone desirable) to have all the menu options permanently visible and the issue is purely the appropriate method of revealing them. However, it is perfectly feasible, and in fact quite common, to have scroll bars permanently associated with each text window. The mechanism used by InterLisp windows is as follows. If the mouse moves off the left-hand edge of a text window a scroll bar appears under the mouse, beside the left edge of the window. If the user continues to move to the left and moves off the edge of the scroll bar (or in fact, anywhere off the scroll bar), it disappears again. How does this measure up to the proportionality tests?

There are several classes of problems so let us consider just a few. One mistake a user might make is to slip off the left-hand edge of a window by accident. The scroll bar appears, but a movement right again back into the window corrects this. Again, the recovery is with the mouse, proportional to the error and in a natural inverse direction. If, on the other hand, the mouse was in the scroll bar and slipped out to the left, the scroll bar would disappear. In order to recover the user would have to move a longer distance to the right to get back over the window, then move left to recover the scroll bar. The recovery is still purely by mouse movement, but the complexity of the correction has grown somewhat, and the movements are less natural. A similar scenario occurs if the mouse is accidentally moved to the right over the left border of a window. Trying to correct by moving the mouse back the way it came would then make the scroll bar appear, incidentally covering up what was probably the focus of attention. Again recovery could eventually be made using mouse movement alone, but the correction is complex and an attempt at a natural recovery leads to further complications.

In the case of InterLisp, the complications arising from trajectory dependence obviously require extra care and fine motor control to avoid mistakes. However, the property of movement as recovery for movement is preserved, and perhaps the sort of mistakes noted are rare. The advantage is that by making the scroll bars appear, they can afford to be larger and thus require less fine hand–eye coordination. There appears to be a trade-off between the two. The appropriate design choice is not obvious from formal arguments alone. What we have done is focus on possible problem areas.

Our final example of trajectory dependence comes from a version of the Gem environment. The system displays a menu bar along the top of the screen. Instead of requiring the user to depress a button over the desired option to obtain the pull-down menu, the system "saves" the user the bother of pressing the button by making the pull-down menu appear as soon as the mouse enters the region of the menu bar. The user can then move the mouse over the desired selection and click on it to make their choice. This is somewhat similar to the previous examples, and might be thought to be no more and no less a problem. If we begin to consider errors and their correction, however, it is seen to have distinct problems.

When the pull-down menus appear, they do not disappear until either one of the menu items has been selected or the mouse is clicked elsewhere. So, if the mouse is accidentally moved over a menu bar title, the mistake cannot be corrected without an event input. Further, the mouse must often be moved quite a way to get it somewhere where extraneous mouse clicks do no harm. That is, it disobeys virtually all the proportionality properties we have mentioned. Is this just a rather extreme example which rarely occurs in practice? I have certainly observed this as a problem myself, and have also watched both children and novice users becoming stuck with a pull-down menu they do not know how to get rid of.

So what did the designers gain by this decision? Apparently they saved the user from pressing a mouse button, and perhaps made their interface slightly more distinctive from similar products. It is fairly obvious as soon as one considers trajectory dependence that this is a potential problem area. By taking this into account, this confusing design error could have been avoided.

10.5.7 Region change events

Just as position dependence can be simplified to region dependence, a similar simplification may occur with trajectory-dependent commands. Both of the examples from the InterLisp and Gem interfaces were dependent only on the regions through which the mouse moved. In the InterLisp case these regions were windows and scroll bars, and in the Gem case they were the menu bar and pull-down menu options. We could capture this form of region–trajectory dependence using the full model something like this:

$$\forall\, h, h' \in H_{Pos} \quad \textbf{if } h \text{ and } h' \text{ are "region equivalent"} \quad \textbf{then } I_{st}(h) = I_{st}(h')$$

where two histories are *region equivalent* if they are the same length and agree exactly at the events and have the same regions at other times:

h region equivalent to h' **if**:

(i) $h = h' = null$

(ii) $h = (c, p) : k$
$h' = (c, p) : k'$
and k and k' are region equivalent

(iii) $h = (\tau, p) : k$
$h' = (\tau, p') : k'$
$region(p)$
and k and k' are region equivalent

In both examples, the regions change during the interaction, complicating matters further; the *region* mapping must take into account the current state of the system. We could go back and add this into the above definition. There is, however, a better way to go about it.

We recall that changes in status are often interpreted as events. The clock hands pass 6 o'clock and I turn on the radio for the news. A continuous, status output of the clock becomes an event for me. In a similar fashion, the continuous movement of a mouse, or the change of any status input, can give rise to system events. In the InterLisp example, as the mouse moves over the window boundary, we can interpret that as an "uncover scroll-bar" event. Similarly, we can regard the movement of the mouse over the Gem menu-bar as a "pull menu" event. The events occur when the mouse moves between regions, and instead of looking at region–trajectory dependence with its obscure definition, we can talk about *region change* events.

These events lie somewhere between the status input device itself and fully fledged user event commands. By regarding these changes in status as events we can discuss the behaviour of systems such as InterLisp and Gem in an "eventy" rather than a status fashion. We do want to retain the distinction between these events and more explicit user events such as keypresses and mouse button clicks. The proportionality properties discussed above become *reachability* requirements on the subdialogues involving region change events, and in general we may want to impose more stringent properties on this class of event. In addition, the regions typically depend on the current state, so the region change events will necessarily have a language, whereas the explicit events are more likely to be unconstrained.

With status change events in mind, we can look again at those windowing systems which direct keyboard input at whatever window the mouse is in. We noted at the time that this makes all commands, both from the mouse buttons and the keyboard, *region dependent*. This is unfortunate as it means that although the keyboard commands may be completely position independent for each application individually, this independence is lost when looking at the system as a whole. If we now regard the movement of the mouse over window boundaries as a "select this window" event, the state of the system then changes in response to the event and the keyboard becomes again position independent. This outlook agrees with our intuition about the system: any oddness is not the "fault" of the keyboard commands, but is because of moving the mouse. Our focus is shifted from the keys towards the region changes, where it belongs.

So, if our focus is on region change events, what questions should we ask about them? One obvious thing to ask is, did the user mean it? The problem with directing keyboard input "at the mouse" is that attention may not be on the mouse, and any movement, and consequent region change event, may be unintentional. Of course, we cannot guard totally against accidents: it is as easy to hit the wrong key as to bump the mouse. However, we can look at a status change event in its expected context of use and ask where the user's attention is likely to be, and thus how often to expect accidents.

Another way to look at this type of error is as a mismatch between the perception of events by the user and the system. If I hit a key, the action of striking it forms an event for me and the reception of that character is an event for the system. On the other hand, if I happen to move my mouse over the pixel at coordinate (237,513) which is totally unmarked, I will not regard this as an event; if the system does so, then a breakdown will occur. The situation is worse if I merely joggle the mouse with my elbow!

In summary, region change events are helpful in describing some types of behaviour. This is recognised at an implementation level in that many window managers will generate events for the programmer when the mouse moves between certain types of screen region. It has also become clear that region change events are a source of potential user problems and should be scrutinised closely during design. In particular, when region change events are contemplated we should ask ourselves whether the events that are recognised by the system agree with the events which are salient for the user.

10.5.8 Complexity and control

We have discussed several classes of model for dealing with different types of positional input system. The complexities of the models correspond to various degrees of hand–eye coordination. The various types of event are summarised below (*fig.* 10.1), set against the degree of control required of the user.

Event class	Motor control required
Position independent	Ability to strike correct keys, possibly touch typing
Region dependent	Ability to retain mouse within target
Position dependent	Fine positioning for duration of "click"
Region change events	Ability to move mouse over target boundaries
Button-down trajectory dependence	Fine positioning for controlled periods
Full trajectory dependence	Continuous fine control
Time–position dependence	....

figure 10.1 *control required for event classes*

An additional category has been added of *time–position dependence*. This is to cover those systems that depend not only on *where* the mouse has moved, but on *how long* it has stayed there. There are probably few examples of this with mouse-based systems outside the games world, but it is more common in other systems. The accelerator pedal of a car has this property: where you get to and how fast depends on how far and for how long the pedal is held down. The same is true for the steering wheel and many other controls.

On the other hand, more complex forms of positional inputs can be used to obtain special and useful effects. The formal distinctions between these classes of systems can be used to warn about possible complexities in the interface, but it is a matter of judgement and perhaps empirical testing how to make the trade-off.

Also, although fine control can be a problem in an interface it can also be an enjoyable feature. Perhaps many people find mouse-based systems more enjoyable to use precisely because of the demands on their skills. Whereas event-based systems have a clerical nature, status input gives more a sense of hand-craft.

10.6 Communication and messages

Virtually all our discussion has been about single-user systems. The models of multiple windows drew on an analogy with multiple users, and this led on to recognising interference between users as one of the sources of non-determinism discussed in Chapter 6; however, this was a minor thrust in both chapters. The emphasis, such as it was, dealt with how to avoid such interference between users. This approach is appropriate in a multi-user system with shared resources but in which each user works independently. For shared or cooperative work a different perspective is required.

There is, of course, a lot one could say about cooperative computing systems and it is one of the current growth areas in HCI research. For the rest of this chapter we deal briefly with some specific aspects which relate to the concepts of event and status.

To begin with, let us think about two contrasting communication mechanisms. First, there is a traditional email system. Each user has a "mailbox". Other users can send messages which should eventually, via various networks, find their way into the recipient's mailbox. When the users look at their own mailbox they find any messages that have been sent to them. Additional facilities offered typically include distribution groups and aliases (as the addressing systems are far from clear) to help send messages, and various forms of visible or audible indicators when mail arrives.

The second type of system to consider is where the model is of a shared information base. At the simplest, users of a shared file system can use this as a communication mechanism. Perhaps two researchers working on a shared paper will read each other's contributions simply by looking at the appropriate files.

Other systems have been designed specifically to encourage information exchange and cooperation. Some are based on hypertext techniques, such as for instance KMS (Yoder *et al.* 1989). The whiteboard (blackboard, chalkboard) metaphor has also been used (Donahue and Widom 1986, Stefik *et al.* 1987), and also simple screen sharing.

These two extremes represent two styles of communication:

- Messaging
- Shared information

The astute reader will have already guessed how these relate to the theme of this chapter, but before discussing this we will examine a few other features of this distinction.

10.6.1 System and user cross-implementation

What a system provides and what users do with those facilities are, of course, very different. Most communication systems are built on top of an underlying messaging protocol between the users' computers. Even where the users share a single file system, their workstations will typically communicate with the file server by network messages. This level of implementation is usually hidden from the users. The network messages may be used to implement an apparently shared data space, and even where the user level model is of messaging this will be built on top of the underlying network in a non-trivial manner. In a similar manner, users are proficient at producing social mechanisms for "implementing" one protocol upon another.

Think about two chess players swapping moves by post. The postal system is a messaging system, but is used to keep their respective boards consistent. That is, they maintain a shared information space. The transfer of files by email is a similar situation. The users want to share the information contained in their respective filing systems and use the email messages to implement this. It is interesting that even where users share a common file system they may still use email for this purpose. This is presumably because of the difficulty of giving the appropriate permissions or specifying the file names.

Social implementation techniques work the other way too. Users of shared information systems make use of drop areas: they set aside parts of the information space for each user and simply add their messages to this area. This mechanism is a direct parallel to the way that many systems implement internal email. Also, bulletin boards may contain a permanent record of all transactions but many users will read only the most recent entries and never consult old entries. They thus become simple broadcast message systems. Again, personal columns in newspapers are messages between individuals which make use of a widely shared information base (the newspaper).

Even though users are quite capable of shifting the communication paradigm supplied by their system, these examples do suggest that support for the actual tasks performed may be beneficial. On the one hand, we may ask whether drop areas in a shared information system form an adequate message system. On the other, it is clear that if we want to transfer information, such as files or more complicated structures, sending messages is not the way to do it.

10.6.2 Attributes of messages

If we want to know how to design communication systems that meet users' needs, we should consider some of the different attributes of messages and shared data. Not all systems need be as polarised as the examples we have given, and the way users implement different social protocols can make it hard to distinguish the crucial attributes that make a system suitable for a task or not. Attempting to uncover the central attributes that differentiate these mechanisms can help us understand this suitability.

If we contrast direct speech with, say, a book or filing cabinet, we see an obvious difference in *persistence*. The spoken message is ephemeral: its permanent effect is in the way it has affected the users' actions and their memory. The shared information in the book and the filing cabinet is much more long lived. This difference in persistence between messages and shared data seems to be fairly characteristic. Email messages tend to be read once only, as compared to a shared database that is intended to be a perpetual resource. However, the distinction is not as clear-cut as all that: both electronic and paper messages are frequently filed permanently, and spoken meetings are made permanent by the use of minutes or notes. In fact, it is a general feature of formal bodies that ephemeral messages become permanent corporate information.

There are two major reasons for recording messages. The first is that they may contain information that is required on a long-term basis. Arguably, if that is the case, then that part of the message was partly a sharing of information that is implemented by the message. A second reason is as a record of a conversation, either to re-establish context when new messages are received, or to serve as a legal or official record in case of dispute. In the former case only the last few messages are required, so the messages are still ephemeral. In the latter case, we have a meta-communication goal, and it is not surprising that it has strange characteristics; the medium through which we communicate has become the object of higher-level tasks.

Other attributes that can help distinguish messages from shared data are *ownership* and *identity*. Shared data tend to remain the property of the originator or may have some sort of common ownership. The *same* data object is available to all. Messages tend to be transferred between parties, and the recipient either gets a copy of the original item, or the sender loses their copy. The typical example of this is the letter. The sender can retain a copy or send a copy, but the letter itself changes hands. Similarly, email messages become the property of the recipient. Again there are exceptions, for instance, if I make lots of photocopies of this chapter to pass to my colleagues to review, but this is a typical example of implementing shared information using messaging.

We can lay out these attributes in a matrix, placing messaging and shared data in it:

		persistence	
		ephemeral	permanent
ownership	transferred	message	(a)
	shared	(b)	shared data

There are two gaps in the matrix, but are there any situations that fit in these gaps? Let us look first at gap (a), that is, transfer of long-lived information. Obvious examples of this in the real world are legal transactions, where the transfer of some written document carries with it the ownership of some asset. This could be seen to be a little stretched because of the link with physical objects. One example of a "pure" information transfer within this category is the passing on of a job. All the relevant information, both electronic and paper, is passed on to the new post-holder. Both these examples seem closer to a data-oriented view of communication. Any messaging that would be associated with the transactions would appear to be additional, even if sometimes within the same enclosure.

The other gap, at (b), is for ephemeral data in some form of shared or common ownership. This is a more difficult gap to fill. One could imagine information in a shared database with limited lifetime, for instance, a meeting announcement. However, the ephemeral nature of this information is related to real time and is very different from the ephemeral nature of messages. Messages are ephemeral because they are no longer needed once they have been read: that is, their timescales are subjective. The only examples of shared data with a subjective timescale that I can think of are items like electronic voting forms, or data items which are "ticked" off when they are read and are removed when ticked by everyone. The former have a quite complex dialogue associated with them, but are essentially message-like. The latter appear to be an example of shared data being used to implement broadcast messaging. A physical example of the latter would be the university lecture, but of course this would lack the guarantee of universal attention.

If anything, the permanence axis seems more characteristic than the ownership one, but all the off-diagonal options are rare or expose failings in the underlying transport mechanisms. Of course, a good shared information store would support many different types of ownership and ownership transfer. Further, it is easier to make permanent data ephemeral than *vice versa*, so the shared data view seems to include messaging as a subcase. This is borne out to some extent by the experience that in shared data systems, the attempts at establishing messaging via drop areas seem more successful than the attempts to establish shared information using email.

So, is that all we need to know about messaging? Should we concentrate on producing effective shared information systems, as these can be configured to include all forms of communication? In fact, there is a vital element missing, and we need to return to the subject of this chapter, events and status.

10.6.3 Messages as events

As we noted earlier, the reader may well have noticed a similarity between the event/status distinction and messages and shared information. Messages happen, information is there. This is exactly the distinction between event and status. The message is ephemeral, because its effect is achieved once it has been received.

One day someone comes into my office and puts a paper to be read on a pile at the back of my desk. It effectively passes into my ownership and my personal workspace. However, let us elaborate on this in two alternative scenarios. In the first, I am away. The paper is left on the pile with a note on top inviting comments. Assuming I notice it there upon my return I will discard the note, but I may not notice it for a while. In the second scenario I am there: "here's a paper you may like to read" ... the paper is flung onto the pile.

In the first scenario, the "message" part of the transaction was ephemeral and transferred ownership, but it failed in its purpose. Because it did not achieve sufficient salience for me, the event of my receiving it was delayed, possibly indefinitely. The difference between the spoken message and the written note is immaterial: if the note had been placed silently on my desk because I was on the phone, it would still have achieved salience and would still have achieved its purpose as a message. To make matters really extreme we could contrast, on the one hand, the note being slipped silently onto my desk while I was in the room but with my back turned, with, on the other hand, it being put there just as silently but in full view. A snapshot of the room just before and just after would be identical, but in the latter case an event would have occurred and a message would have been passed.

10.6.4 Why send messages?

Why is the salience and event of reception so important? We must look at the reasons for the sending of messages.

The first reason is imperative. The knock on the door and cry "Fire!", and the Poll Tax demand are both intended to prompt action. The salience is necessary if we want the action to occur in time; however, the immediacy required in the two cases is different. Putting a letter through the door with the word "Fire!" on it would not be appropriate, neither would knocking at the door and shouting "Poll Tax".

Often, combined with this desire for action is a more subtle reason. If you know that your message was salient, you *know* that the recipient has read it. In addition, the recipient knows that you know, etc. That is, it achieves its effect by creating and transferring meta-knowledge. Examples of where this is important would include informing interested parties about a decision even when they have no opportunity of affecting it.

When combined, messages may achieve quite complex social purposes. For instance, not only does the Poll Tax demand prompt you to pay, but the fact that *you have received* the demand and have not paid makes you liable to prosecution. Letters sent to those who do not hold television licences which request confirmation of receipt have a similar legal entrapment function.

Speech-act theory treats a conversation rather like a game, with various moves open to the participants at different stages. Breakdowns can occur during face-to-face conversation due to misunderstandings, but remote conversations open up further possibilities. If I make a move in the conversation game and the other participant does not know about the move, then she may make a move which is perfectly valid from her view of the conversation but may be confusing or meaningless from my point of view. Thus the salience of my messages forms an important role in synchronising our conversation.

10.6.5 Appropriate events and user salience

Thus messages must give rise to events salient to the recipient. There are various ways by which users perceive events:

(i) An event-type system output, e.g. a beep or sudden flash of the screen.

(ii) An indicator on the display, e.g. a mail ready flag.

(iii) A change in the display, e.g. a mailbox size increase.

(iv) An indicator which is uncovered during subsequent interaction.

(v) A change in status which is uncovered during subsequent interaction.

Notice that (iii) and (v) are both cases where a change in status is interpreted as an event. There is, of course, the possibility that such status change events are *never* noticed by the user: perhaps by the time the user's attention is on the appropriate part of the display or database, they have forgotten what the previous state was. The eye is very good at noticing changes, so display changes will be noticed as long as they occur reasonably close to the user's focus of attention. However, changes that are not immediately visible are very likely to be missed and so it would seem good policy always to have some persistent indicator rather than to rely solely on status change.

The same is true for the event outputs of class (i). These have the greatest immediate salience but, once missed, will never be noticed. Again, a permanent indicator will usually be required. These requirements for status records of events were discussed earlier in this chapter. A good example of an event indicator which combines both immediate salience with persistence is the Mac "Alert". These dialogue boxes disable the system entirely until they are responded to, hence they cannot be missed, yet they remain on the screen if the user is not present. They have other less fortunate properties, however. They interrupt other work, but the same could be said for a fire alarm. More seriously, they often demand that the user takes some action such as a decision before allowing the user to continue. Thus their purpose as an event, and a connected requirement for user intervention are conflated.

Status indicators of classes (ii) and (iv) only become events when the user notices them. Again, as we discussed early in this chapter, they may require some pattern of use to be sure of being noticed. For instance, my workstation has a flag on the mailbox which is raised when new mail has arrived (class ii). Periodically I look at the flag to see if there is any mail. If the flag were not there I would periodically have to use a mail command to examine the contents of the mailbox (class iv).

If the sender of a message wants to know that the message will be received then something must be known about the way the arrival of the message is signalled, and possibly about the habits of the recipient. If I know that someone looks at their mailbox only once a day when they log on to the system, I would not use email to ask them whether they can meet over lunch. When we discussed this issue earlier we noted that the granularity of events would be related to the pace of the task. Then we were thinking of automated tasks with a single user. A similar observation holds here though, that the pace of the conversation will influence the acceptable granularity of perceived events. This is why it is unacceptable to put a letter through the door warning of fire, whereas the task of paying taxes is measured in weeks (at least!), and therefore my infrequent checks on my doorstep are sufficient to make the letter an appropriate message.

We see then that in order to deal with a range of types of communication, we must not only be able to deal with information transfer and associated issues of ownership and persistence, but must also account for events. This is the key feature which mail systems possess but shared information lacks. Where users implement messaging using shared data they have to develop protocols for examining the appropriate parts of the data space. Adding features to a shared data model to raise appropriate events would be a major step towards producing a common framework for communication.

10.6.6 User status and system status

Finally, after considering the central role of events in communication, we return to status. The shared information entered by users as part of their communication is part of the system. A crucial part of face-to-face communication is the status of the other participants: whether they are attending, whether they want to speak, etc. Without such information it can be hard to establish effective communication. Once users are separated by a computer system they may not even know whether the other participants are at their terminals, let alone whether they are attending.

In Chapter 4 we considered techniques for capturing user task knowledge about the interdependence of windows. There is a similar issue here. Obviously, only information within the system can be presented to the user. However, the crucial status is about the users themselves: their attention, attitude, presence, etc. This is not immediately available to the system and cannot therefore be passed on to the other users. There are several possible ways of dealing with this problem.

First, we might ask the users to tell us what they are doing. This is likely to be an imposition and may be neglected. Still, if users know that the information they supply will be useful to other participants in a conversation, they may cooperate.

Alternatively, there may be information within the system which gives a reasonable indicator of the user's status. For instance, if the user has been typing recently then it is reasonable to assume that the user will be attentive to new messages, especially if they are accompanied by an audible or strong visual signal. Screen-savers work on precisely this principle, and thus the relevant mechanism may already exist on the system. A similar indicator is a screen lock, which implies that the user is away, but will return. These are both partial indicators of the user's status: the user may leave the screen locked (or turn the machine off) whilst still in the office. Similarly, the user may be attending to the screen but not actually typing.

It is possible to go further. Not only can existing system information be used, but also dialogues can be deliberately designed which expose aspects of the user's status. For example, in an experiment at York, each participant was required to select windows as part of their normal use of the system. These selections were designed to reveal information about the user's current activity. This information was then made available to the other participants.

10.7 Discussion

I have discussed status and events in many detailed circumstances, and it is time to recap on some of the properties that have been encountered. However, before I do that I would like to reflect on one last facet.

10.7.1 Events and dynamism

When we consider status it is (almost by definition) about *static* properties. It can be derived from a snapshot of the system at a particular moment. Furthermore, if we look at a time when there is no event, the status a little before and a little after is likely to be very similar. Hence status also emphasises continuity. Events, on the other hand, are at moments of change: they punctuate the dialogue, marking discontinuity and *dynamism*.

We can see intimations of this in previous chapters. The distinction between static and dynamic invariants has a similar feel. Static invariants relate various aspects of the status of a system. Dynamic invariants still talk about status, but limit the change in status when events occur. However, dynamic invariants do not tell us about what will happen at specific events, but provide only general restrictions on the sort of thing that can happen.

Both pointer spaces and complementary views told us something about the dynamism associated with particular events, but in rather different ways. For complementary views, the concentration on process-independent updates meant that we virtually ignored the *method* of change at an event, and instead looked only at *what* change occurred. This simplified matters considerably in what was still a tricky area; however, when we got to the end of the chapter and took a quick look at dynamic views a different perspective was required. We discovered that different sequences of updates could take the underlying database from the same start point to the same final value, and yet a different view was required. That is, the method of change became crucial. This led us to see the similarities with dynamic pointers and pointer spaces.

The pull function in a pointer space is a rich construct as it focuses on the *manner* of change. We went on to mention the concept of change information in general, of which the pull function and locality blocks were examples. Unlike the result and the display, which are status information, this information is associated with the event itself.

It is easy when thinking about concrete interfaces to concentrate on static aspects: at the surface level, screen mockups and perhaps at the semantic level, data structures. Dynamics are difficult to describe, both in written language and in the visual language of screen dumps. It is, however, the dynamics that are often crucial to the feel and effectiveness of a system. One of the points made in the pointer spaces chapter was that the dynamics of positional information are typically not considered carefully enough in design. For instance, it would be easy to approach the design of a novel graphical hypertext with beautiful screen displays and subtle representations but not discover the problems of updating the links until it was implemented.

Events are obviously regarded as important in design and, in particular, dialogue descriptions are almost purely event oriented. It is, however, in linking the worlds of events and status that we are likely to come unstuck. The common approach is to look at the event as a state transition. That is, we reduce the event to its effect on status, rather than looking at it as important in its own right. Bornat and Thimbleby (1986), in their editor ded, went to great lengths to make the manner in which the screen is updated emphasise the direction of movement through the text. This is possible on character terminals which frequently have fast methods of scrolling upwards and downwards. Unfortunately, these insights have been all but lost in bit-mapped systems. Bit-map scrolling is often far too slow and the tendency is simply to repaint the screen. So, even at the raw interface, the event of movement is reduced to a change in status. This loss of dynamics is sad, and it is particularly unfortunate that an apparent improvement in technology should lead to a reduction in other aspects of interface quality.

10.7.2 Summary – properties of events and status

We shall now briefly review the various properties of status and events. The fundamental distinction is perhaps that events *happen* whereas status simply *is*. This was exactly the same distinction that we noted between messages and shared data, and the same issues tended to arise in both the single-user and multi-user cases.

Granularity and context

The way in which we regarded the ringing of the alarm at the beginning of this chapter depended on the timescale with which we looked at it. If we look at the ringing at a fine timescale it is a status, yet at a coarse timescale it is an event. The appropriate granularity depends on the task and the context: hence we must establish a contextual framework for statements about status and events. In a similar way, the context of communication determines what would be regarded as an effective message.

Status change

An important way in which events and status interact is that changes in status may be perceived as events. Just as events, when looked at closely, can often be interpreted as status, so can apparently continuously changing status, when examined, be seen as many change events. More important, however, even at the scale whereby the individual changes are not salient, certain changes of status through critical values become, for the recipient, events. Examples of this include the hands of a clock passing a certain time, a mouse moving over a window and seeing a postcard appear on a noticeboard.

Salience

The "may" at the beginning of the previous paragraph was essential: changes in status are interpreted as events only when and if the recipient notices them. Similarly, objective events become subjective events for the user (or the system) only when they are noticed. Thus, in determining what are to be seen as salient events, we must have regard to the focus of attention of the recipient (and grab that attention if necessary). Further, for status change events it is often necessary that the recipient engages in some periodic procedure to scan the appropriate status. Sending an effective message requires not only that it should arrive, but also that the recipient will notice that it has arrived and thus act upon it.

Synchronisation

As events are partly subjective, there is the possibility that different partners in an interaction, whether a system and a user, or people in a conversation, will differ in their interpretation of an event. It is crucial that a message I send is not only an event for me, but that it becomes an event for you also. Our individual perception of the state of our conversation depends on mutual recognition of events; without this, we may diverge in our respective interpretations of where we are. This is especially important when silence is interpreted as a speech act, or where a message is intended to terminate a conversation. Synchronisation was also important when we discussed the way that the system may regard some changes in mouse position as events which the user considers inconsequential.

Dynamism

Finally, we have only just discussed the role of events in injecting a sense of dynamism. Talking via a pin-board is a far cry from a face-to-face conversation.

10.7.3 Conclusions

We have examined the role of status and events both in single-user interaction, and in the form of messages and shared data for multi-user cooperation. It is not surprising that similar features arise in the two areas, as we can, of course, regard the single-user case as that of a user in cooperation with the machine. In fact, it was a consideration of the parity between the two which began this chapter.

The most interesting thing about status and events is the interplay between the two. Unfortunately, it is hard to find models which adequately describe both. As we have noted, the models in this book largely fail in this respect, as do other formal models of interaction. This failing is not confined to HCI; similar problems are apparent in general computing. For instance, models of concurrency tend to be totally event-based, ignoring status completely. On the practical side, database systems are almost entirely status-based and this is a major shortcoming when they are considered as vehicles for cooperative work. The only computing paradigm that could handle both to any extent would be *access-oriented* programming.

This lack of allowance for the range of status and event behaviours seems particularly unfortunate since our real-life interactions are typically mixed. An office worker will respond to memos and phone calls, consult paper files and databases, have conversations and smell lunch cooking in the canteen. A cyclist must be aware of road signs and road markings, traffic lights, the movements of other road users, and car horns. In addition, the cyclist will control the cycle's road position, use hand signals, and perhaps a bell. A factory controller will watch various status indicators, temperatures, pressures, and stock levels, whilst being ready for breakdowns, the arrival of bulk deliveries and knocking-off time.

In order to comprehend the complexities of the world, it is often necessary to simplify. However, this wealth of variety suggests that there is much to be gained from a richer understanding of event and status in interactive systems design.

CHAPTER 11

Applying formal models

11.1 Introduction

We have seen a wide range of formal models for different circumstances and classes of systems. Formal models can be applied not only in many circumstances, but also in many different ways.

The most obvious way is as part of a rigorous formal development. The models are related to a specification and properties proved of it. The specification is then refined to a working system. The last two sections of this chapter follow this path. The first of these considers an example specification of a text editor using layered design and the dynamic pointer paradigm as proposed in Chapter 8. The full specification that this section discusses is given in Appendix II.

The last section looks at the issue of refining specifications for interactive systems. It uncovers conflicts that are are latent in the very nature of linear formal development, but are particularly important when weighed against the requirements for rapid turnover of reasonably fast-running prototypes. The section's title: "Interactive systems design and formal development are incompatible?" is a little ominous. However, the conflict arises largely because users and computers are incompatible. The job of HCI is addressing that incompatibility and the lessons from formal development are that this job is never easy.

This very rigorous approach is not the only, nor always the most useful way of applying formal models. All the way through this book we have applied the models in more informal ways. The definitions and analyses may have been formal, but the way these are related back to the real world is partly a matter of metaphor. In fact, the very nature of the formality gap means that there must always be a step between the formalisms and the things they represent. An over-emphasis on the formal steps can make us forget this, the most important part of the development process.

There is thus a whole spectrum of ways we can use formal models, ranging from the vague use of concepts and vocabulary to the full-blown formal process described above and at the end of the chapter. So, before we move on to these more formal uses, we look at two alternative approaches. The first is a "back of the envelope" analysis of a lift system, and is similar in tone to a lot of the stories and examples that have arisen throughout this book. The second is a notation, action–effect rules, developed by Andrew Monk, a member of the Psychology Department at York. It is a deliberate attempt to link the models described earlier into a framework more digestible to the typical interface designer.

These different approaches are not exhaustive, nor are they mutually exclusive. Even where the fully formal approach is used, more informal, creative input is required. This may be independent of the models, or may be an application of the more subtle nuances of the models that are not fully captured by the formal process. Real systems tend to be like, but slightly different from, some formal model. The properties we want are close to, but with some exceptions from, the principles defined using that model.

11.2 Analysis of a lift system

Some years ago I attended an HCI conference on a university campus. The delegates were housed in several tower blocks. My room was on the 11th floor, so I often used the lifts. I am not sure whether the lifts were hard-wired or used a microcomputer, but the interface was typical of a large class of automatic systems with restricted interfaces, from electric drills to cash registers.

There were, in fact, two major interfaces to each lift. On each floor (except the basement and top floor) there were two buttons to call the lift, one if you wanted to go up and one if you wanted to go down. Now that in itself seems fairly simple, but there is often some confusion even over that. On this lift the button with the upward arrow meant "I want to go up" and the downward arrow meant "I want to go down". However, recently I was on the top floor of an hotel which had a single lift call button with an upward pointing arrow, which meant "bring the lift up". This is somewhat reminiscent of the problems with arrows for scroll boxes. Does an upward arrow mean move the window up over the text, or move the text up over the window? Fortunately, this confusion is less widespread in lift systems and the standard convention on the campus lift caused no problems.

The interface inside the lift was more complex (see *fig.* 11.1). There were a set of buttons, one for each floor. The floors were lettered and there were two lifts which served alternate floors. In addition, some letters were omitted: hence the

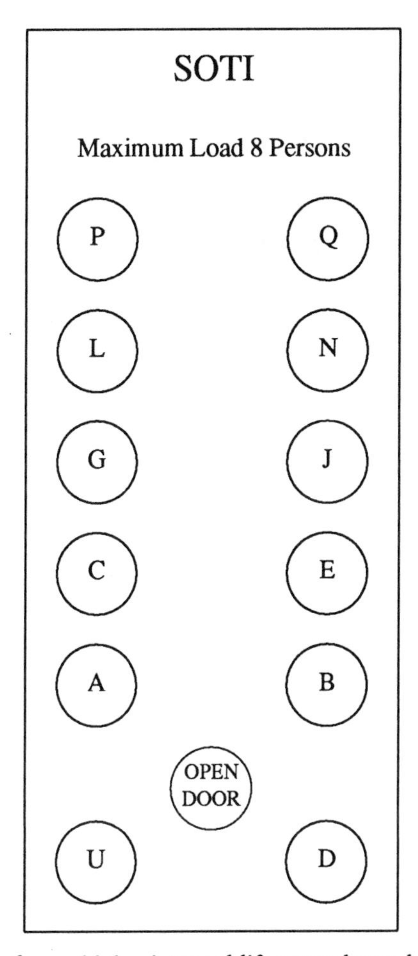

figure 11.1 *internal lift control panel*

odd labelling. Above the door was a strip with illuminated letters, which indicated the current floor, and up and down arrows indicating the lift's current direction of travel. It was this internal interface which had interesting effects.

11.2.1 The ritual

After a while I began to notice an interesting ritual I performed upon entering the lift. Other people seemed to follow the same ritual, which went something like this:

- I get into the lift, and press the button for my floor.
- I wait ...
- and wait ...
- I press the button again and at once,
  clunk – the doors close, and whoosh – up goes the lift.

When several people got into the lift the ritual became a little more complex. It began in the same way, except that buttons were pushed for each floor. The lift reached the first selected floor, one passenger got out leaving the rest of us, and then:

- We wait ...
- and wait ...
- Someone presses the button for the next floor and at once,
  clunk – the doors close, and whoosh – up goes the lift.

Now, I am sure that each of the passengers knew that the lift would have gone up anyway whether or not the button was pressed again, but we also knew that it seemed to make a difference. Analytic science versus empiricism? I am a mathematician and I did it: what chance for the experimental psychologists?

On the other hand, lifts have a reputation for unreliability and I have certainly found several times that lifts do not take me where I want them to. Computer systems share this reputation for unreliability and lack of control, though fortunately lifts rarely "crash". The compulsive pressing of the floor button is rather like the habit of repetitively striking an ENTER key.

11.2.2 Analysing the problem

Now, however well the interface had been designed the passengers may have still been impatient and engaged in compulsive behaviour, but this interface had an important flaw which is obvious as soon as the problem is described: there was no feedback from pressing the floor selection buttons. The only way the passenger could know whether the lift had registered their request was if they waited for it to go to their floor. That is rather like using a word processor with no display: you know if you have mistyped or not when page 73 gets printed. In the terms introduced earlier, the lift system is not predictable.

Once we have noticed the flaw, we can imagine simple fixes. Many lifts have a light behind each floor button, which is illuminated if that floor is due to be visited. Alternatively, if we preferred direct manipulation we could have buttons that stayed down until the lift passed the floor, but it is doubtful whether that would do much for the lift's reliability.

Analysing the problem in retrospect and classifying it in terms of generic issues is quite interesting in its own right. More importantly, we could easily have predicted the problem beforehand, using largely automatic analysis. The internal state of the lift system may be quite complex, including oil levels, maintenance indicators and the like, but the state insofar as it affects the passenger has four major components. Three of these are:

- The current location.
- The direction of travel.
- The set of floors to be visited.

If we had wanted to demonstrate the lift's predictability early in its design we would have had to show that each of these components could be obtained from the display. As the state and the system were simple, we would probably have required the simplest form of predictability, and demanded that all the relevant state was derivable from the current display. The first two of the components were permanently displayed above the door, so they were catered for. However, there was no way of observing the third component, and in this lies the heart of the problem.

I did say there was a fourth component of the state, and, to be honest, I did not even think of this component till some time after I had written out the rest. This final component is:

- The time left before the lift starts.

Now this is not just an observation of the lift, such as how long it will take to get to the third floor, which depends partly on the future of the interaction with other passengers and partly on physical factors, but is a necessary part of the state. When I get into the lift and press the button for my floor, it does not immediately start up, but instead waits for a while in case other people want to get in. Somewhere in the electronics of the lift a timer will be ticking away the seconds until the lift starts again. It is interesting, especially in the light of the discussion of dynamics in the previous chapter, that I should neglect this component, and perhaps when writing out a description of the state of a system one should be particularly careful not to miss out time dependencies.

Now, unlike the floors to visit, it did not even occur to me that we should think of displaying this component until I wrote the state out in full. Having noticed this important component, the designer might have added some feature such as an hour glass above the open-door button which gradually lost its sand until the doors closed, or alternatively a dial, bar strip or digital indicator that ticked the seconds away. The sand (or seconds) could be topped up every time the open-door button was pressed. It is interesting that although many lifts have lights on the floor request buttons, I have never seen a delay indicator. The reader may like to imagine (as have previous readers) ways in which the passenger might be given some control over this aspect of the state.

Control panels like the lift system's are common. The underlying systems tend to have simple states. Often they are finite-state machines, or at most have a few simple additional components like strings or numbers. The displays tend to be quite simple too. There are various CAD-type packages around that help a designer layout control panels; it would be very simple to add facilities to check automatically for properties such as observability. Putting such an interface design package together with a simulator of the functionality would guard against the designer missing out components of the state, such as the door delay.

11.3 Semiformal notations – action–effect rules

There are a large number of semiformal notations used in interface design, from the psychologically oriented methods like GOMS (Card *et al.* 1983) to the systems-oriented techniques such as JSP (Jackson 1983). Although they may contain textual descriptions or are vague in places, a significant part of these notations may be put into forms suitable for formal or automatic analyses. Unfortunately, the task-oriented techniques frequently miss the output side of a system, concentrating on the task–action mapping, which means that they are less well matched to the principles in this book than the systems description techniques.

It is important to forge a link between semiformal techniques that may be more comprehensible to a typical human factors expert or systems engineer, and the full formalisms that are capable of rigorous analysis. This section describes a notation developed by Andrew Monk, a psychologist at York, as a first attempt at bridging the gap between the types of formal model presented in this book and these semiformal notations (Monk and Dix 1987).

11.3.1 Action–effect rules

Action–effect rules consist of three elements:

- A description of the possible actions available to the user; this is equivalent to the command set of the PIE.
- A textual description or snapshot of typical displays, one for each visual "mode" or major state of the system.
- For each visual mode, a set of rules that describes how the system responds to each possible user action. This may either cause a change to the display within a mode, or cause a mode switch.

For the text editor which I am using at the moment, the possible actions would include all the keys on my keyboard, including function keys and cursor keys, plus combinations of these with various shifts. The visual modes would include:

(M1) The start-up page with copyright notice

(M2) The main editing mode with text covering most of the display and some context information at the top.

(M3) Several help screens

(M4) Setting the find/replace strings

Typical rules for the main editing screen (M2) might include:

(R1)	Printable key	⇒	character inserted in text before cursor.
(R2)	UP ARROW	⇒	**if** cursor not at the top of the screen **then** move cursor up one line on screen **else** scroll screen down one line leaving cursor at the top.
(R3)	DELETE	⇒	remove character before cursor.
(R4)	F1	⇒	**goto** Help screen (M3).

11.3.2 Analysis

These rules can be subjected to various forms of analysis. We can check them for *completeness*, and if there are any gaps in our knowledge go back and try out the various combinations. This test is fairly automatic, as we can simply see if there are any actions for which no rule is specified. We may, however, miss out some condition that would complicate a rule, making it dependent on context. For instance, rule R3 says delete the character before the cursor; however, I should really have added extra context conditions depending on whether the cursor is at the beginning of a line, or even at the beginning of the whole document. Similar problems occur with R1, and rule R2 should have yet more boundary conditions.

We can also test for *predictability*, first by checking that the modes are visually distinct, then by examining any rule with a conditional right-hand side and asking whether the user could tell from the display which of the possible responses would be taken. Designers could perform this analysis using their own professional judgement or by asking potential users. This could be done using a "live" prototype based on the description, or simply with paper mock-ups.

Consistency can be assessed by looking at the similarity (or lack of it) between rules for the same action in different modes. We would looking for several things here, most obviously, for a general consistency throughout the interface with the standard benefits of improved learning. Also, we might be especially wary where we see two modes where almost all the rules are identical, but where just one or two differ. This would be an obvious point where confusion could occur and we would have to decide whether the visual distinctiveness and semantic necessity were strong enough to overcome the propensity for error.

We can also analyse *reachability*: whether the chains of actions and modes form a completely connected net, or whether there are some blind-alley modes. This is not a totally automatic procedure, as the movement between modes may depend upon context. Also, the designer would have to analyse as a separate exercise the reachability within each mode. However, this separation of concerns would be no bad thing.

Further, we can ask about the *reversibility* of the rules: if we follow a rule, how many actions do we have to do to get back to the mode we have left? This gives us some measure of the difficulty of undoing an action. Again, this analysis would require the designer to look at the movement between modes, and also the ability to recover within (or after re-entering) a mode.

These analyses are done by eye, with the formality of named rules and modes helping the designer to book-keep. There are various subjective choices to be made by the designer during analysis. For instance, when assessing predictability, there are places where the user would not expect to know the exact form of the next display: for instance, if the user had just read in a new file. The designer must distinguish these cases from those where predictability has been violated in a serious manner. It is also partly a matter of opinion whether the effects of actions are similar in different modes. For instance, in an integrated system with word processor and spreadsheet it might not be clear what the consistent spreadsheet equivalent to the word processor's carriage return would be. The rules therefore give the designer a structure within which to perform analyses, but do not take over from the designer's professional role.

11.3.3 Layout choices

There are also choices to be made during the laying out of the rules. By introducing more visual modes the number of conditional rules can be reduced. For example, we could introduce 24 modes, one for each line the cursor could be on. We would then have a copy of rule R2 one for each of the 24 modes. Each rule would be simple, as there would be no conditional, but there would be lot of rules! In addition, the conditional would reappear as a violation of consistency. Similarly, there are several Help screens that can be scrolled through using the cursor keys. These could be regarded either as one mode with a rather vague rule:

mode M3:	(R5)	RIGHT ARROW	⇒	change Help information field to next subject.

or as lots of modes with rules of the form:

mode M3a:	(R5a)	RIGHT ARROW	⇒	**goto** second help screen (M3b).
mode M3b:	(R5b)	RIGHT ARROW	⇒	**goto** third help screen (M3c).
...			...	
mode M3i:	(R5i)	RIGHT ARROW	⇒	**goto** first help screen (M3a).

This ambiguity could be addressed by extending the rule system to include some sort of inheritance hierarchy, but this would significantly complicate the notation. The designer can also choose to be more or less precise about the descriptions used. Obviously, a very vague rule will not be very useful for analysis.

11.3.4 Comparison with task–action mappings

Most of the psychologically founded notations are task-oriented: the focus is on how the user's internal goals are translated into actions upon the system. Typical of these would be GOMS (Card *et al.* 1983) or TAG (Payne 1984). These are, in effect, designer's models of the user's actions. We shall refer to the whole genre as task–action mappings during the following discussion. Action–effect rules look at a different part of the user-system feedback loop:

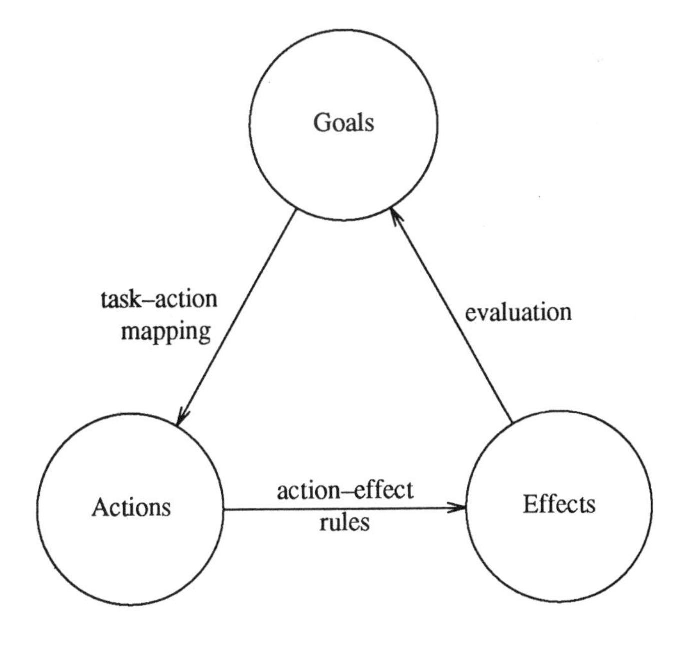

The final side of the loop is labeled *evaluation*, from Norman's analysis of interaction. It is a pity that psychologically founded notations are poor on this side, as this would nicely complete the picture.

It is clear that task–action mappings and action–effect rules occupy different design niches and can therefore complement each other in the design process. To get an idea of when we should be using one approach rather than the other, we shall contrast their respective strengths and weaknesses.

The task–action approach seems at first glance more user-oriented; the action–effect rules merely describe the system. They do, however, describe the system from the *user's* perspective. We could contrast the two approaches as looking *inside* the user, compared with standing *alongside*.

Concentrating on task is clearly important in optimising systems for typical behaviour. It enables the designer to match the form of the dialogue with the structure of the tasks the user will perform. The danger of this approach is that, by optimising for particular tasks, a system may deal badly with those that are missed. New systems are always used in ways that the designers did not consider. Good task analysis may help to predict these, but will never discover all possible uses. In addition, we have to consider error recovery, which massively increases the set of "tasks" that should be allowed for.

Action–effect rules have completely opposite strengths and weaknesses. They favour no particular task and, without being paired with a task-oriented input, would allow systems that were totally unsuited to their eventual use. However, by being uncommitted they favour systems that can be used in new unforeseen ways. In particular, by focusing on the effects that a user can achieve and how to achieve them, it allows users to set their own goals.

A similar contrast occurs in descriptions of direct manipulation. On the one hand, there is the concept of cognitive directness (Hutchins *et al.* 1986), the natural flow of user's mental models and goals into the system. On the other, is the concentration on exploration, recognition and goal seeking (Shneiderman 1982). The former gives us the clues to what metaphors are relevant within a particular system, but does seem to assume a blind and deaf user. The characteristic user of the latter would, I suppose, be extremely sensually acute, but with little foresight.

11.3.5 Prototyping and automatic analysis

Some of the rules given above were very explicit, in particular those that sent the system into new modes. Others were almost entirely textual. With a little effort the major mode changes could be captured entirely, and the notation rendered in a machine-readable and implementable form. This would allow various levels of automatic design support.

First, it would form a rather crude prototype. The resulting system would behave in a gross manner like the target, the user's inputs taking the prototype through its visual modes. More detailed behaviour might be met with a dialogue box saying "the character X is inserted before the cursor", but with no actual change to the display. This would be rather similar to the slide-show prototypes built using hypertext systems.

The same description could be used for some automatic checking. Some of the analysis presupposes a level of interpretation, but some of the more tedious checking and book-keeping could be taken over by the computer. Further, when a detailed specification is produced this could be compared with the action–effect description, and checked for consistency.

11.4 Specifying an editor using dynamic pointers

In Appendix II a specification is given of a simple editor. This section gives an overview of this example specification. It is a simplification of an editor specified and implemented as part of an experimental hypertext system at York several years ago. The original included a folding scheme that allowed disparate parts of a document to be viewed at the same time, a more sophisticated pretty-printing mechanism and hypertext links to other documents. The editor sat within individual windows of a multi-windowed system including a structural view of the hypertext.

The specification is an example of a layered design. It is specified in three layers:

- *Display* – The physical display as seen by the user. It represents a standard 80 by 25 character screen.
- *Text* – A formatted representation of the application objects, but of unbounded extent.
- *Strings* – The application object itself, a simple stream of characters.

Each layer has an associated pointer space, and the layers are linked using projections (*fig.* 11.2).

The specification proceeds in two major stages. First, the major data types are defined, *Disp*, *Text* and *String*, these being related by appropriate projections. These data types are then used to define states, with appropriate projection and interaction states being added to the various layers.

Proceeding in this two-step fashion has various advantages. The operations defined over the basic data types are "pure": that is, there is no concept of update, new values are merely constructed from old. It is often argued that it is easier to

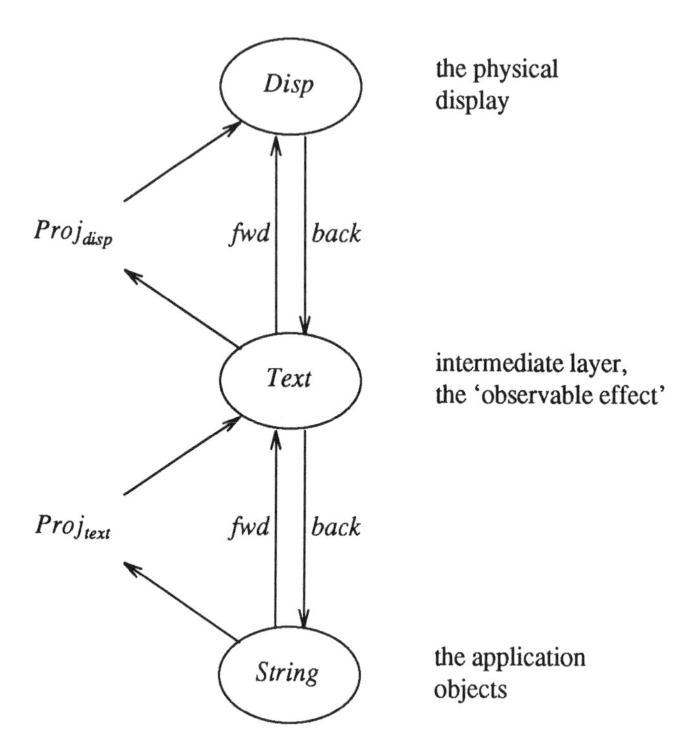

figure 11.2 *pointer spaces and their projections*

reason about pure functions. Thus we can perform some proofs or analysis of the specification at this level.

The operations will typically not correspond directly to simple keystrokes. For instance, the insertion operation on a string requires both a character and a position at which to insert it. Similarly, a find/replace command requires two strings as parameters. These additional parameters are part of the "interaction state" of the system. The former would typically be an insertion point, and the latter may reside as some permenant part of the state of the system or be demanded interactively as part of a find/replace dialogue. These decisions can be left to the second phase of the specification, thus simplifying both stages.

The role of the projection distinguishes the specification given here from similar specifications given elsewhere (Ehrig and Mahr 1985, Chi 1985, Sufrin 1982). The information in the projection would typically be part of the state definition phase and would be likely to be dispersed throughout the specification. The use of pointer spaces means that this information is both separate from the data types themselves, and part of the functional data-type phase. This both eases tractablity and factors the design.

11.4.1 Data types and projections

The display and text are very simple, the display being merely an 80 by 25 array of characters and the text a sequence of lines. The pointers for each of these are merely number pairs giving row–column offsets. The projection between *Text* and *Disp* is just a window into the text at a position specified by a text pointer. The *fwd* and *back* maps associated with this projection are simple enough in the "normal" case; one merely adds or subtracts the relevant offset. There are, however, numerous special cases. For example, when we wish to take a display pointer *back* to a text pointer, if the display pointer is over blank space we want it to be mapped to the "last" sensible text position. We cannot simply say that such mappings are out of range, as we will want to translate mouse clicks after the end of lines and treat them as selecting the appropriate line end. The important point to note is that these special cases are isolated in the description of the *back* and *fwd* mappings, rather than being spread throughout the definition of each operation on the objects.

The actual objects being manipulated are simple streams of characters including new-line characters. The pointers to these are single integer offsets into the strings. The string data type has two operations defined which yield new strings and pull functions:

$$insert: \ P_{string} \times Char \times String \ \rightarrow \ String \times (P_{string} \rightarrow P_{string})$$
$$delete: \ P_{string} \times String \ \rightarrow \ String \times (P_{string} \rightarrow P_{string})$$

These are simply character insertion and deletion at the appropriate positions. In addition, there are *succ* and *pred* functions which take a pointer and give the succeeding or preceding pointer. Each of these requires not only the pointer but also the string on which it is acting. This is so that the operation can do "range checking" and thus yield only valid pointers.

The projection between strings and text is merely the breaking up of the string at new-line characters. The line breaks are defined using a sequence of string pointers, *line_starts*, which give a pointer to the beginning of each line. Thus *line_starts* (3) is a pointer to the beginning of the third line. A more sophisticated approach would have been to have block pointers encompassing each line, but would be essentially similar.

Allthough not part of the example, it is possible to update this information in a highly efficient manner using locality information (§8.4). If we know that only a small part of the underlying string is affected by an operation, then we can simply use the *pull* function to modify parts of this *line_starts* structure outside the locality. This seems a little over the top for such a simple projection function, but forms the pattern for more sophisticated treatments. In the full version of the system a more complicated pretty-printing scheme was used (Dix 1987b). With the addition of contexts for the projection as described in §8.4.7, the structure of the system was very similar.

Although the action of splitting a string at line breaks is very simple, the *fwd* and *back* functions require a surprising amount of care at the boundary cases. As with the text–display projection this area of potential mistakes is localised to the definition of the projection, rather than being spread around the specification. It is thus more likely to be both correct and consistent.

11.4.2 String state

The lowest level of state in the system is the stream of characters being manipulated. It is just the current value of the string being manipulated, and thus the "state" is the same as the data type. This is so simple that it is not defined as a state as such in the specification, but is merely a component of the text state. The operations on strings as a data type can be regarded as operations on the underlying string state. The character required for the *insert* operation is likely to come directly from the keyboard and will be passed in as part of the user's command. The point at which to put it and the point required by the *delete* operation will be the cursor or insertion point. This will not be supplied by the user on a character-by-character basis, but will be part of the interaction state of the system. This is the additional information required to form the next level of the state of the system.

11.4.3 Text state

The state at the text level consists of a string (the lower level of state) plus a string pointer which will be the insertion point. The text is obtained by using the string-to-text projection. This projection has no parameter information and thus nothing additional is needed in the text-level state. Operations on this level of the state are of three classes:

> *Update* – *insert*(*c*) and *delete* are inherited from the string layer and "passed" on after suitable conversion.
>
> *Movement* – *move_right* and *move_left* are operations which move the insertion point. These use the *succ* and *pred* operations at the string level to update the insertion-point component of the state, and hence automatically ensure that the pointer is valid.
>
> *Selection* – *select*(*tp*) is an entirely new operation which sets the current insertion point to the text pointer *tp*.

The update operations are very similar to one another: they simply add the insertion point as an extra parameter and update the string part of the state using the appropriate string operations. The new insertion point is obtained by applying the *pull* function from the string operation. The display level will want an equivalent of the *pull* function on the text pointers in order to update its own state. This is obtained in a canonical way, by first translating text pointers to string pointers using the *back* function on the original string. The string pointer can then be updated using the *pull* function before being translated back to a text pointer using the *fwd* pointer of the updated string:

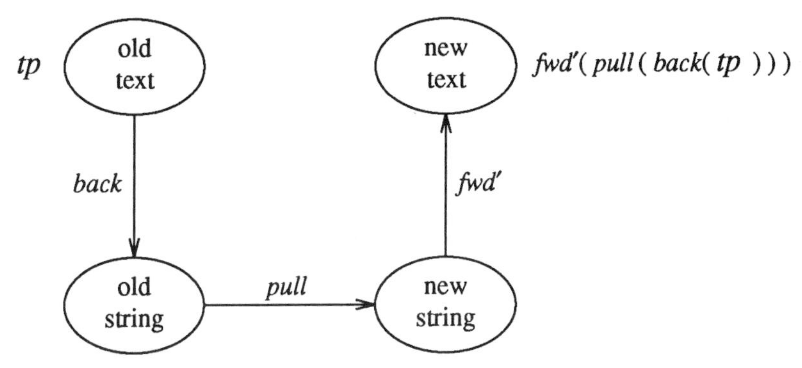

This construction is the same no matter how complicated the projection, and forms part of the pointer-space "tool-kit" for constructing manipulative systems.

The update operations are very simple, applying *succ* or *pred* to the insertion-point component of the state. Selection is almost as simple, it sets the insertion point to the supplied text pointer. However, as the insertion point is held as a string pointer, it must first be translated with the *back* function. An alternative design would have the insertion point as a text pointer. The cursor would then be able to range all over the semi-infinite text space, and not be constrained to the formatted region. The *back* function would then have been applied before passing the insertion point to the update operations.

The text state has a "display" mapping $display_{text}$ which uses the projection to obtain a text and translate the insertion point to a text pointer. This is not a physical display, but would correspond to the virtual display that most users would perceive as lying behind their screens. The result mapping of the system would be either the actual string or the text portion of the projection. The former would be the case for a text editor in a filing system, the latter for a printed document.

11.4.4 Display-level state

At the physical level, the user sees a screen and presses actual keys. The projection from text to display requires a text pointer as a parameter. This text pointer is added to the text state to obtain the complete system state. This can be though of as the display "map" component of the state.

The commands at this level are all inherited directly from the text level. Update commands are simply the appropriate keystrokes, as would be the movement commands. The selection command would be obtained from a suitable pointer and thus be a display pointer.

Selection is the simplest operation, which is not surprising as the whole dynamic-pointer paradigm revolves around the issue. The display pointer is simply translated into a text pointer using the *back* function from the text-to-display projection, and the selection operation passed on to the text level. The text level will, of course, go on to translate the pointer yet again! However, this does not concern the display level.

The esential structure of update and movement is quite simple. The relevant operations are performed on the text state and, in the case of update, the text pointer "map" component is updated using the *pull* function. However, both update and movement are complicated by the existence of a *static invariant*.

A standard observability condition is that the insertion point be visible whenever operations like insert and delete are performed. There are several ways of accomplishing this:

- The operations can be made illegal when the insertion point is not visible.
- The insertion point can be scrolled into view when an update operation is performed.
- The insertion point can always be kept on screen.

I will not go into the various arguments now, but the specification chooses the last option. This then becomes an invariant of the system. It can be expressed by demanding that the insertion point (treated as a text pointer) is obtainable by applying *back* to a display pointer. Because it relates the state and display at each instant, it is a static invariant.

As we discussed in Chapter 7, it is usually the case that to maintain static invariants some sort of adjustment must occassionally be made to the display mapping. An *adjust* function is defined in the specification for this purpose. It takes an updated text state and the *pull*ed map component. If the pair satisfy the static invariant they become the new display state. This is a simple form of display inertia. If, however, the new insertion point lies outside the display, a new text pointer is chosen for the map component. The choice of this position includes various cases depending on where the insertion point lies relative to the beginning and end of the text, but essentially centres it in the screen. Again, a lot of complexity and boundary issues that might otherwise be distributed about the system are collected within the *adjust* function. As with the projection functions, this both simplifies the rest of the system specification and ensures consistency.

The actual display of the system is obtained by using the text-to-display projection and using the *fwd* function to take the text-level insertion point into a display pointer. The final display then consists of the character map together with a cursor position. No new commands were added at the display level, but commands such as scroll-up or scroll-down could be be added here, updating the map component appropriately.

11.4.5 Summary architecture – adding new functionality

In retrospect, we can look back on the state architecture and see how commands are passed consistently back from level to level, and the display obtained (*fig.* 11.3):

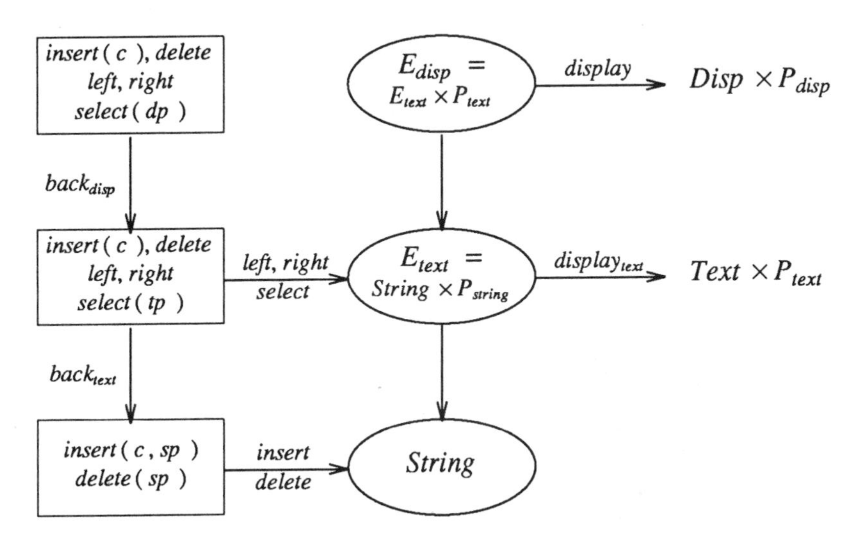

figure 11.3 *summary architecture*

The *back* arrows on the left-hand side show how the user-level display pointers are translated back through text pointers and, where appropriate, to string pointers. Of course, this is not the only aspect of translation. Pointers may be added (as with the insertion point to the update operations), or other data types. Also, the translation scheme may be far less one-to-one than this simple system.

The arrows down from E_{disp} to E_{text} and from E_{text} to *String* remind us that they are related by an abstraction relation. Although the dynamic-pointer paradigm has allowed us to build the various layers from one another, the essential feature is still one of abstraction.

A lot of the complexity of the system is wrapped up in the definition of the projection mappings and the display adjustment function. New commands are quite easy to add. As an example of this, the Appendix shows how a find/replace command would be added. We assume we have such a command for strings. It takes two parameters, the search string and its replacement, and updates the string returning also a *pull* function. It is, of course, this pull function which then makes the rest of the process easy. The find/replace command is plugged in in exactly the same way as the other update commands. The text layer simply updates its insertion point using the *pull* function, and produces a *pull* function for text pointers by using the *back–pull–fwd* trick. The display level again updates its mapping component, a text pointer, using this pull function, then applies the *adjust* function. The only extra mechanism that is required is choosing somewhere to add the string parameters. For a self-contained editor the appropriate place would probably be at the text level, with commands to select editing of text or these strings. In a more integrated environment these may be supplied as part of a high-level dialogue.

The mechanisms described above might seem a trifle heavy for a simple text editor. However, they factor nicely even this design. More importantly, the basic architecture using pointer spaces is the same for more complex systems. A general design strategy using dynamic pointers goes something like this:

- Define application objects and add pointers for them.
- Choose approppropriate user-level layers corresponding to the physical screen and the "observable effect" and add pointers to these (the screen pointers will be character or pixel coordinates).
- Generate the projections between these and the application objects.
- For each operation on the application objects, add a *pull* function.

The rest fits effortlessly together... well almost!

11.5 Interactive systems design and formal development are incompatible?

The most rigorous use of formal models would be as part of a strict formal refinement process. However, when we consider the relationship between the requirements of the *process* of interactive system design, and the *process* of formal development, several conflicts occur. These problems are not unique to interactive systems design, but are inherent in the concept of formal development; it is just that the rigours of interactive systems intensify and bring these problems to light.

After considering the differing requirements of the two domains of interactive systems design and formal refinement, we will proceed in a dialectic style. Conflicts will become apparent between the sets of requirements, which we will attempt to resolve, eventually leading to the need for structural transformation at the module level during refinement.

We will find that although interactive systems design and formal development are not completely incompatible, their combination is both complex and instructive.

11.5.1 The requirements of interactive systems design

Interactive systems design gives rise to three problems:

- Formality gap.
- Rapid turnaround – iterative design.
- Fast prototypes.

These are described below:

Formality gap

We discussed this in the first chapter. This is the gap between the informal requirements for a system in the designer's or client's head, and their first formal statement. The steps within the formal plane are relatively simple compared to this big step between the informal and the formal. If the formal statement of requirements is wrong, then no amount of formal manipulation will make it right. Correctness in the formal domain means preserving the badness of the first formal statement.

We have noted that if an abstract model is designed well for a particular class of principles, it can help bridge the formality gap. To achieve this, there must be a close correspondence of structure between the abstract model and the informal concepts. However, the abstract model will capture only some of the requirements, and the same principle of *structural correlation* must then apply to the entire interface specification.

Rapid turnaround

Because of the formality gap, we will never entirely capture the requirements for an interactive system (Monk *et al.* 1988) and thus some form of iterative design cycle is usually suggested. That is, a prototype system is built based on a first guess at the interface requirements; this is then evaluated, and new requirements are formed. The turnaround of prototypes must be fast for this process to be effective, perhaps days or even hours. Frequently this is done using mock-ups that have the immediate appearance of the finished product, but lack the internal functionality. However, many of the interface properties we have studied permeate the whole design of the system, and hence we would see it as necessary to have some reasonable proportion of the functionality in this prototype.

Fast prototypes

Not only does the turnaround of prototypes have to be rapid, but the prototypes themselves must execute reasonably fast to be usable. A slow interface has a very different feel from a fast one, and hence wrong decisions can be made if the pace of the interaction is unreal. The evaluator may be able to make some allowance, but this ability is limited. It would be very hard to evaluate an interactive system where you type for a few seconds, then have to go away for ten minutes and have a cup of tea before seeing what you have typed appear. The only thing to be said in its favour, is that it might encourage predictability, as a result of the "gone for a cup of tea" problem! Not only is it difficult to appreciate the system at all, but also very poor performance can encourage the wrong decisions to be taken, negating the benefits of prototyping. For example, imagine designing a word processor. One design is a full-screen, "what you see is what you get" editor, the other a line editor with cryptic single-character commands. At the speed envisaged in the production version, the full-screen editor would be preferable, but when executed a hundred times slower, the line editor, requiring fewer keystrokes and not relying so much on screen feedback, would appear better. It was for just such situations that line editors were developed! Of course, such a major shift would be obvious for the designer; however, there may be many more subtle decisions wrongly taken because of poor performance.

Contrast with non-interactive systems

We can contrast the above requirements with the design of non-interactive systems such as data-processing or numerical applications:

Requirements contrast	
Interactive systems	DP or numerical applications
wide formality gap rapid turnaround – iterative design fast prototypes – for usability	well-understood requirements slow turnaround functionality sufficient

Whereas the requirements of interactive systems are very difficult to formalise, for a DP application like a payroll, this may not be too much of a problem. The requirements are already in a semiformal form (e.g. pay scales, tax laws) and they are inherently formalisible. Similarly, because these requirements are well understood, there is less need for iterative design: changes in requirements may be over periods of months or years, with corresponding turnaround times. Finally, with such applications, it is sufficient to prototype the functionality only, with little regard for speed. It is easy to produce a set of test data and then run a numerical algorithm overnight to check in the morning for correctness. Even complex distributed systems may be able to be simulated at a slow pace.

11.5.2 Formal and classical development compared

In this section I compare formal development with classical development. By classical I mean a non-formal approach to development (hacker rather than Homer). I assume that the requirements are known and consider how these requirements are used to produce a first implementation, which is then optimised to create a version that is suitable for use (perhaps as a prototype or even as a product). Finally, I examine how these two processes respond to changing requirements, and the implications this has for the development process.

Initial development

Classical

Given loosely stated requirements, the programmer will, possibly using some intermediate graphical or textual plan, produce a first implementation of the program. Typically this will already be structured with efficiency in mind. However, if this is not fast enough, several stages of optimisation may ensue before reaching a version suitable for release (*fig.* 11.4):

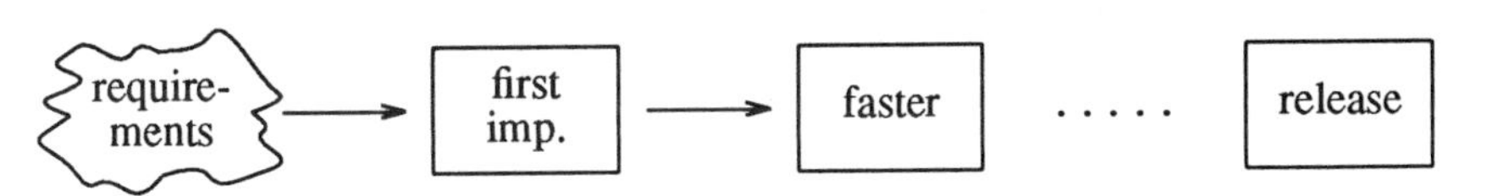

figure 11.4 *classical development*

Interspersed with this optimisation process will be debugging. In principle, one might debug each version in turn until one is sure of correctness. In practice, this debugging will be distributed, errors present from early implementations being corrected only later in the process. These changes will probably never be reflected in those early points, as the early versions are likely to be overwritten, or at best stored in a source control database.

Formal

At this stage the formal development process is quite similar. The requirements will be used first to generate a specification. This first specification will then go through several levels of refinement within the formal notation, both to make it sufficiently constructive for implementation and possibly as a first stage in moving towards an efficient formulation. The final product of these specifications will be used to derive the first implementation, which itself may then be optimised through several versions before a releasable version is produced (*fig.* 11.5):

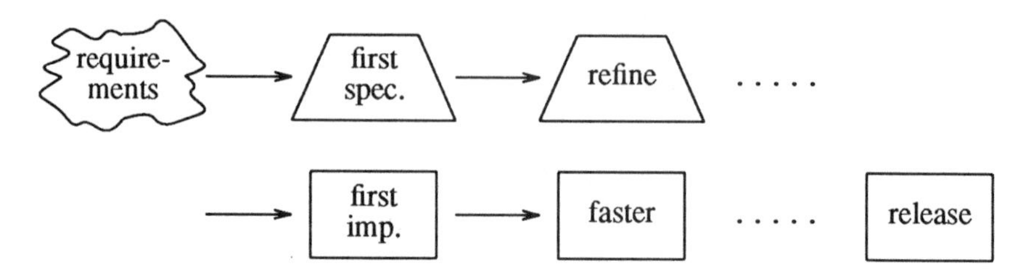

figure 11.5 *formal development*

Again there will be debugging steps, both in the specification as each refinement is checked against the previous specification for correctness, and in the implementation as this also is checked for correctness. The only difference here from classical development is that debugging is less likely to involve errors from previous versions. In particular, because the initial specification is derived much more directly from the requirements (it may in fact be a formal statement of the requirements), there are likely to be less changes needed to make the final product match the informal requirements. With some formal development paradigms, such as refinement by transformation, it could be argued that there is *never* any debugging, as all versions are guaranteed to be correct. However, even here backtracking in the transformation process is a form of debugging. From now on we will largely ignore these debugging steps, as they form a development process at a finer granularity than we are considering.

Changing requirements

Classical

If the changes in requirements are extreme, the programmer may be tempted to throw away all previous work and start from scratch. More usually, however, the most developed version of the system will be used and altered to fit the new needs (*fig.* 11.6). Because the fastest, optimised version is used there is unlikely to be much need for optimisation steps, except where radically new algorithms have been introduced. Note especially that the changing requirements are not seen to affect at all the early unoptimised versions of the program; they are just history.

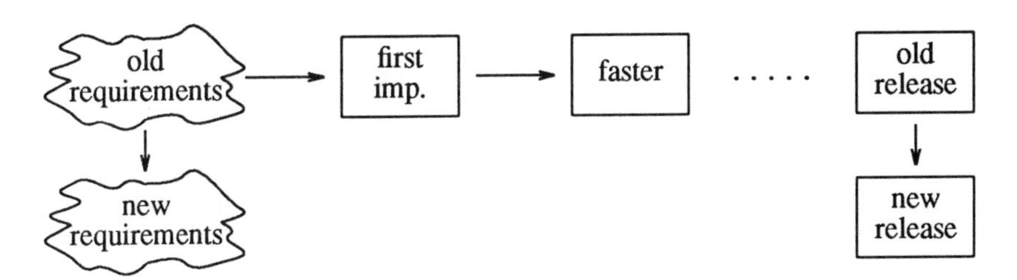

figure 11.6 *classical development, change in requirements*

Formal

The formal situation is very different. The intermediate versions, and especially the first specification, are the *proof* that the final version really does satisfy the requirements. There may well be testing and validation as well, but it is the process of development itself which is the major source of confidence in correctness. This is even (and especially) true when the process does not employ automatic checking. It is insufficient (although tempting) to change the specification a bit, change the final version a bit, and say the process is still formal. No, for formal correctness a change in requirements demands a complete rework of the whole development process, from initial specification to final optimised system (*fig.* 11.7). Again, in the worst case, this may involve a complete rewrite, but will usually be obtained by propagating small changes through the stages:

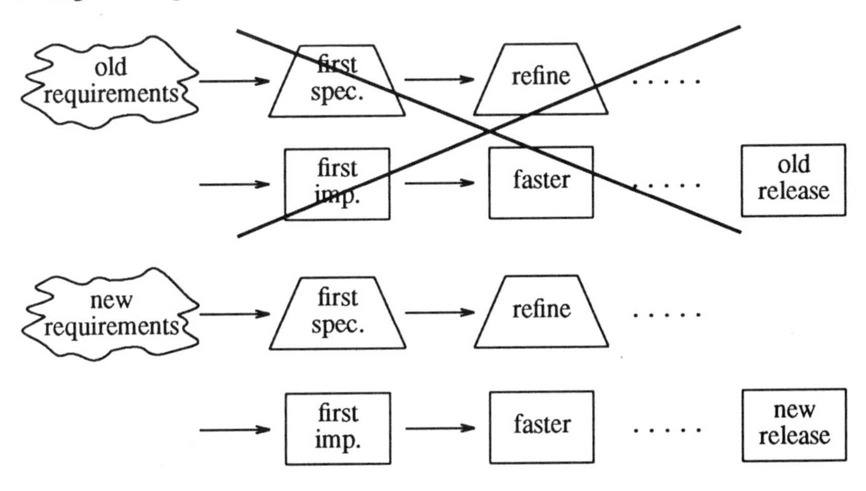

figure 11.7 *formal development, change in requirements*

11.5.3 Conflict – rapid turnaround and refinement

How do the requirements for interactive systems design fit into this picture?

- Rapid turnaround $\Rightarrow$ lots of changes in requirements, say m in all.
- Fast prototypes $\Rightarrow$ lots of steps in the refinement process, say n steps.

That is, the length (n) of the development chains, and the number of them (m), are far greater for interactive systems compared with non-interactive. These interact with the cost figures:

- *Classical* – This takes n steps to produce the first version, and one further step for each requirements change, so $m+n$ versions will have finally been produced.
- *Formal* – This too takes n steps to produce the first version, but then all requirements changes also require n bits of work, so the number of separate specifications and programs eventually produced will be $m \times n$.

At first glance the difference between these two cost figures, $m+n$ for classical development and $m \times n$ for formal development, is astounding. One wonders if formal development can ever be a practical proposition. A one-off cost can be acceptable, even if high, but a recurring cost like this could never be.

Happily, the situation is not as bad as it seems. First, the number of "changes in requirements" may be vastly different. For a system using classical methods, many of the requests coming in for changes will not be true changes in requirements, but restatements of parts of the original requirements the system does not satisfy: that is, long-term debugging. As we've noted, formal development should reduce this to a minimum (or even zero if the requirements are a formal document themselves). Unfortunately, this argument does not hold too well for interactive systems design, as we've noted that here formal methods will not perfectly capture informal requirements. Again, the number of development steps may differ. Formal development may require the additional refinement steps, but the costs of each step will be much reduced, as there will be less debugging.

Orthogonal development

It is clearly critical just how costly the changes to the intermediate specifications are when requirements change. In the best case, the n changes needed for formal development may be a distribution of the effort of the single change in the final implementation for classical development. We would hope that a small change in requirements, would only require a small change in each of the intermediate specifications.

Modularisation is the standard vehicle for such localisation. A small change in requirements is likely to require changes in only one or two modules of the specification. If the refinement and the implementation preserve the specification's module structure, then we can alter the relevant modules in the refined specifications and implementation and re-use all other modules. That is, the refinement process must obey the principle of *structural correlation*.

Cedar (Donahue 1985) uses such a system, carefully maintaining dependency information so that recompilation upon the change in a module is minimised. (Compilation is, of course, a form of refinement.) However, the idea of locality in Cedar, as in most traditional languages such as Ada (Ichbiah *et al.* 1983), is based on type correctness. The semantic repercussions of changes are not considered, and it is not obvious therefore that the rest of the system will behave as expected.

Happily, the situation in formal development can be much better. Modularisation based upon semantically defined interfaces leads to full semantic independence between modules. One module cannot affect another module without a change in the interface. This leads to an *orthogonal* development idiom. The specification consists of many modules, as do each of the intermediate stages to the final implementation. Development within any one of those modules is independent of all the rest. If a change is made in one specification module then an early, perhaps very inefficient, version of the corresponding implemented module can be used with highly optimised existing modules. This means that we can get a reasonably fast prototype back into the hands of the interface designer very rapidly.

11.5.4 Conflict – structural correlation and orthogonal development

To summarise where we have got to: the formality gap forces us to match the structure of interaction in the original specification. The conflict between the need for rapid turnaround and fast prototypes, and formal refinement forced us to use orthogonal development, with structural correlation throughout the refinement process. Thus we will arrive at an implementation with a structure still matching closely the structure of interaction. However, at the end of the day, in order to obtain a fast prototype we will need an efficient implementation structure. Unfortunately, the interaction structure and the structure required for efficiency do not usually agree.

It is worth noting that the experience of formal development is likely to be very different here from classical development. In the latter, the structural design will from the beginning be oriented towards efficient implementation. In fact, the same can be done with formal specifications. However, we argue that this is not what a specification is for, and such trends are likely to yield errors in translating requirements to specification (Jones 1980). In short, a good formal specification is likely to have problems with orthogonal development.

Non-orthogonal development – interface drift

However, despite large gains through intra-module optimisation, the time may come when the system is still too slow and inter-module optimisation is needed. Design steps which involve several modules are *non-orthogonal*, since subsequent changes to the specification of one module will require the other modules to be redone from the non-orthogonal design step onwards. The challenge is therefore to achieve this in as well-structured a way as possible, preserving correctness, and making easy the re-use of existing modules as requirements change.

The simplest situation involves just two modules. We can think of one as a *server* providing some functionality, and the other a *client* which uses this service. The need for intra-module refinement shows up in two ways:

- *Information* – The client module could perform further optimisation if the server could give it some more information: for instance, the use of locality information to tell the client about where the server has made changes. This locality information is not usually present in the original specification interface and is therefore not normally available to the client, although the server "knows" it. This type of information may enable the client to optimise changes in its own data structures.
- *Services* – There may be computations in the client module involving many calls across the interface to the server. Such operations may well be performed far more efficiently by the server, reducing function call

overhead at very least, but quite likely being more efficient in general due to the server's knowledge of representation and (again) information not available across the interface.

As we can see, the two are not entirely independent, and both require *interface drift*, a movement of some information or functionality across the interface between two modules. In general, the information category requires a drift of functionality across the interface *upwards* from server to client, as the server "opens itself up". Similarly, the services category requires a *downwards* drift of functionality, as the server "takes over" jobs previously performed by the client.

A technique for managing the latter form of interface drift was developed at York as we struggled with the refinement of an interactive system (Dix and Harrison 1989). This technique involves the encapsulation of the functionality that is to be transferred into a separate module. The functionality is then transferred in two steps. The first separates the encapsulated functionality from the client module. The second step then merges this with the server. The existence of the additional module both simplifies proofs and also allows substantial independent development of the two modules despite the non-orthogonal step.

This technique for interface drift has been cast into a general framework for formal manipulation of specifications at the module level (Dix 1989). Important issues that arise when non-orthogonal steps are taken include the ability to trace when they occur and maintaining control if the process involves more than one person. Interestingly, it is the formal *interface* between the modules that is crucial in the approach and, in particular, promoting it to a "first-class" object in the development environment.

11.5.5 Human interfaces for development environments

The discussion has focused on the rigours placed on formal development when the subject of that development is an interactive system. However, it leads to important questions about the cognitive demands of the resulting process, and the user interface to any support system. Recalling the comparison between classical and formal development processes, the classical took $m+n$ steps to the formal's $m \times n$. Although we argued that the actual effort expended would not be as extreme as this suggests, it is likely that the number of *documents* (formal and informal) produced will approach this level. That is, the complexity of formal specification "in the small" is likely to approach that of classical programming "in the large".

Large-scale programming has in the past been supported by well-documented analysis and management structures, and currently project support environments are being developed to aid this process. However, the timescales involved in interactive systems design preclude these approaches, requiring instead environments more akin to exploratory programming (Goldberg 1984). Marrying these conflicting styles will require exceptional organisation and ingenuity in the environment and its user interface.

Whilst the development process is totally orthogonal, the complexity is unlikely to be too bad as the documents can be located in a matrix, the dimensions of which are reasonably small. However, non-orthogonal development steps significantly complicate this picture. They make the structure more complex, introducing problems of naming and representation. Whilst such issues are apparently insignificant formally, they have a major effect on the usability of formal development environments.

11.6 Summary

This chapter started by placing the various sections on a gradation based on formality. We began with the least formal and moved on to more and more rigorous approaches. However, there is a different way of classifying the examples based on *professional* skills.

All the examples required different skills and expert input of various kinds. Let us begin with the full formal refinement process (§11.5). The specification, which is the starting point, captures the user interface requirements (and most other requirements); hence there is little if any room for further human factors input. This is an advantage, as we need only look for someone with software engineering skills to manage this part of the design process. If we use the analogy to the building trade, this stage corresponds to working out the order and method of construction given detailed plans. The building may not take shape in the order it was drawn, but it will end up as the architect planned it. (Well almost: in both interface and construction there is likely to be some feedback.)

The production of a formal specification of a specific system (§11.4) will inevitably involve design decisions which affect the interface. Thus the practitioner will require skills both in formal methods and in human factors (or at least, have ready access to appropriate advice). This may well be a two-stage process, starting out with a well-fleshed-out but not rigorous specification of the interface produced by someone with human factors expertise and a smattering of the appropriate formal knowledge, which is then continued by someone with complementary skills. If we look at the use of the formal models in this stage, the important activity is selecting the appropriate models and interface architectures to use and then applying them. We will discuss this selection process in more detail in the next chapter. Once the appropriate models are chosen their application is more or less mechanical. The building equivalent would be the drawing up of detailed plans, using formal techniques such as stress analysis where appropriate. The skills required are draughting, understanding of human features in the building not captured by formal analyses, and knowing just where the formal analyses are appropriate.

The action–effect rules (§11.3) capture some of the properties of formal models in a less formal (and perhaps less frightening) framework. The appeal is to someone with reasonable human factors skills but little if any formal expertise. Certainly, no knowledge of formal modelling as expounded in this book is required over and above that within the framework itself. Thus the formal models are not applied directly but packaged up. In designing a large dam an engineer might use soil mechanics to calculate the effects of the water and dam on the surrounding ground. However, if the job were building a two-storey house, soil mechanics would rarely be applied directly. Instead the builder would use packaged principles: perhaps six inches of concrete are needed over clay, but two feet on sand. The builder needs neither an understanding of soil mechanics nor a long experience in trying different foundations to apply the principles. The comparison between the dam and the house foundations is similar to that between a full formal specification and the use of action–effect rules.

Finally, we come to the lift example (§11.2). This is really a sort of "back of the envelope" application of formal models. Formal techniques are applied in a similar style elsewhere. A professional engineer would know from previous formal calculations where stress is likely to be greatest on a beam, and could estimate the likely failure modes of a structure without performing all the intermediate calculations. Some years ago, I used to work on electrically charged fluids. Frequently the question arose: could we ignore say surface tension, or viscosity in a particular circumstance? A quick order-of-magnitude calculation based on an average flow rates, surface radius, etc. was usually sufficient to rule out one or more effect. It was not necessary to calculate the precise shape of the fluid surface, or the flow characteristics. It is precisely these "back of the envelope" calculations which mark the *professional* application of formal techniques.

So let us finish with this paradox: the more informal the application, the more professional expertise in formal models is required.

CHAPTER 12

Conclusions – mathematics and the art of abstraction

12.1 Introduction

In Chapter 1 I said that to formalise is to *abstract*. Although formal statements are often couched in mathematical notation, the heart of the formal statement exists independently of the notation and may often be captured just as well in ordinary English.

Since I first wrote that, several years ago as part of the introduction to my thesis, I have heard several people define formalism as precisely the manipulation of notations independent of meaning. That is, a manipulation of the *form* only. The proposer of such a view then immediately throws away any pretension of dealing with real people or real things. Now that sort of statement immediately sets teeth on edge, but by a dictionary definition I suppose it is correct. In which case I don't really want to be a formalist at all!

In retrospect what I should have written was:

> In essence, *mathematics* is to abstract ...
>
> In, short I use formal notation because it is useful, but the mathematics is not restricted to the formalism ...

The trouble is, if I had been writing about *mathematical* models of interactive systems I would probably have frightened off even more readers! However, as you are all now reading the conclusions I can safely nail my true colours to the mast and become a card-carrying mathematician.

The interplay between mathematics and formalism is subtle. Although a mathematical argument may begin with an imaginary inner tube, or something equally non-formal, it is usually assumed that it will eventually be transformed into a rigorous formal notation. When applying mathematics we may either formalise our problem and attack it using formal mathematical techniques, or alternatively "lift" a formal result into the problem domain and apply it at the informal level. Probably, a mixture of approaches is most common.

Looking at the previous chapter, the first approach, of formalising the problem, corresponds to the specification of a system and relating the abstract models to this specification. The second approach, lifting the formalism, corresponds to the "professional" application of the models as exemplified by the analysis of the lift (*sic*) system.

The major problem we are faced with when dealing with a situation mathematically is precisely what to put into the formal notation. What is the appropriate abstraction? In Chapter 4, we began the formal discussion by focusing on windows as having *content* and *identity*. We then went on to produce a model which encompassed these two features. There was mathematical work needed at both these stages. Choosing the appropriate way to represent content and identity of windows is not easy; there are many equivalent representations, and depending on which we choose the statement of various properties becomes more or less hard. However, I am never too worried about transformations between equivalent mathematical formalisms, and the most important step seems to be the identification of content and identity as central features. That is, a mathematical step of *abstraction*, which is, at that point, entirely in the informal domain.

This example shows us that the job of "bridging the formality gap" requires work on both sides of the divide. We choose to abstract, to look at things from a particular perspective, on the informal side; and on the formal side, we choose an appropriate model or notation which matches the abstraction. There are various heuristics which can be applied to the latter task: identifying entities and operations, etc. However, the mathematical, but informal, task requires more creativity. If abstraction is the essence of mathematics, then developing abstractions is the *art* of mathematics.

12.2 Abstractions we have used

Let us now look again at some of the issues in the previous chapters focusing on abstraction. We will begin by looking at abstraction in the sense used in describing models as abstract. This is, in essence, abstraction by the designer of multiple systems. First (§12.2.1) we will look at how the abstraction chosen can limit the *domain* of our models, that is, the types of statements and systems that the model supports. Then (§12.2.2) we will see how abstractions introduce a *grain* into our analysis. That is, we are led along certain directions by the nature of the abstraction.

Any particular system can be viewed at various levels of abstraction. This is considered in §12.2.3. Both the designer and the user will have such abstract ideas about the system which may or may not coincide. These abstraction are basically intellectual ones, and in a sense dig into the system *away* from the interface. They lose detailed information about the interface and tell us about the inner functionality.

The views we have of a system are, in a sense, the opposite: they are abstractions of the system *towards* the interface. The most abstract such view is that of the concrete pixels of the display, and we move inwards towards less abstract views of the system such as the *monotone closure*. These abstractions of state and display are discussed in §12.2.4.

Finally, in this section we look at the ways abstract models are applied to specific systems (§12.2.5). We see that the most important decision to be made is the level of abstraction at which to apply our models.

12.2.1 Abstract models – domain

Of course, we have laid great emphasis on the abstract nature of the models presented. This was to help achieve generality and to focus on the relevant aspects of the systems. Of course, the generality is never complete and each model has its own *domain* of applicability. We started out with the PIE model. This was claimed to be very general, encompassing nearly all interactive systems. However, as the book developed we have seen many more models addressing areas which were inexpressible within the PIE model.

Some of the restrictions in the domain of applicability were because of the *level* of abstraction. That is, the PIE model was just too abstract to express certain important features. For example, we were unable to address adequately display-mediated interaction and had to look at more architectural models such as dynamic pointers (Chapter 8). Earlier than that, we developed a model of windowing (Chapter 4). The original PIE model was an abstraction of this, as the user's mouse and keyboard commands could be related to the screen without explicitly talking about the windows. However, although the PIE model *could* describe the external view of such systems, a more refined model was required to discuss the more detailed properties of windows as entities in their own right.

Now, this limitation due to level is only to be expected: the purpose of abstraction is to lose some detail and if we later decide we want to talk about that detail we must clearly use a different (more refined) abstraction. The new abstraction may be a direct refinement of the original (as with windowing) or have a more complex relationship (such as dynamic pointers). It may even be that two completely different abstractions are used to address different aspects of the system: for instance, we may use an abstract model of interaction whereas someone else may be using an abstract model of system security or safety.

Limitations due to the level of abstraction restrict what we can say about a system. The other limitation in domain comes about when the model chosen implicitly restricts the *range* of possible systems. That is, the model is only capable of describing a subset of possible systems. If we again start with the PIE model and then look at the temporal model of Chapter 5, we see that the PIE model was an abstraction of the τ-PIE but only under certain circumstances, when the system admitted a steady-state functionality. Thus not only did the PIE model abstract away from the detailed interleaving and timing of system inputs and outputs, but it *implicitly* limited discussion to systems with a stable steady-state behaviour. For instance, the PIE model is incapable of describing most computer games. We see a similar example in Chapter 10: a general model was given of status input, and then a second model which implicitly required trajectory independence. In fact, it was this second model that I first wrote down, and later, when I realised its limitations, I began to frame the concept of trajectory dependence.

In both cases the formulation of the abstract model limited the range of systems that could be considered. Whereas the limitations due to level reduce *expressiveness*, these limitations of range reduce the *generality* of the models. Now since one of the declared purposes of abstract models was to make generic statements, this loss of generality is rather worrying. The picture is not quite as bad as it seems; there are good reasons for restricting the range of models. Abstract models can be simpler because they abstract away some features of the system; in the same way, models of limited range may be simpler than more comprehensive models. If we are *not interested* in the systems that are not covered by a particular model, then it is more appropriate to use that simpler limited model than a more complicated but complete one.

A more serious problem is that the limitations in range may not be obvious. The level of abstraction may hide the limitations and effort may be wasted on a model that misses the very systems of interest. Further, if the model is used as an architectural framework for developing the specification of a particular system, the system will be limited in scope by the assumptions implicit in the model. Suppose we chose to develop a system using the pipeline architecture (§7.2.3), where input parsing and display generation are seen as separate activities. We saw that the architecture does not support display-mediated interaction (§7.4) and hence the system developed would lack any sense of direct manipulation. More importantly, this central design decision may never have been explicitly made; instead, the structure of the model will have led the design in that direction. This danger of implicit design decisions is a problem not just of abstract modelling; any design system we use will have such implicit assumptions. Application generators will usually produce very stereotyped systems, and interface design methods will only support certain styles of interaction. Even programming languages will limit the sort of interfaces that can be produced; for example, they may lack real-time programming facilities. The important thing is to be aware of the range of the abstract model (or whatever else) one is using.

Limitations in range can be turned to advantage. If we choose to develop a system based around a model, we do not have to maintain explicitly the properties which are implicit in the model. So if we are designing a mouse-based system and want all commands to be trajectory independent, we can use the model in Chapter 10 which supports only trajectory-independent systems. Similarly, we may enforce appropriate steady-state behaviour by designing the major part of a system using a model (such as the PIE) which does not allow the expression of more complex real-time behaviour. As a separate stage we can map this model into a fully temporal one with due regard for the additional features. This is precisely the design approach suggested at the end of Chapter 5.

It is non-determinism which enables abstract models to describe complete systems whilst ignoring detail. That is, it allows us to increase *expressiveness* without too great an increase in complexity. It is not so obvious that we can increase the *range* of a model by use of non-determinism. However, this is precisely what happened towards the end of Chapter 7. We had a linear model of interaction that was (as we have already noted above) incapable of expressing mediated interaction. However, we were able to increase its range and generality by using *oracles*. So, by adding non-determinism, effectively the same model could describe not only the original range of systems but also a whole range of display-mediated ones.

12.2.2 Abstract models – grain

We saw above that models are restricted in *domain* in terms of both *level* and *range*. We also noted that the most dangerous part of such limitations is when we are not aware they exist. However, there is a far more subtle way that a model may influence the sort of systems that are produced. It may be *possible* to model a large range of systems within a given framework, but the framework naturally steers us towards a certain class.

I pick up a piece of wood to smooth with a plane, look along it, perhaps feel it with my fingers. I try to work it in one direction; the plane sticks, then jumps, taking an unsightly chip off the surface of the wood. So, I turn the wood around and begin to plane in the opposite direction; shavings curve gracefully off the wood and pile as a rich carpet around my feet. I have found the *grain* of the wood. Of course, it is possible to plane the wood against the grain, if you are careful, if the tools are sharp; but it is not easy.

Any material, set of tools, or language will have such a grain. It restricts what can be produced or expressed, not by what *can* be achieved, but simply by what is easy. Perhaps there are exceptions, like carving candle-wax, but they are rare. In particular, of course, the models and concepts presented in this book and formalisms in general induce a grain into the design process.

There is a much greater awareness of the effects of grain when using physical materials than with software. Indeed, part of the aesthetics of many crafts is the way that the intended function of an artifact reflects the inherent dynamics of the material. A Windsor chair produced out of steel would be both inappropriate and possibly structurally weak.

If we question the proponents of computer notations or formalisms there is far less awareness of these issues. We are frequently told that the particular method is all-embracing and general. If the proponent is of a formal nature, this may take the form of a proof of Turing completeness. Alternatively, if we suggest a particular system for which the technique seems ill-suited we get a reply of the form "ah but – you can get the effect by...". The fact that it is *possible* to achieve the desired effect clouds the fact that we *would* not develop such a system using the notation. Only recently, I read a remark that we could do without a range of different programming languages, Cobol, Fortran, Ada etc. because C++ could do it all. I can only assume that the author had never worked at a data-processing site!

How is this reflected in the models in this book? In Chapter 10, I said that it was possible to describe status input behaviour using the PIE or τ-PIE models. These models did not, however, immediately suggest status input, and we would not tend to design systems with status input if we had these models in mind. So they have a definite *grain* to them.

Where does this grain come from? Well, it is partly in the formal expression: by not separating out status and event inputs in the model, the possibility is not suggested by the formalism. Also, aspects of the formulation such as the use of sequences suggest discrete events, and the types of principles stated tend to be appropriate for events rather than status inputs. Moreover, the examples used were of event input systems which tended to limit the way we thought of the system.

In Chapter 8, we can see another example of the way examples influence the way we think of a model. All the earliest examples of dynamic pointers were of atomic pointers. Only later in the chapter did we get onto block pointers. Now partly because of this, and partly because of the connotations of the word *point*er, I have found that readers assume that block pointers are somehow different, rather then just a special case of dynamic pointers. The formal expression of pointer spaces with *pull* maps and projections with their *fwd* and *back* maps apply equally well to the atomic and block cases, but the examples and the nomenclature bias the reader.

If we were looking for a piece of wood from which to make the prow of a ship, we could just take any piece and steam-bend it. However, steam-bending tree trunks is rather hard work. Alternatively, we might look for a piece of wood (or a tree) that already had the desired shape. Of course, it wouldn't be exactly right, but we would be working with the grain of the material, not against it.

We can do the same with interactive systems. We saw that we can deliberately choose models which constrain a design to have certain features. In a similar way, we can choose a model or notation so that its grain encourages a style of design. This happens in various contexts: an interface toolbox may in principle allow the designer total freedom but actually encourages a particular style of interaction. Many programming languages are largely sequential in character, and I have shown elsewhere (Dix 1987c) that this leads to systems where the computer has excessive control over the dialogue. As a deliberate attempt to counter this, I showed how a non-sequential computer language could be augmented with input/output primitives in a non-standard way deliberately to encourage user-controlled dialogue.

The red-PIE model was always intended to be applied at many different levels of abstraction, and to various parts of the system. Now although this message largely got across at the level of overall abstraction, so that people felt comfortable with the display being a bit-map or simply an abstract description, it has never really been successful when applied to facets of a system. Other workers at York were interested in the way that certain parts of a screen and of the result were salient to the user at different times (Harrison *et al.* 1989). They emphasised this by having functions called *display templates* and *result templates*, which abstracted from the current display and result the features that were important for a particular task. Now arguably these are unnecessary; we could just apply the red-PIE model choosing the display and result to be the appropriate bits, and relate it back to the complete system by way of an abstraction relation. However, although the templates may not be strictly necessary in terms of expressiveness, they make explicit a feature of the system and therefore bring it to attention. That is, they have created a model with a certain *grain* which encourages systems where the subdisplays and subresults corresponding to tasks are well thought-out and discriminated.

Now this issue of grain is interwoven with the concept of *structural correlation* that I have talked about previously. One of the most obvious forms of grain is where the structure of the model influences the structure of the system. Indeed, the paper I cite above, where a language was designed to encourage user-controlled dialogue, was written in precisely these terms. The critical point was the inherent structural correlation between the programs and the ensuing dialogues, and hence between the programming notations and the interface. So, when we cross the formality gap between what we feel we want and formally expressing it, we must be careful that the grain of the formalism flows with us as we produce our description.

12.2.3 Abstractions of the system

The issues we have discussed, the expressiveness and generality of a model, are about the abstraction of the model. However, we can look at a particular system at various levels of abstraction. The models may be appropriate at various levels and may even help us to describe the levels of abstraction within a system. This is essentially what was going on in Chapter 7. We wanted to view a system at the physical level and at an inner, logical, application level. These layers of abstraction were actually represented by a pair of PIEs with functions between them. That is, we used an abstract model to talk about abstraction!

It is common in both HCI and software engineering to discuss systems at various levels. In the latter, the levels correspond to layers of the software system that isolate the true heart of the system from the physical input and output. In the former, the levels are more interesting as they may correspond to the way the user construes structure from the interaction. I say *may* as they may equally well be a re-expression of the system side; however, to be generous we should be thinking of levels of abstraction *for the user*. On the input side (computer's input, that is), this ranges from some level of tasks or goals through various levels to the actual movements of the user's fingers on a keyboard. On the output side, this includes the actual seeing of a screen through to the recognition of the content of the display as pertinent to the user's intentions.

The user is often unaware of the lower levels of abstraction of the interaction. Much of the time I am "writing a book", only partly aware that I am also "using a word processor". I am not thinking that I am "typing at keys" or "reading characters", and certainly not "moving muscles" or "interpreting light patterns". In fact, I am probably incapable of even performing these last actions at a conscious level. The times I become aware of the lower levels of abstraction are regarded as "breakdown", often an exposure of faults in the system.

There is a general hope that levels of abstraction for the user will correspond reasonably well to the layers within the system. My understanding of the system matches what is there. Now in §11.5 we saw that this was in general impossible. The pragmatics of producing reasonably efficient systems means that the systems must match structurally the demands of the machine. If the layers understood by user and system are to agree, then the user must adopt an implementationally efficient view of the system (or buy a supercomputer). Sadly, this is the state of much software: the user is taught to interact as a machine.

Happily, the analysis of refinement led us out of this impasse. The *specification* of the system can be layered in a way which presents a natural structure for the user, whilst the *implementation* of the system may adopt a different structure. The techniques for managing this structural change are especially important given that the implementation layers refer to the way the system is built. In Chapter 7 we saw that the abstraction of a system which we could regard as the functionality is in general *not* a component of the system as a whole, but an *abstraction* of the system. By appropriate transformation we can retain this abstraction relationship within the early specification and move to the constructional relationship as development proceeds.

Movement between levels may occur not just because of breakdowns: we intersperse high-level, more abstract, planning with more concrete, less abstract actions. It is frequently the case that these movements in level are associated with syntactic units in the grammar of interaction.

Examples abound: in my particular case I decided to edit a particular section of this book, so I invoked the editor; when I have finished I will exit it. These shifts in level between the abstract "edit the section" and the more concrete actual writing are reflected in the interaction, and could easily be found by anyone with a trace of the dialogue. Within my editing various subtasks occur, perhaps a global replacement of a habitual misspelling. Again the shift in level is reflected in the dialogue.

Now the levels of abstraction by which we understand a system need not agree with the levels by which it is implemented (nor even necessarily specified), so long as the models are consistent. Of course, the place that this consistency must be found is in the physical interaction itself. So the system need not interpret the structure of our interaction in the same way as I do, but it must at least be consistent with my interpretation. (Or to be precise, I suppose I must be consistent with it!)

Assuming principles of structural correlation hold and the designer's model of the system agrees with the user's, we can relate the movements in the user's level of abstraction to the layering in Chapter 7. The points in the dialogue when the *parse* function yields abstract commands correspond to the shifts in level. Other workers in York have formalised this in the concept of a *cycle* (Harrison *et al.* 1989). The whole interaction is looked at as a sequence of cycles. Each cycle affects the result of the interaction, but no major changes occur within the cycle. At one level of abstraction, each cycle can be thought of as one abstract command. This corresponds to the inner level in §7.2.1. The cycles can have subcycles which themselves can have subcycles. This exactly parallels the stacking of layers of abstraction within a system. They are particularly interested in the ways by which the user can tell when a cycle has finished: that is, how the user and the system *synchronise* their level of abstraction. They focus on the clues in the display which are repeated when cycles are complete, such as main menus or prompts. At this point they can begin to make prescriptive statements about the necessity of such clues to interaction.

To summarise: users and designers must agree statically on the abstractions by which they understand the system, but they must also agree dynamically as the interaction proceeds as to which level is current.

12.2.4 Abstractions of the state and display

In several places, particularly in Chapters 3 and 9, we have been interested primarily in the output side of the system. Arguably, Chapter 9 on views was about input as well: as we were interested in the way we could update the state through the view. However, we were concerned not with the dialogue, just the relationship between the internal state and views of that state. The treatment of different forms of predictability and observability in Chapter 3 was of a similar form. There were issues of dialogue hidden within the discussion, in particular the *strategy* used by the user to view the system. However, these were effectively hidden by packaging up this dialogue in the *observable effect*, a static view of the system's state.

In both places the prevailing ethos was of the output as an *abstraction* of the state. In user-oriented terms, we could say that the user is aware of an abstraction of the state of the system at any time, and these models capture this awareness in different ways.

Note that the "state" that the red-PIE and the views model capture are very different. In the red-PIE model the state was the *entire* internal state of the system. Now it is a truism that the display and results of a system are abstractions of the internal state, as the state of the system includes all the memory locations that hold the screen display, etc. The red-PIE discussions therefore focused on the relationship between various system outputs.

In the views model the basic state we dealt with was very different. We called it a "database", partly because of the interrelationship with database theory, but partly because this was not the entire internal state but merely the state of the "objects of interest". The additional state due to the dialogue, physical display, etc., was ignored. Note that the very fact we were talking about the user seeing *views* of this state was prescriptive, and is an example of a implicit restriction of the *range* of the model.

State and monotone closure

Of course, we may look at many abstractions of the internal state of the system. For instance, the programmer will have access to some of that state, but may not know about some aspects such as the precise screen bit-map, or partly buffered user input. From the user's point of view the *monotone closure* introduced in Chapter 2 is particularly important. For any viewpoint, it gives precisely the most abstract state of the system which is still capable of telling us everything about the future behaviour from that viewpoint. I know that this concept has caused difficulty with many people, but it is incredibly useful in expressing and understanding interface behaviour. So we shall spend a short time with monotone closure and abstractions of the state before discussing displays and views in more detail.

Part of the difficulty of the monotone closure is that it does not belong to any one or any part of the system. The display and result of the system are things that the user can find out about. In a sense the user *has* these abstractions. Similarly, the levels of abstraction in the system *belong* to the designer, in that they are explicit parts of the specification. The system itself may also directly express levels of abstraction, even if these are different from those expressed in the specification. The monotone closure is different. In general, it is uncomputable. If this is the case, then it is impossible to look just at a bit of the system and get it as abstraction. It is there, but inaccessible. The designer may be able to talk about it, but may not be able to say what is in it. Similarly, it affects the behaviour of the system as seen from the user's viewpoint but may not be directly viewable from that viewpoint. It expresses all the *potentiality* of the viewpoint, but no amount of exploration need uncover all that potentiality. Remember, this is not philosophising, just the outcome of a formal procedure, but it does perhaps have a message that we can apply to other domains.

Now although mystery is a normal and important part of everday life, it is an element that we usually try to minimise in interfaces. So, *in general* it may not be possible for a designer to elaborate the monotone closure for a particular view of a system; however, we would expect as a normative requirement that most systems would be susceptible of such elaboration. That is, we require as a part of the design process that the monotone closure of each view of the system is given. At a formal level this sounds a little heavy, but it really comes down to asking the fairly basic question: "what parts of the state can possibly affect this viewpoint?". The question is not as easy to answer as it seems because we must consider all possible future inputs to the system. However, if the designer does not know the answer to the above question, there is probably some deficiency in the design.

We would relax this requirement a little for the inner functionality of the application itself. For instance, we might have a system that took as inputs triples of numbers, (x,y,z). At the nth input it would print a one or a zero depending on whether:

$$x^n + y^n = z^n$$

Now assuming Fermat's Last Theorem (Hardy and Wright 1954) is true, the monotone closure of the state after the second input is void: all future outputs are zero, thus no state is required. If, however, Fermat's Last Theorem is false, we need to keep a count of the number of inputs so we know the value of n to compute the formula.

So, we would look for computable monotone closures for the *interaction* state of the system but not necessarily for the *application* state. Of course, this distinction is particularly hard to maintain (Cockton 1986); to be totally formal we would have to talk about a non-deterministic system state where the non-determinism corresponds to the results of the application. In the example above, the truth of the expression would be regarded as non-deterministic. This, of course, requires an extension of the definition of monotone closure to the non-deterministic systems as described in Chapter 6. Such a definition was not given; it is fairly easy to produce a generalisation for the non-deterministic case, but a little care is needed. However, I would imagine that total rigour is unnecessary; the formal analysis has told us what questions to ask, and the designer will have enough wit to decide which elements of the state are part of the application and unknowable, and which are the domain of the interaction.

Display and result

We can move on to the display and result as abstractions of the state. Because they were abstractions of the state, they had corresponding monotone closures $D^{\dagger}$ and $R^{\dagger}$ (§3.2.3). The result's closure, $R^{\dagger}$, constitutes *all* we need know about the state in order to predict the future results of the system. In particular, if we had an operating system with a "browse" facility for viewing files, while we were using the browser its internal state would be completely ignored as part of $R^{\dagger}$. No changes could be made to the result of the system (presumably the file system) and hence it would not be part of $R^{\dagger}$. The only thing we would require would be sufficient state to mimic the exiting of the browser.

From a task point of view, $R^{\dagger}$ could be seen as *the* important abstraction of the state, as it predicts exactly the future behaviour of the result. However, the result may be the endpoint of the task, but does not constitute the whole *interactive* task. The monotone closure captures only how the system will behave *if* the user enters any particular inputs. Of course, the behaviour of the interaction depends on which inputs the user enters, and hence on the other outputs the user receives. This same point was made when we considered window independence in Chapter 4. This is clear from the browser example: the future behaviour of the system may not depend on what the user does with the browser, but the future behaviour of the user will.

The display's closure, $D^{\dagger}$, is all we need to know about the system in order to predict future displays. We are on much safer ground in asserting that this is precisely the important abstraction of the state for considering the interaction. The behaviour of the user and computer together is completely determined (from the computer's side) by this. Elements that may be in it, but not in $R^{\dagger}$, include the state associated with browsing, as above, elements such as clocks or pop-up calculators, and less desirable features like mistakes in display update.

The closure $D^{\dagger}$ is what affects the display, but may in general be inaccessible to the user. The *observable effect* (§3.3) was precisely the abstraction that the user was able to obtain from the system. It is an output abstraction of the state: it is not an abstraction of the actual outputs at any point in time, but, like the monotone closure, expresses a form of potentiality. Unlike the closure, which expresses the potential effects of the system on the user, the observable effect expresses the user's potential for finding out about the system. It is thus an encapsulation of the user's control over the disclosure of internal details. It emphasises the user as an active participant rather than as a passive recipient.

It is precisely because the potential expressed in the observable effect captures the user's control, that we sought to have principles that showed that those features that might affect the user were observable from it. Predictability and observability properties assert the user's mastery over the system.

In practical design we would not want to take a system and then try to prove properties about it. Instead, we would look at the system and seek to make it satisfy the predictability properties. We have already thought about how the designer can ask about which parts of the system may effect the user. Having done this, the next stage is to ask: "how can the user know about this component of the state?". The display and command repertoire can then be designed (or altered if this is *post hoc*) to allow the user access to the relevant parts of the state. Of course, in designing this display and dialogue more components of system state will have been added, requiring further analysis...

The above discussion talks about *the* display and result. In fact, the display and result depend on the level of detail at which the system is viewed. For instance, at the operating system level, the result might be the file system and the display an entire workstation bit-map. If we consider an editor used within that operating system, the result is a single file and the display one window within the screen. At an even greater level of detail, we may look at the specification of search/replace strings. The result is the strings themselves, and the display perhaps a dialogue box within the editor window.

The red-PIE model can be applied individually to these cases; however, as mentioned above (§12.2.2), other workers at York have found it useful to apply additional abstraction functions to the display and result. The display D and result R are then chosen to be maximal for the range of tasks of interest, and then for particular subtasks different functions $template_D$ and $template_R$ are chosen to capture what the user is focused on in both domains:

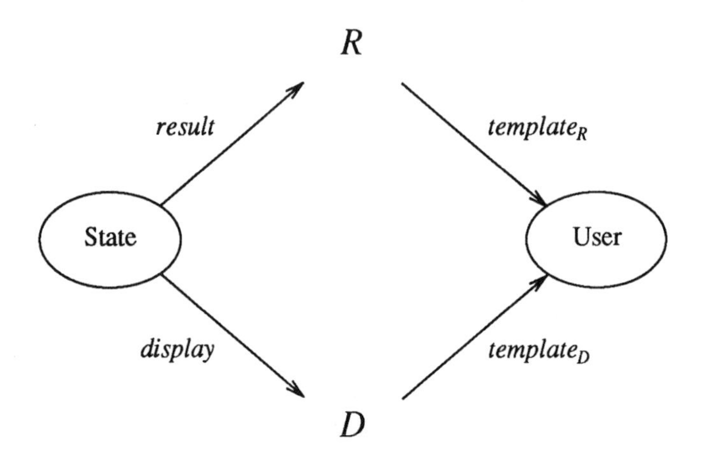

Note the similarity to the perception functions δ and ρ in the WYSIWYG model in §1.5. It is appropriate to add these additional abstraction functions because they make *explicit* within the model the additional levels of abstraction from these domains.

Views

We noted that the state that we considered in Chapter 9 was already an abstraction of the full internal state and represented the objects of interest within the system. In particular, we ignored the interaction state. In fact, the process of producing complementary views and update strategies was part of the process of constructing such an interaction. Thus this process sits logically before the analysis of display and state, although some of the insights have been applied to a PIE-like model by Michael Harrison and myself (Harrison and Dix 1990).

The views are simple functional abstractions of the underlying database, but the sorts of properties that arise give some insight into more complex forms of abstraction. For example, the search for a complementary view is a particular example of the general problem of framing: knowing the unchanging context against which we are to set the additional information we are given. Framing problems occur in many different contexts and functional views are surely one of the simplest.

When we move from the views themselves to the problems of update and complementarity between updates to the view and database, we are extending our abstractions to be abstractions of the dynamics of the system. This is somewhat like the situation in Chapter 7 of layered systems, except there the abstraction was away from the physical system as the user experiences it inwards towards the system. The abstraction of the system given by views is one where the abstraction is outwards, from the system towards the user. The abstraction is capturing the fact that the user does not see all of the system.

The contrast between views in general and product databases as a special case is a second reminder of the distinction between components and abstractions. There is often a tendency for users of formalisms to miss this distinction and make assumptions based on a component-wise view, when the practical reality is a far more rich abstraction. The change in perspective between a product database with constraints and a totally view-based approach is not simple. They are equivalent, but have a very different feel. When constraints are few, the product formulation is often easiest, and even where views of interest do not fall out as simple components, they are likely to be expressed as functions of components. The major "extra" given by product databases is that the complement to choose for a particular component is obvious. There may be many complements that we can choose, but the fact that a Cartesian product formulation is used *implicitly* encourages us to assume that unnamed components are unchanged. This is a form of *meta-framing* principle, and, thinking back to the previous discussion (§12.2.2), represents a *grain* to the product database model.

12.2.5 Using abstract models

If we want to apply existing abstract models to a particular system, how do we proceed? There are several decisions to be made: which models to use, which principles should hold for this system. The choice of model depends partly on the sort of system we have and partly on the sort of things we want to say about it. So, if we have a windowed system we would instantly think of the model from Chapter 4, but if we are interested only in the temporal behaviour of the system we may not need to talk about the windows explicitly and may be able to use a model such as that in Chapter 5 directly. Of course, we may want to know about both. We could use a model encompassing both windowing and temporal aspects, but most likely we would apply each model to a different *abstraction* of the entire system. When we apply the windowing model we would be considering the steady-state behaviour (assuming it exists!), ignoring detailed timing, and when we applied the temporal model we would consider the relation between steady-state and full temporal behaviour ignoring windows.

So, in general we proceed by associating a model with an abstraction of the system in which we are interested. Each element of the model must be associated with an element of the abstracted system. For instance, if we consider the PIE model and a syntax tree editor, we may want to look at an abstraction that ignores the bit-map display but instead regards the command set C as including operations like "delete subtree", and the display D to be the raw syntax tree. Alternatively, we may want to look at the physical level and talk about the command set C as mouse clicks and keystrokes, the display being lighted and unlit pixels.

Typically, any model may be applied in several different ways to the same system. Part of the purpose of the discussion of layered systems in Chapter 7 was to capture these simultaneous abstractions *within* the modelling process. However, the *choice* of abstraction and the *mapping* between model and system are not in themselves formalised. Further, for each application of the models we may want different principles to apply. These design choices represent the formality gap in reverse, moving from an abstract formal statement back to the real world. The movement over this gap is not, and cannot be, part of the formal process itself, but is a step that must be accomplished for any formalism to be useful. Throughout this book, with each model and especially in Chapter 11, are examples of applying models to various situations. There are heuristics and guidelines, but the final step is always one of creativity.

For an example of this in another domain, we look at simple mechanics. In principle, given the moduli of elasticity of a material it is possible to calculate its deformity under various loads. One standard problem is to predict the displacement of a bar when a small load is applied to its end. I remember in school struggling for a long time with this calculation, but in vain, as the equations I generated had no easy solution. My problem lay in the approximations I had made. For small peturbations there are many different equivalent ways of describing the system, and only by selecting the correct one does the solution fall out easily. That is, the solution is obtained by applying the formal model of elasticity to the right abstraction of the system.

As I said at the introduction, developing general abstractions is an art, perhaps the art of mathematics, and it is by this means we develop formal understanding. The corresponding art of applying mathematics in general, and abstract models in particular, is choosing the correct abstraction to which to apply this formal understanding.

12.3 About abstraction

Abstractions can be tricky things. The further we get from the concrete, the easier it is to make silly mistakes, but there again the more powerful the things we can say. We have seen how abstraction is a common theme running through this book, and so in this (nearly) final section, we shall look at some of the misunderstandings and dangers of abstraction and formalism.

As well as being central to mathematics, abstraction is central to all language, both natural and esoteric. We can only communicate by abstractions, ignoring what is irrelevant in order to talk about what is relevant.

So, although abstraction and formalism have problems, they are often problems shared by more "informal" approaches. On the other hand, abstractions, both in language and in formalisms, have the potential of richness and power.

We begin by looking at the belief that mathematics and formalisms are precise and therefore form a useful means of communication. Both beliefs are seen to be false: formalism is by its nature ambiguous and, unless that ambiguity is addressed, may block communication.

We then go on to look at problems due to an uncritical acceptance of abstractions and formalisms. Both are powerful and useful in developing our understanding of the world, but must be used with due consideration for their limitations.

Finally, we look at relationships and analogies. We see that some of the richest uses of formalism are when the formal model is seen as an analogy of the real world, rather than drawing out the precise formal relationship.

. . . . .

Before moving on here is a short story: its relevance will become clear further on. Read it through once and see if you can answer the question at the end. The correct answer is given later, but be careful as you read – don't be distracted by irrelevant details.

Tom Tipper the tipper truck driver went out one day, delivering sand. In the back of his tipper truck were six piles of sand. He stopped first at Miss Jones' house, where she was building a barn to house her collection of vintage leeks. She had ordered three piles of sand, but decided she really needed an extra one as well. Tom shovelled out four piles of sand and went on to old Mr Cobble who wanted one pile of sand to make concrete gnomes with. After that Tom called back at the yard to pick up another four piles of sand as his tipper truck was looking a bit empty.

Tom Tipper had a quick sandwich and drove out along the bumpy track to Farmer Field who was making a new patio for her husband. Tom shovelled out three piles of sand for her and then went for his final call of the day. "I really wanted five piles of sand for my new cell," said Mr Plod, "but I'll just take what you've got." So Tom tipped out the rest of the sand and went home for his kippers and ice cream.

How many piles of sand did Mr Plod have to build his new prison cell with? Remember your answer for later.

12.3.1 Ambiguity and precision – the myth of formalism

If you asked a software engineer who was "into" formal methods to list their advantages, the list would probably include a statement like: formal methods provide a precise means of describing systems and communicating to others. Although there is some truth in this, it is deceptive, and potentially dangerous – *formal descriptions are inherently ambiguous*.

Some readers may have already thought through these issues and may find this fact self-evident; I think many will not. To the convinced formalist it may seem almost heretical, and to others plainly false: surely the whole nature of mathematics is precision. It is one reason why many with a background in the "soft" sciences or the humanities dislike formalism: there is too much precision to allow for the complexities of real life.

However, mathematics *is* ambiguous precisely because it is founded on *abstraction*. The nature of abstraction is to ignore detail, to focus on one aspect of the world at the expense of others. So it trades total precision about some aspects at the cost of utter ambiguity of others. This is no bad thing – it is the power and strength of mathematics – but it is a wise thing to bear in mind when you use it.

Consider the simplest form of mathematics. You have two oranges and get one more orange. You now have three oranges. You have two apples and get one more apple. You now have three apples. Abstract:

$$2 + 1 = 3$$

We now know something general that can be applied to all sorts of things: oranges, apples, atoms of hydrogen, sheep, greater-spotted aardvarks, bank balances, piles of sand... oh yes, piles of sand.

If you have forgotten your answer to Tom Tipper's quiz, you could have a quick look back now.

How many piles of sand did Mr. Plod get? Two? The correct answer is one. Tom Tipper tipped out the remaining sand. It fell, of course, in one (big) pile.

It's Christmas day, there are two plates, on one are two roast parsnips and on the other, three. Tom Tipper likes roast parsnips a lot, which plate do you give him? I hope you are prepared now, it depends how big the parsnips are.

The laws of arithmetic depend on the items being regarded as "units" which retain their identity and are each to some degree equivalent. But two piles of sand may become one pile of sand, and one roast parsnip may be bigger than another. All oranges are created equal...

It is not just the mathematics which causes problems, even terms like "roast parsnip" and "orange" are themselves abstractions. The use of numbers helped to steer us away from the reality of the precise parsnips involved, but as soon as we said "parsnip" the actual golden steaming roots on the plate were lost. To be utterly precise we must be rooted in the individual items and moments of existence, but that rules out all communication and understanding.

So if we want to communicate we need to use words like "orange". The word "orange" is ambiguous. There are many different oranges, of different sizes, shapes and tastes. But when I say "orange" it means something to you. In a book I cannot gesture at my fruit bowl. My understanding of, and ability to discuss, oranges in general inevitably carry the possibility of misunderstanding about precise oranges. Abstraction and ambiguity go hand in hand.

To go from parsnips to programs, we are bound to lose something as soon as we move from the precise system to talking about systems in general. Indeed, even to talk about the system without regard to a single use in an explicit environment is an abstraction. Some workers in HCI would take precisely this view, that interactive systems cannot be discussed in general but must be experienced in context. In the light of the above discussion, this is undoubtedly true, but I'll bet they add up their change in the supermarket.

The level to which we abstract depends partly on the situation, partly on our individual temperament. One thing that we must be aware of is that we always abstract, and that we are always ambiguous. What is essential is to know that the ambiguity is there, to know just what is being abstracted away, and what the limits to that abstraction are.

Neither formalist nor non-formalist can be smug on this point. We all abstract and we all forget at times that we have done so. The special danger for formalists and for those listening to them is when either side believes the bald statement "mathematics is precise".

This brings us back to communication. We need abstractions to communicate, but forgetting about the inherent ambiguity can destroy that communication entirely. It has frequently been the case, even when working relatively closely with colleagues, that a new model has caused considerable trouble. Take the windowing model from Chapter 4. When I first showed this to a colleague I wrote down the formal model, the sets and mappings, and then started to discuss formulation of principles. As the discussion proceeded there was obviously a growing level of misunderstanding. We agreed about the formal model, it was written there before us. But we differed on the *interpretation* of that model. To me, the "handles" in the model were merely place-holders to denote the identity of the windows. My colleague was interpreting them as functions which extracted the contents of the windows (like the views in Chapter 4). The difference was not important for the static properties of the windows, but (as we saw at the end of Chapter 4) it became so when we wanted to consider dynamic properties.

Now if we were engaged in Pure Mathematics this might not have mattered. We could both have seen the correctness (or agreed about the errors) of the formulation. Internal consistency would have been sufficient. (In fact, this is rather a caricature, even of Pure Mathematics.) However, as soon as we start to say things like "this would be a good principle" or "an obvious extension of the model would be...", things become more complicated. What makes a good principle depends on meaning. External consistency is paramount. The formal model captured denotation but lacked connotation. Although there are limits to shared understanding, we must have some level of common meaning to the words we use.

Now if we were talking in everyday English, we would perhaps have been more prepared for problems of language. It is well known that many arguments boil down to a different interpretation of words. The problem with the formal description was not that the ambiguity occurred, but that the myth of formal precision made us unprepared for such misunderstandings.

On the other hand, common knowledge of the ambiguity of language does not seem to stop it being a problem in many situations. One advantage that formal models seem to have is that the ambiguity, *once we realise it is there,* is more easy to pin down.

The problem my colleague and I faced could also be described as *meta-aliasing*. Simple aliasing happens when two things (windows, views, parts of documents) look the same but are different: or, in the language of Chapter 4, content does not determine identity. Here we have a similar problem, but at the level of talking about systems. Two abstractions of the system have the same formal expression, but are different. In common with other types of aliasing, the similarity will break down sooner or later.

So formalisms capture some properties of interest, but divorce themselves so much from the things they represent that ambiguity arises, not just ambiguity about what is ignored, but even ambiguity about which abstraction is being used. This is very similar to the sort of problems we had with views in Chapter 9:

Aliasing

which view – which abstraction

Complementary views

what don't we see – what is abstracted away

Views deal with abstractions at the concrete level of what the user can see of a system. Formal models are abstractions at the level of talking about systems. Both share many properties and, having seen the problems at the concrete level first, we have been perhaps a bit better prepared for distinguishing and discussing them here.

If formal notations can lead so easily into misunderstanding, why do some people claim to find them such useful tools for communication? My guess is that in all the situations where this is claimed the notation is shared by a closely knit design team. The meanings of each part are known by all, as the formal language is built in a community. Similarly at York, although there may have been problems as described above, once a common understanding has been gained the models form a rich common language. However, I deeply doubt the scenario painted by some software engineers, of a formal analyst producing formal specifications which are signed over to some lowly programmer to implement.

Unfortunately, you, the reader, were not part of this group wherein the common understanding of these models has grown. How have you managed to get this far in a book littered with formal models? Of course, if I wanted to learn Urdu I would not just grab a few books written in Urdu and stare at them hopefully. I would want either to see Urdu juxtaposed with English, or to hear it spoken in context. Just so when we deal with formal notations: the model or specification must be developed in context; meaning must be attached to the formal symbols.

Although many formal notations pay lip service to well-documented specifications, as far as I am aware only Z has included this as part of the standard language. A Z specification is always part of a document which mirrors in natural language the progress of the formal specification. So formal notation and meaning are acquired in parallel.

With specifications of concrete systems, the natural language descriptions are concerned chiefly with explaining complex formulae. The basic symbols often inherit their names from the real systems that are described. So if we have a specification that talks about a *screen* being an array of 80×25 characters and a *cursor* being a pair of integers, we will probably know what is meant by these symbols with little further explanation.

The more abstract the formal description is, the less easy it is to attach meaning to the symbols. Even if we know what is referred to by a term such as "window", we do not know what level and way it is abstracted. Any sort of communication of formal models must therefore be full of examples and textual or verbal description.

Looking back over this section, it all seems almost too obvious to bother writing. However, the myth of formal precision persists and so some reminder is obviously necessary.

12.3.2 Uncritical dependence on abstraction

We have already seen how some problems can arise if we forget about the ambiguity inherent in abstraction, but there are other problems associated with abstraction and with formalisation. Many of the problems can be recognised both in the application of formal methods to interaction and in everyday life. Perhaps recognising problems in the relatively simple world of formal models may help us to understand more complex issues.

One of the reasons for introducing abstract models was to *generalise*, to state properties of whole classes of interactive systems (§1.2). This means we can recognise a property in a particular situation, see how this is represented in the abstract model, and then apply this generalised property to all the systems described by the model.

The obvious potential danger is, of course, *overgeneralisation*. For example, many problems in older text editors are due to modes: the same keystroke has a different meaning dependent on the current editor mode. From this we might conclude that modiness is a bad thing, and frame principles of mode freedom in the abstract model. Menu and mouse-based systems are incredibly mode-ridden: virtually everything we do depends on what is being displayed. If we blindly applied principles of mode freedom, our mouse-based systems would become very boring indeed.

Why then did the generalisation fail? Well, even in the process of describing modes above, I began to abstract. I talked about the problem as one of different meanings at different times. The problem is more rooted in the particular situation than that. Modes are a problem, partly because the user may not know the current mode, partly because of remembering different meanings. In the menu or mouse-based system, the user's attention is on a screen which indicates clearly (sometimes) what the meaning of a mouse click or a particular key will be. The presence of visual cues removes ambiguity and prompts recall. In retrospect, we should have talked about the need for visually distinct modes, the user's focus of attention, etc. By framing the problem in terms of meaning of keystrokes, the range of appropriate generalisation is limited to those systems for which these other factors are similar.

Now I hope it has always been clear that the various properties and principles framed throughout this book are putative properties that we may want to demand of a particular system. By the nature of abstraction, these principles ignore factors which are abstracted away, When the principles are applied, these additional facets need to be brought into account.

Note also that the *grain* of the abstraction we are using will tend to make us view a problem or property of a system in a particular way. We are driven by the abstraction to look at certain classes of problems. Choosing an abstraction colours irrevocably what we can say and think. This is at least part of the problem of prejudice: as soon as we decide to use colour, sex or race as the principal attributes by which we describe people, we inevitably use them to discriminate. Alternative abstractions, such as friendliness or generosity, generate completely different world views. Of course, any abstraction tends to hide the individual.

Another related problem is "nothing but"-ism: we use an abstraction to make sense of the world, but then forget that it is an abstraction. Part of my reticence in including the user explicitly in models stems from this. We could describe the user as a non-deterministic function from displays to keystrokes. This would make clear the feedback nature of an interactive system, whilst making no assumptions about the user's behaviour. It does not worry me too much if someone should mistake the PIE model as saying that an interactive system is *nothing but* a function from inputs to outputs. However, I am loath to allow even the suggestion that a user is *nothing but* a non-deterministic function from displays to keystrokes. Nothing but-ism abounds in everyday life. The sun is nothing but a ball of hydrogen and helium plasma – but it is also a God-given sustainer of life. A factory's future is determined solely (or on nothing but) economic grounds – but it is also a sense of community, the livelihoods of workers. Abstractions encourage nothing but-ism, by focusing on certain aspects and ignoring others. When we use abstractions, whether formal or informal, we must make a compensating effort to look at the world beyond the abstraction.

12.3.3 Uncritical dependence on formalism

When I print rough drafts of chapters for this book, they are always typeset on a laser printer. The quality of the earliest copies will be very similar to the finished book. It is amazing how strong an aura of authority well-printed text has, no matter how bad or good the content. The same text printed on a dot-matrix printer (or in my hand-writing!) would carry far less weight. Now formal notations can give a similar appearance of weight and credibility whilst not necessarily adding anything in terms of content.

It is all too common (more so in software engineering than HCI) to see papers where a "formal bit" is included which adds little to the content and amounts to using the notation of sets and functions to carry the same meaning as a box diagram. If the authors had used the equivalent box diagram the text would have been clearer and just as rigorous. Of course, the presence of the equations (especially the odd λ and special symbols) makes the paper look very formal and impressive. The authors probably worked hard to get the correct formal syntax, but if you look closer you find deeper problems. A common mistake is that the functions are not really functions: vital parameters are left out. With a box diagram, this sort of imprecision is accepted, and the use of such a diagram would have been *more* rigorous and accurate. There is clearly no intention to fool the reader, just a half-hearted attempt to use formalism. At the best this merely wastes a little of the author's time and effort, but at the worst can hide the fact that the author has not really developed an understanding of the domain.

Not only can formal notations be convincing in print, an incorrect formal argument may be believed, *however much it conflicts with reality*. The belief that, once something is in mathematics it is correct, is common even in academia.

Some years ago I used to work on the mathematical modelling of electrostatically charged sprays. The basic equations were easy enough to write down, but under most conditions the only way to solve them was by computer simulation. In such cases it is always useful to have analytic solutions to simple cases, as these give one a far better "feel" for the general behaviour, especially when there are lots of different variables affecting the solution.

One obvious special case is to look at spray eminating from the centre of a circle or sphere, with the circumference held at some constant voltage. This would be a good approximation to the centre of a cone of spray. One day, I came upon some papers written by two Japanese *physicists* who gave solutions for precisely these cases. I was interested; I had got stuck each time I had tried to solve these particular situations, and they had even used more complex forms for air drag on the spray. I tried to work out how they had got their solution, but could not reconstruct it. This bugged me for some time, then one day I looked at the sort of results they were obtaining. Some of the outputs from their equations seemed wrong. If you had designed the equivalent physical experiment they would have been *inputs*. So the *mathematical* results did not correspond to *physical* reality. Having noted that something was wrong I went back over the calculations, looking for an error. Sure enough, you could obtain their results if you made a standard school-book mistake.

Now, the mistakes were not complex: I am sure the authors were quite capable of understanding that level of mathematics. Why then did these *physicists* not notice that their results were unphysical, whereas I, a mathematician, could see this? (remember I did not spot the mistake from the mathematics alone). The answer is that once they had set out their problem mathematically they trusted the mathematics totally and threw away their physical understanding of the system. Even if they had noticed some problem with the results, they would not have dreamt of doubting the mathematics.

Now computers generate the same sort of awe in many people. The combination of formal methods in computing can leave many simply dazed. On the other hand, the opposite attitude of complete distrust of computers and formalism (or of formalism by computer scientists) is equally unhelpful. If formalisms are to be useful one must develop a respectful, but not irrational, distrust of the results. If you used a calculator to add up a supermarket bill and got an answer of £2,073.50, you would guess that something was wrong. Formal models must similarly be constantly checked against one's intuitions about the system. If they differ, both the formal reasoning and intuition can be reassessed. I frequently make mistakes in both realms.

12.3.4 Abstraction and analogy – mathematics and poetry

We have been concentrating on abstraction, both as a unifying theme for the models in this book, and because of its central role in mathematics. However, in any mathematical study there is a stage before full formal abstraction which gives a pointer towards the meaningful application of formalisms.

Earlier, we went through the first stages of developing addition: two oranges and another orange, two apples and another apple... Then apparently the magic step: $2 + 1 = 3$. How did this abstraction arise? We saw there was a similarity between the situations we encountered in concrete. The two oranges were similar in a way to the two apples, the extra orange was similar to the extra apple, the resulting piles of fruit were similar. Some things were different in the two scenarios: one talked about oranges, the other talked about apples. The appropriate abstraction is therefore one which abstracts away the points of difference and retains the similarity of structure. Our earlier discussion would also remind us that the abstraction will usually have side conditions attached to it; in this case the distinctness of the objects (no piles of sand) and their equivalence (all parsnips are equal).

The jump from recognising this similarity to developing the abstraction is not usually immediate: in many branches of mathematics this has taken hundreds of years. Usually the development of an appropriate abstraction is closely linked to the development of a suitable notation, although this may be misleading as the individuals may have used an abstraction, but not have been able to record the fact. In natural language we also see that the naming of a concept is closely linked to the understanding of it. Although it is possible to see similar situations and grasp those situations somehow as one, it is the naming of that new concept which marks a watershed. Naming allows communication and is traditionally associated with power.

However, it is a mistake in searching for an understanding of a class of systems to jump into an abstraction prematurely. We have seen that there are many dangers associated with abstraction, and these are obviously heightened if we choose the wrong abstraction to start with. A better way to begin is to immerse oneself in the individual examples and look for the similarities. If something is a problem in one situation, rather than trying to generalise too soon it is often better to look at similar situations and ask what is the equivalent to the problem there. Looking at similar situations is a common experience: if we are faced with some new problem we will instantly think, "something happened just like that the other day". The analogy helps us to understand the new situation in terms of the old.

Abstraction and analogy can go hand in hand. If I say "the steel tube was as as cold as ice", the reference to "cold" tells us that it is the coldness (an abstraction) of the steel and the ice that is important; I would not expect there to be any similarity concerning the roundness of the tube or the wetness of the ice. In Chapter 6, the use of non-determinism helped us to see similarities between disparate types of problem. It focused us on certain aspects and thus acted as an abstraction of the different domains (sharing, real-time, uncertainty). The abstracted situations were not identical, but by losing some of the detail we were able to see that there were similar problems and thus similar strategies for dealing with them.

Analogies differ in their precision. Some are so precise that we can draw inferences about one situation from the analogous one. For instance, suppose I have two piles, one of oranges, the other of baked bean tins. I find I can draw a one-to-one correspondence between them perhaps by lining them up opposite each other; or possibly by painting coloured spots on them: one purple spotted orange, one purple spotted tin. I then take away one orange and one tin, and count the tins. If I have 57 baked bean tins left, I know I also have exactly 57 oranges.

Any similarity which is going to give rise to a formal abstraction has to be this precise. Mathematics is littered with words which describe such correspondences: translations, equivalences, congruences and morphisms of all hues (automorphisms, homeomorphisms, endo-, epi- and isomorphisms). Computing formalisms have added a few of their own, including bisimulation and observational equivalence. In music too (often allied with mathematics) we find transposition and counterpoint; however, the interest is often in the way that some theme is repeated in a similar, but not quite identical form.

Such looser similarities turn up in mathematics as well. Two mathematical structures may have a roughly similar structure, but differ in details. If some property is known to be true of the first structure, it is natural to wonder whether a similar property holds for the latter. We may even try to follow the proof of the first in proving the other; these will differ from each other, and at some stage the analogy will break down and a different and more complex procedure will be required. However, we often find a similarity in gross structure.

This latter form of analogy is much more like the analogies found in literature and poetry. Simile and metaphor are some of the principal means by which poets convey meaning to their readers. Sometimes these can be pinned down, in a fairly tight manner, but sometimes are more suggestive. Let us look at probably the most well-known poem in the English language. Wordsworth says of the daffodils along the lake shore (Hutchinson 1926):

> *Continuous as the stars that shine*
> *And twinkle on the milky way,*
> *They stretched in never-ending line*
> *Along the margin of the bay:*

Notice the level of detail in this simile. The milky way is continuous yet composed of distinct stars. The daffodils will similarly give the simultaneous appearance of a continuous mass and yet be composed of distinct flowers. The distinctiveness is especially obvious in their movement (they are "*Fluttering and dancing in the breeze*") and this is mirrored by the twinkling of the stars. Even the sweep of the milky way suggests the curvature of the bay. Of course, there is far more in such an analogy, subtle nuances, the cosmic nature of the milky way suffused with the magnificence of the panorama, but it is not too dissimilar to a mathematical analogy.

Of course, the PIE morphisms in §2.10 and the layered models of Chapter 7 are mathematical-style relationships in the context of particular models. When applying models, the most obvious "formal" method, direct refinement, is equally strong. We take a system, draw up a precise mapping to the model (this data type is the result R, this one the display D, these are the possible commands C, etc.) and then apply principles defined over the model to the system by direct translation.

In other places, we have dealt with slightly less precise analogies, more like the weaker mathematical analogies and perhaps the sort of simile above. We use the reader's understanding to draw relations between things that are not 100% precise, but yet which carry an obvious meaning. The reason that we went into so much detail about the simple PIE model in Chapter 2 was to perform various analyses on this simple model, so that they would not need to be repeated later. The implication was that although the models were slightly different there would be a roughly similar behaviour. For instance, consider the definition and construction of the monotone closure (§2.5). This cannot be applied directly to the windowing model, or the non-deterministic PIE, but something like it would apply. We would certainly require a more complex definition in each of these cases, but would retain the general notion of a state which captures just what affects the user, but no more. Further, the formal definitions would be very like the definition for the PIE, but have extra bits and small differences.

In Chapter 6, we developed a non-deterministic version of the PIE model. It is fairly clear that this process could be applied to other models. Similarly, we have seen properties of different models labelled as forms of predictability, observability or reachability, and aliasing has cropped up in different forms. In each of these cases the analogy is not quite precise, but allows us to organise and understand the relevant formal models.

Think back to the lift example in §11.2. The analysis there *could* be seen as completely formal: the command set is the lift buttons, the display consists of the various lights inside, etc. However, my *initial* analysis was more one of simile. It was the "likeness" to the model rather than the precise relationship which sprung to mind. Indeed, to complete the formal analysis of the lift system I would have needed a model which dealt adequately with the external aspects of the system such as the actual movement of the lift and other passengers. Note then that I was able to draw the formal relationship only because I had first seen the informal analogy.

Let's go back to the poem. It begins:

I wandered lonely as a cloud
That floats on high o'er vales and hills,

At first sight this is like the "cold as ice" analogy. We are comparing Wordsworth to a cloud and it is the abstraction of loneliness which is of interest. However, a moment's thought and we realise that clouds are not lonely. The cloud is alone surely, but lonely? It is only by personifying clouds that the analogy makes sense. We see Wordsworth in the cloud, then feel the feelings he would feel. The second line unfolds this a little. The cloud floats high above the earth; Wordsworth, although physically upon the ground; is wandering and abstracted. The cloud's floating is like the floating of a lonely person, like Wordsworth himself. The simile tells us as much about clouds as it does about Wordsworth.

A moment later Wordsworth sees

..... a crowd,
A host of golden daffodils;

This is terse metaphor. We could read it as a very precise analogy: a "crowd of daffodils" means a lot of daffodils close together. However, because Wordsworth was lonely and alone, the word crowd evokes all sorts of additional feelings: companionship, togetherness. The contrast is especially pertinent when we compare the floating clouds that do not touch the ground, to the daffodils which are gaily rooted and part of the lakeland landscape. Again, the analogy has a richness far beyond the simple matching of attributes.

Let's look again at abstract modelling. In Chapter 6, we developed a non-deterministic model to help us deal with some formal properties of systems. We wondered what this signified in the real world. Now some of the implications were direct formal analogies between the models and the real systems they denoted. For instance, non-determinism due to timing corresponds exactly to the non-determinism that arises in the formal model. However, the discussion ranged far wider than that. The analogy between formal models and interactive systems had a richness, which although by no means as beautiful is not so dissimilar from the poetic analogy.

The discussion of events and status in Chapter 10, although rooted at various places in formal models, ranged far wider than these models. Formal and informal concepts were counterpoised, and, like Wordsworth and the cloud, we ended up knowing more about both.

This stretching of understanding generated by the use of formal models is possibly their greatest benefit. The application by strict refinement is necessary and perhaps desirable in many situations, but is rather utilitarian compared to rich and exciting formal metaphors. Not only can I do more, I know more. Nowadays when I see a system, one of the things I look for is aliasing. Note I do not match the system to a model and then look for aliasing via the model. Aliasing has become part of my understanding. The same could be said for other properties I have mentioned, such as predictability, dynamism and structural change.

I am not sure how many authors would admit it, but I believe that many uses of formalism are of this analogous nature. We have already discussed the way that formal statements often serve much the same purpose as box diagrams, and often less clearly. There are circumstances, especially where temporal reasoning and change is involved, where diagrams are not so useful. In these cases a formal model may be enlightening even if it is not an accurate reflection of the real system. This is because it serves as an analogy. Problems arise if the author mistakenly believes that the formal statement is in precise mathematical correspondence rather than analogous to the real system. There is nothing wrong in telling a reader exactly what are the limits of a particular statement, but go on to use it to guide an analysis of a wider class. However, to go on in ignorance may be dangerous.

Of course, this discussion of poetic analogy and formal analogy is an analogy in itself. It would be unwise to be too precise in looking for a correspondence between Wordsworth's poems and abstract models. However, as a looser, more poetic analogy itself, it suggests some of the richness that we can gain through analogous use of formalisms.

12.4 Other themes

This chapter has been concentrating on abstraction as a unifying theme. However, a few other concepts have repeatedly arisen during its course which deserve reminders.

Some of the earliest properties we discussed concerned the *predictability* and *reachability* of the system; or to put it in other words, what you can see and what you can do. Similar properties have recurred in subsequent chapters, especially various forms of predictability. For instance, in Chapter 5 we wanted to be able to observe the current state of systems liable to delayed responses or buffered inputs. Later, in Chapter 9, we were particularly interested in predicting the effect of view updates and whether or not an attempted view update would succeed. The ability to know what is and will happen is an obviously desirable property of a system. Chapter 6 reminded us that what is or is not predictable depends on who you are. Some response may be predictable to the program's designer, but not to the user; to the expert, but not the novice. If we detect such non-determinism, we can begin to think about ways of making such a system more predictable.

The way predictability was defined for PIEs in Chapter 2 emphasised the possible future interactions with the user. In Chapter 3, we found that predictability and observability properties were expressed most easily as functions denoting what could be deduced from one view about another. The fullest form of predictability was when we could tell the whole current state from the display or collection of displays. It was important that this state was not the explicit internal state, but the *monotone closure*, the state as it may affect the user. This same insistence on not looking at the explicit state as given by the implementation arose again in Chapter 4. It was important there that any definition of sharing or interference dealt with the implicit linkage as perceived by the user, rather than any explicit data model.

Connected to, or perhaps a special case of, predictability was the issue of *aliasing*. This has cropped up already in this chapter, and we saw examples of aliasing in Chapter 1, where the concept was first introduced, Chapter 4, where we considered aliasing of windows, and Chapter 9, when we asked whether users were aware of what views they were seeing. Aliasing is an important special case of predictability because it is easy to overlook. When assessing a design, we will more easily notice that certain attributes of an object are not visible, than that the objects are not themselves uniquely identifiable.

In several places there has been a stress on *dynamism*. This was most obvious in Chapter 8, where we discussed dynamic pointers as compared to their static counterparts. Also it was a key feature in the distinction between status and events in Chapter 10. Indeed, the relationship between events and dynamism was the focus of part of the summing up of that chapter. Because of the major differences between static and dynamic properties of an interface, Chapters 6 and 7 attempted to divide the consistency properties into static and dynamic invariants. However, this was mainly in order to emphasise the special role of static invariants which are applied before the dynamic ones. This seems to contradict the stress on dynamism, but in fact highlights a nice counterpoint between the two. Static properties tend to be what make an interface correct, whereas dynamic ones tend to make it interesting.

One form of dynamism which is particularly important is *structural change*. This underlies the requirement for dynamic pointers; if the data structures were structurally static the pointers to them could be, and hence we would require no *pull* functions, and everything would be a lot simpler (but less fun). Although we discussed some properties of windowed systems when windows were being created and destroyed (§4.8), the required properties are far less obvious. Similarly, when we thought about dynamic views at the end of Chapter 9, these clearly had much more complex properties than the static case. Note, we had several levels of dynamism going on here. The simplest case would be simple structurally static views: the views are simply what we see and we never try to *do* anything through them. The chapter was principally about updating through views. We were thus interested in an aspect of dynamism: the contents of the database changed. However, in so far as it affected the view we assumed that the database was structurally static. We then considered the case where the views "moved" when the database was altered. We noted how dynamic block pointers could form a descriptive or implementation mechanism, underlying the importance of dynamic pointers. The final case, which we considered only briefly was when we wanted to update the fundamental structure of the data through views. Each level of dynamism added complexity. In the introduction to Chapter 5, we noted how the temporal properties of interfaces are often poorly specified and documented. One simple reason for this is that it is easy to draw a sketch of a snapshot of a system, but more difficult to describe its evolution. Precisely because it is easy to overlook and complex to handle, we must be especially vigilant when considering aspects of structural change.

Having talked about structural change we move on to *structural correlation*. Precisely because structural change is difficult to handle we want to avoid it in the design process. As far as possible we want the design notations, models and specifications to match the designer's intuitions. There is an obvious tendency to specify systems in an easily implementable fashion. This is correct up to a point: it is no good designing a wonderful interface only to find that it is totally impractical. However, this process of checking the realisability of a specification should not dominate the structure. As I pointed out right at the beginning of this book, the greatest barrier to a successful interface is the *formality gap*. Our major effort in design must go towards narrowing that gap as much as possible.

This might mean defining new models or notations which match the domain, or simply choosing appropriate structures within the notations we have. Structural transformation within the formal domain might be a bit of a pain, but is fundamentally manageable. Indeed, we saw how *interface drift* allowed us to perform structural transformation even in the presence of conflicting goals.

12.5 Ongoing work and future directions

The work described in this book is not static. I will now describe briefly some more recent related work and possible future avenues for development.

I have shied away rather from making the user an explicit part of the model, preferring the psychological insight to be in the choice of level of abstraction. I have given some reasons for this reticence above (§12.3.2), but within these constraints there is room to make the cognitive elements more precise. I have already noted the work on display and result templates (Harrison *et al.* 1989). These try to add a more psychological perspective to the display and result by adding an idea of focus. This still leaves the defining of such areas of focus to the human factors expert or to experiment, and therefore does not run the risk of reducing the user to a model. Perhaps other cognitive aspects could be included explicitly into models without compromising the user's humanity.

Another feature of these studies is the way that they developed a specific model, the *cycle* model, to capture aspects of a particular system. They were studying a specific bibliographic database system, but found it useful to abstract away from the specifics of the system in order to understand it better. This model was not as wide in its generality as say the PIE model, but was better able to express specific properties of interest, in the same way as the model in Chapter 4 addressed windowed systems. This process of finding domain-specific *interaction models* has been promoted to a specific design and analysis heuristic.

Most of the models in this book are largely *declarative* in nature. They are intended to describe systems in such a way as to make it easy to define properties but not to determine how those systems work. This is the correct first step: the discussion of the formality gap told us that we must concentrate first and foremost on *what* we want to be true of the interface. However, we saw in §11.5 that we cannot expect the structure of such a model to match the structure of the implemented system. More important, the definitional models need not even be of suitable structure for the specification. The models address specific issues, and ignore others, but a complete interface specification must cover all areas. There is thus a need for parallel *constructive* methods which are tied into the individual methods. We could just take an existing specification notation and map the model into it, but this would not be too successful without some sort of methodological guide; the gap is too big. The specification in §11.4 used layered models and pointer spaces to generate an architectural plan. These two formulations are themselves examples of intermediate formalisms. I regard dynamic pointers as a central area for study and experiment, partly because they crop up so often, and partly because they are useful in implementation as well as definition. Sufrin and He's (1989) interactive processes and Abowd's (1990) agent architectures are also aimed at this gap.

The big gain from developing generic architectures is that a lot of the work of checking properties can be done as general proofs at the abstract level; specific uses of the architecture would only require the verification of simpler properties. In practice, however, I think that the complexity of real systems means that more domain-specific models are often required. These are able to have more of the usability prepackaged, and come with simple rules for building, but are limited in their scope.

In the spirit of this chapter, I don't want to finish on the formal application of formal systems. Some while ago Harold Thimbleby and I were discussing consistency in interfaces. Metaphors like the desktop were clearly very useful, but tended to break down when pushed. In general, we concluded, you could not expect powerful complex systems to be expressed entirely by simple direct-manipulation interfaces. However, we began to use the analogy of differential geometry (the underlying mathematics of Einstein's General Relativity). We imagined systems which were too complex to be represented by a single simple interface but which were composed of simple overlapping parts; where each part had a direct manipulation interface and where the boundaries between these parts were consistent. I don't vouch for the practicality of the idea, but we found that the formalism was a tool for thinking, a spark for the imagination, and fun.

Appendix I

Notation

Specific notations are introduced in the course of the book for different purposes. These are usually of local significance, and it would not help to list them out of context. The notations used for general concepts, sets, sequences, tuples and functions are given below. There is also a short description of the notation for structures used in Chapter 8.

I.1 Sets

Standard mathematical notation is used for these: {...} to enclose enumerated sets, $\cup$ and $\cap$ for set union and intersection, $\times$ for Cartesian product and $\in$ and $\subset$ for membership and subset. The following notation applies:

$\mathbb{P}X$ – the power set of X, that is, the set of all subsets of X

$\mathbb{F}X$ – the set of finite subsets of X

$\| X \|$ – the cardinality (number of elements in) X

$\mathbb{N}$ – the set of natural numbers, $\{0,1,2,...\}$

I.2 Sequences

Many of the models deal with sequences, or lists of values. These are usually finite but of arbitrary length. The following notation is used for sequences:

X^* – the set of all sequences of elements from the set X

X^+ – the set of all non-empty sequences of elements from the set X

null – the empty sequence

[...] – encloses a sequence, so that $[\ a\ ,\ b\ ,\ c\ ]$ is the sequence with a as its first element, b as its second and c as its third and last, [] on its own also represents an empty sequence

(...) – used on occasions to enclose sequences: this is more normal in mathematics, and less obtrusive where there is no confusion

"..." – quotes enclose character sequences

The sequences are glued together using three operations:

: – adds an element to the front of a sequence, e.g.:

$$x : [\,a,b,c\,] \;=\; [\,x,a,b,c\,]$$

:: – adds an element to the end of a sequence, e.g.:

$$[\,a,b,c\,] :: x \;=\; [\,a,b,c,x\,]$$

; – concatenates sequences, e.g.:

$$[\,a,b,c\,] ; [\,x,y,z\,] \;=\; [\,a,b,c,x,y,z\,]$$

Where the meaning is clear juxtaposition is sometimes used to mean any of the above three, so that if a is an element and p and q are sequences:

$$\begin{aligned} a\,p &= a : p \\ p\,a &= p :: a \\ p\,q &= p \,;\, q \end{aligned}$$

The more specific notation is used where there is likely to be confusion lexically (because of the complexity of expressions) or semantically, in particular, when dealing with sequences of sequences.

Sequences are intrinsically ordered:

≤ – denotes *initial* subsequence, so that:

"ab" ≤ "abc"
but **not** "bc" ≤ "abc"

This same symbol is also used with its normal meaning for numbers and for information ordering on lattices and on view spaces (Chapter 9).

I.3 Tuples

Round brackets are often used to enclose tuples, leaving it to context to disambiguate them from sequences. Angle brackets <...> are also used. In particular, they are always used for major tuples denoting complete models. These occur in free text and the angle brackets emphasise the formal nature of the contents, in contrast to textual parentheses.

I.4 Functions

The bracketing notation $f(x)$ is used for function application rather than lambda calculus juxtaposition. The only exception to this is the translation *Tf* of f in Chapter 9, which follows the notation of Bancilhon and Spyratos (1981) whose work it extends. In addition, note the following:

$f : X \rightarrow Y$
– f is a function from the set X to the set Y; if f is partial its domain is defined close to its definition, otherwise it should be assumed to be total

dom f
– the domain of f, that is, the set of elements for which f is defined

range f
– the range of f, that is, the set of elements that are possible values of $f(x)$

id_X – the identity function from the set X to itself

$f|_X$ – the function f restricted to X, that is:

$$f|_X(x) = \begin{cases} f(x) & \textbf{if } x \in X \\ \textit{undefined} & \textbf{otherwise} \end{cases}$$

I.5 Structures – functors

Chapter 8 deals with general parameter structures. For example, if X is a set of interest we might want to talk about the following functions in a uniform manner:

$$a : X \times \mathbb{N} \times X \rightarrow X$$
$$b : Char \times X \rightarrow X$$

We will use the notation that both are of type $F[X] \rightarrow X$ where $F[X]$ represents the structure with elements from X. So for a and b the structures are:

$$F_a[X] = X \times \mathbb{N} \times X$$
$$F_b[X] = Char \times X$$

The structure $F[..]$ can be applied to any set, so that if Y is a set:

$$F_a[Y] = Y \times \mathbb{N} \times Y$$

Given any function $f : X \rightarrow Y$ we can generalise this to a function from $F[X]$ to $F[Y]$, mapping the corresponding elements of the structure. To be strict perhaps one ought to write $F[f]$ for this function, but the meaning is always obvious from context. In category theory $F[\,..\,]$ would be called a functor. This description is partly repeated when it is first used in Chapter 8 to remind the reader.

I.6 Reading function diagrams

In many places diagrams are used to show the interaction of various functions. In these diagrams, sets are the nodes and arrows the functions between them. The simplest such diagram has only two nodes and one arrow, e.g:

$$X \xrightarrow{f} Y$$

This simply says that f is a function from the set X to the set Y. That is:

$$f : X \rightarrow Y$$

A more complex example is found in Chapter 3:

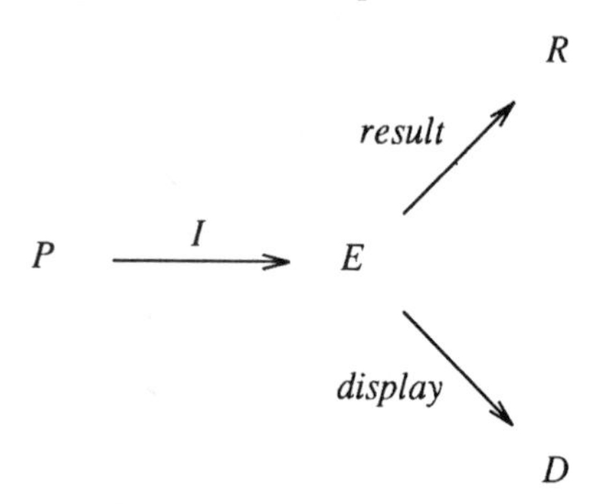

This says that there are four sets, P, E, R, and D. I is a function from P to E, *result* is a function from E to R and *display* is a function from E to D. In other words:

$$\begin{array}{lll} I: & P & \rightarrow E \\ result: & E & \rightarrow R \\ display: & E & \rightarrow D \end{array}$$

Neither of these diagrams has more then one path from place to place. However, this need not be the case. A special sort of function diagram is a commuting diagram. In these diagrams there is more than one way to get from place to place, but it "doesn't matter" which path you take.

Consider the natural numbers $\mathbb{N}$ and the three functions *add* 1, *add* 2 and *add* 3, which add 1, 2 or 3 to their arguments, respectively. For example, *add* 2 is defined as:

$$add2 : \mathbb{N} \to \mathbb{N}$$

$$add2(n) = n + 2$$

We can draw the following diagram:

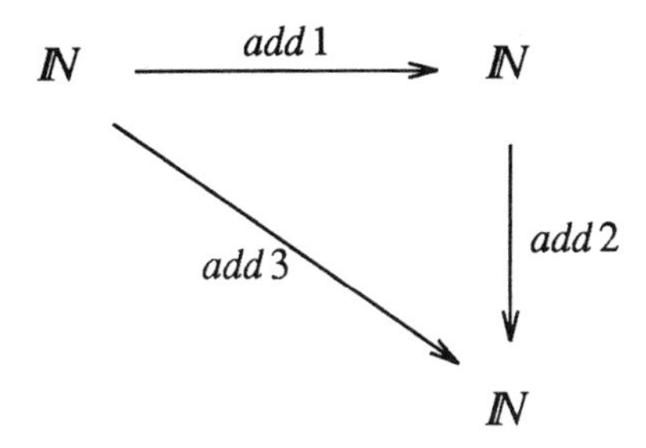

This diagram commutes because it doesn't matter whether we first add 1 and then add 2 (i.e. left and then down) or whether we just add 3 (take the diagonal path). In terms of the functions:

$$\forall n \in \mathbb{N} \quad add2(add1(n)) = add3(n)$$

Typically, the sets are different and the diagrams may have many more paths through them.

Appendix II

A specification of a simple editor using dynamic pointers

Pointer spaces can be used to produce an extensible specification for a layered display editor. The display editor consists of three layers:

- *Display* – The actual display as seen by the user. This includes a representation of the cursor, and commands at this level include mouse selection.
- *Text* – A textual representation of the objects of interest, bridging the gap between the objects and the bounded display.
- *Simple strings* – The underlying objects being edited.

In this simple design, the extra text level might be seen as a trifle excessive, but even here it helps to factor the design. This layering becomes essential in more complex systems, where the mapping between the application objects and their textual representation is less straightforward. Further, the layering is useful in proving user interface properties. The text representation is what the user is likely to obtain as an observable effect, and thus a proof of observability can be factored into a proof first that the whole of the text can indeed be observed, and second that the objects can be obtained from some parsing of the text.

We will address the design in several stages. First the three pointer spaces, and the projections between them will be introduced. Then in §II.2 we will see how these can be formed into an editor state. In the next section we will see how the design is extensible by adding a find/replace operation. Finally, we will summarise the insight gained from this example.

II.1 The pointer spaces and projections

In this section we describe the pointer spaces and projections necessary for the three layers of the editor design. At the surface is a simple grid of characters, representing the user's screen. This display is assumed to also have a single cursor *between* characters. This would be presented on most terminals as a cursor over a character, and the last column would have to be left blank in order to represent the last cursor position:

$$Disp = Array\ [\ height, width\ ]\ of\ Char$$
$$P_{disp} = \{\ 1,...,height\ \} \times \{\ 0,...,width\ \}$$
$$\forall\ disp \in Disp \quad vptrs(\ disp\ ) = P_{disp}$$

The intermediate level is two-dimensional text, a ragged-edged sequence of lines of printable characters:

$$Text = list\ of\ list\ of\ Char - \{\ [\,]\ \}$$
$$P_{text} = \mathbb{N} \times \mathbb{N}$$
$$(\ n, m\ ) \in vptrs(\ text\ ) \Leftrightarrow n \in \{\ 1,...,length(\ text\ )\}\ \textbf{and}\ m \in \{\ 0,...,length(\ text[\ n\ ]\ )\ \}$$

Note that the empty text is represented by "[[]]" (the sequence consisting of a single empty line), and in general the last line of a text is *not* assumed to have an implicit trailing new-line. The disallowing of "[]" (the empty sequence) as a text is thus quite consistent.

The projection between text and display is simple framing, and requires only a single text pointer as parameter specifying the offset of the "window" into the text:

$$Disp\_struct[\ P_{text}\ ] = P_{text}$$

$$proj_{disp} : P_{text} \times Text \rightarrow Disp \times (\ P_{text} \rightarrow P_{disp}\ ) \times (\ P_{disp} \rightarrow P_{text}\ )$$

$$proj_{disp}(\ (\ n, m\ ), text\ ) = disp, fwd, back$$

The display *disp* is constructed as the relevent section of *text*, with the locations that don't have corresponding *text* locations filled with blanks:

$$\begin{aligned} disp[\ i, j\ ] &= text[\ i{+}n{-}1\ ][\ j{+}m\ ] \quad \textbf{if}\ i{+}n{-}1 \le length(\ text\ ) \\ &\qquad \textbf{and}\ j{+}m \le length(\ text[\ i{+}n{-}1\ ]\ ) \\ &= space \quad \textbf{otherwise} \end{aligned}$$

The *fwd* map normally merely subtracts the window offset (n, m), but has to return extremal values when the text pointer is outside the window. Note that for this and the *back* map, the last clause is the "normal" case after all exceptions have been dealt with:

$fwd((n', m')) = i, j$
where
  if $n' < n$ **then** $(i, j) = (1, 0)$
  else if $n' \geq n+height$ **then** $(i, j) = (height, width)$
  else if $m' < m$ **then** $(i, j) = (n'-n+1, 0)$
  else if $m' > m+width$ **then** $(i, j) = (n'-n+1, width)$
  else $(i, j) = (n'-n+1, m'-m)$

The *back* map has similar problems with "out of range" display pointers, this time ones which represent blank screen locations:

$back((i, j)) = n', m'$
where
  if $n+i > length(text)$
  then $(n', m') = (length(text), length(text[length(text)]))$
  else if $m+j > length(text[n+i])$
  then $(n', m') = (n+i, length(text[n+i]))$
  else $(n', m') = (n+i, m+j)$

The text, in turn, is represented by a projection from a single-dimensional string of characters with embedded new-lines:

$String = list\ of(Char+\{nl\})$
$P_{string} = \mathbb{N}$

$vptrs(string) = \{0, \ldots, length(string)\}$

This has various operations defined on it, character insertion and deletion, and incrementing and decrementing of pointers:

$insert: P_{string} \times Char \times String \rightarrow String \times (P_{string} \rightarrow P_{string})$

$insert(n, c, string) = string', pull$
where
  $string' = string[1, \ldots, n] :: [c]$
    $:: string[n+1, \ldots, length(string)]$
  $pull(n') = n'$ **if** $n' < n$
    $= n'+1$ **if** $n' \geq n$

$delete: P_{string} \times String \rightarrow String \times (P_{string} \rightarrow P_{string})$

$delete(n, string) = string', pull$
where
  $string' = string[1, \ldots, n-1] :: string[n+1, \ldots, length(string)]$
  $pull(n') = n'$ **if** $n' < n$
    $= n'-1$ **if** $n' \geq n$

$$succ: P_{string} \times String \rightarrow P_{string}$$
$$pred: P_{string} \times String \rightarrow P_{string}$$

$$\begin{aligned} succ(n, string) &= n+1 \quad \textbf{if } n < length(string) \\ &= n \quad \textbf{otherwise} \\ pred(n, string) &= n-1 \quad \textbf{if } n > 0 \\ &= n \quad \textbf{otherwise} \end{aligned}$$

Note that the successor and predecessor functions both require the string as an argument, so that they can do range checking. In fact, we can see that the predecessor function could do without; however, it is better always to require an object context for pointer operations, as it leaves room for alternative representations – we could have had pointers representing distance back from the end of the string!

We can now construct the obvious projection from strings to texts:

$Proj\_struct$ is empty

$$proj: String \rightarrow Text \times (P_{string} \rightarrow P_{text}) \rightarrow (P_{text} \rightarrow P_{string})$$

$$proj(s) = text, fwd, back$$

The maps *fwd* and *back* and the result *text* are defined using two arrays of string pointers, *line_starts* and *line_ends*, both consisting of valid pointers for s:

$$line\_starts: \ list\ of\ vptrs(s)$$
$$line\_endss: \ list\ of\ vptrs(s)$$

The former is taken to have a pointer to the beginning of each line in the string, and the latter to the end of the line. Thus $length(line\_starts)$ is the number of lines in the string and $line\_starts[n]$ is a pointer to the beginning of the n th line. We can then define *text* by the following properties:

$$length(text) = length(line\_starts) = length(line\_ends)$$
$$text[i] = s[line\_starts[i]+1,...,line\_ends[i]]$$

and we obtain *fwd* and *back* fairly directly from *line_starts* and *line_ends*:

$$fwd(p) = (n, m)$$
st
$$line\_starts[n] \leq p \leq line\_ends[n]$$
$$m = p - line\_starts[n]$$
$$back((n, m)) = line\_starts[n] + m$$

Finally, *line_starts* and *line_ends* themselves are defined by:

$$
\begin{aligned}
&\textbf{let}\ \ len\ =\ length(\ line\_starts\ )\\
&line\_starts[\ 1\ ] = 0\\
&line\_ends[\ len\ ] = length(\ s\ )\\
&\forall\ i \in \{\ 1,\ldots,len\ \}\\
&\qquad nl \notin s[\ line\_starts[\ i\ ]{+}1,\ldots,line\_ends[\ i\ ]\ ]\\
&\forall\ i < len\\
&\qquad line\_starts[\ i{+}1\ ] = line\_ends[\ i\ ] + 1\\
&\qquad \textbf{and}\ \ s[\ line\_ends[\ i\ ]{+}1\ ] = nl
\end{aligned}
$$

The definition given is not constructive; however, one can easily produce an effective procedure for calculating *text*, *line_starts* and *line_ends* from a string.

II.2 Construction as state model

We can now use the preceding pointer spaces and projections to define an editor state. Again we will do this in two stages, first defining an editor for texts, with no finite display, then adding state to define the display.

The text editor state simply consists of a string and a string pointer to act as an insertion point:

$$E_{text} = String \times P_{string}$$

Its interface is a text and a text pointer (the insertion point), yielded by application of the projection:

$$display_{text} :\ E_{text} \rightarrow Text \times P_{text}$$

$$
\begin{aligned}
&display_{text}(\ (\ s\ ,p\ )\ ) = text\ ,\ tptr\\
&\textbf{where}\\
&\qquad text\ ,fwd,\ back = proj(\ s\ )\\
&\qquad tptr = fwd(\ p\ )
\end{aligned}
$$

There are five operations defined. The first two are simple insertion-point moves and the third sets the insertion point to a given text position (for use later with mouse selection). These three have no effect on the pointers, so the *pull* function is just the identity and is omitted from their definitions. The others, insertion and deletion of characters, alter the object and yield *pull* functions so that anything using this can convert its own text pointers. Note that the user of the package has no knowledge of the implementation using strings; the interface is entirely in terms of texts and text pointers:

$\{\ movement\ \}$:

$$move\_right_{text}:\ E_{text} \rightarrow E_{text}$$
$$move\_left_{text}:\ E_{text} \rightarrow E_{text}$$

$$move\_right_{text}((s,p)) = s, succ(p,s)$$
$$move\_left_{text}((s,p)) = s, pred(p,s)$$

$\{\ selection\ \}$:

$$select_{text}:\ P_{text} \times E_{text} \rightarrow E_{text}$$

$$select_{text}(p_{new},(s,p_{old})) = s, fwd(back(p_{new}))$$

$\{\ update\ \}$:

$$insert_{text}:\ (Char + \{\ nl\ \}) \times E_{text} \rightarrow E_{text} \times (P_{text} \rightarrow P_{text})$$
$$delete_{text}:\ E_{text} \rightarrow E_{text} \times (P_{text} \rightarrow P_{text})$$

$$insert_{text}(c,(s,p)) = (s',p'), pull_{text}$$

where

$$p' = pull(p)$$
$$s', pull = insert(p,c,s)$$
$$pull_{text} = fwd' \circ pull \circ back$$
$$text, fwd, back = proj(s)$$
$$text', fwd', back' = proj(s')$$

$$delete_{text}((s,p)) = (s',p'), pull_{text}$$

where

$$p' = pull(p)$$
$$s', pull = delete(p,s)$$
$$pull_{text} = fwd' \circ pull \circ back$$
$$text, fwd, back = proj(s)$$
$$text', fwd', back' = proj(s')$$

Note the similarity between the definitions of $insert_{text}$ and $delete_{text}$. It is therefore very easy to prove properties of E_{text} and its operations from those of *String* and its operations, for instance, $delete_{text}(insert_{text}(c,e)) = e$.

We construct the full editor by adding the display structure, which is a single text pointer:

$$E_{disp} = E_{text} \times P_{text}$$

$$\forall\ (e_{text}, p) \in E_{text} \qquad p \in vptrs(text)$$

where

$$text, ip = display_{text}(e_{text})$$

This has as its interface a display with cursor, obtained by using the display projection on the text obtained from e_{text}:

$$display_{disp} : E_{disp} \rightarrow Disp \times P_{disp}$$

$$display_{disp}((e_{text}, p)) = disp , ip_{disp}$$

where

$$ip_{disp} = fwd(ip_{text})$$
$$disp , fwd, back = proj_{disp}(p , text)$$
$$text , ip_{text} = display_{text}(e_{text})$$

As an observability condition, we would always like the text insertion point to be on screen:

$$\forall \ (e_{text}, p) \in E_{disp} \quad ip_{text} \in back(P_{disp})$$

We recognise this as a *static invariant*, relating the display and state at an instant. In order to maintain this, when the normal course of operations breaks this invariant, we need an internal adjustment function to be used after each operation:

$$adjust : E_{text} \times P_{text} \rightarrow P_{text}$$

$$adjust(e , (p\_line , p\_col)) = (p\_line', p\_col')$$

where

$$\textbf{if } ip\_line - p\_line + 1 \in \{ 1,...,height \} \textbf{ then } p\_line' = p\_line$$
$$\textbf{else } p\_line' = \max (1, ip\_line - height/2)$$
$$\textbf{if } ip\_col - p\_col + 1 \in \{ 1,...,width \} \textbf{ then } p\_col' = p\_col$$
$$\textbf{else } p\_col' = \max (1, ip\_col - width/2)$$
$$text , (ip\_line , ip\_col) = display_{text}(e)$$

We can now use this to define the final display editor operations. These are simply the operations on E_{text}, followed by *adjust*:

{ *movement* }:

$$move\_right_{disp} : E_{disp} \rightarrow E_{disp}$$
$$move\_left_{disp} : E_{disp} \rightarrow E_{disp}$$

$$move\_right_{disp}((e_{text}, p)) = (e'_{text}, p')$$

where

$$p' = adjust(e'_{text}, p)$$
$$e'_{text} = move\_right_{text}(e_{text})$$

$$move\_left_{disp}((e_{text}, p)) = (e'_{text}, p')$$
where
$$p' = adjust(e'_{text}, p)$$
$$e'_{text} = move\_left_{text}(e_{text})$$

{ *selection* }:
$$select_{disp} : P_{disp} \times E_{disp} \rightarrow E_{disp}$$

$$select_{disp}(p_{new}, (e_{text}, p)) = (e'_{text}, p)$$
where
$$e'_{text} = e, back(p_{new})$$
$$e_{text} = e, p_{old}$$
$$back = proj_{disp}(p, text).back$$
$$text = proj_{text}(e).obj$$

N.B. No adjustment is necessary, as $back(p_{new})$ is in $back(P_{disp})$ by definition.

{ *update* }:
$$insert_{disp} : (Char + \{ nl \}) \times E_{disp} \rightarrow E_{disp}$$
$$delete_{disp} : E_{disp} \rightarrow E_{disp}$$

$$insert_{disp}(c, (e_{text}, p)) = (e'_{text}, p')$$
where
$$p' = adjust(e'_{text}, pull(p))$$
$$e'_{text}, pull = insert_{text}(c, e_{text})$$

$$delete_{disp}((e_{text}, p)) = (e'_{text}, p')$$
where
$$p' = adjust(e'_{text}, pull(p))$$
$$e'_{text}, pull = delete_{text}(e_{text})$$

II.3 Adding features

As well as dividing concerns very clearly between the various levels, the design is very easily extensible. For instance, imagine we designed an additional operation *find_replace* for strings:

$$find\_replace : String \times String \times String \rightarrow String \times (P_{string} \rightarrow P_{string})$$

where $find\_replace(f, r, s)$ replaces all occurrences of f in s and puts r instead. We can trivially extend the rest of the editor to include this:

$$find\_replace_{text} : String \times String \times E_{text} \rightarrow E_{text} \times (P_{text} \rightarrow P_{text})$$

$$find\_replace_{text}(f, r, (s, p)) = (s', p'), where pull_{text}$$

where

$$p' = pull(p)$$
$$s', pull = find\_replace(f, r, s)$$
$$pull_{text} = fwd' \circ pull \circ back$$
$$text, fwd, back = proj(s)$$
$$text', fwd', back' = proj(s')$$

$$find\_replace_{disp} : String \times String \times E_{disp} \rightarrow E_{disp}$$

$$find\_replace_{disp}(f, r, (e_{text}, p)) = (e'_{text}, p_{text})$$

where

$$p' = adjust(e'_{text}, pull(p))$$
$$e'_{text}, pull = find\_replace_{text}(f, r, e_{text})$$

The final lexical level of the editor has not been included here. Some commands are simple: insert, delete and movement would map onto single keystrokes and selection would be a mouse click. The find/replace operation would require extra mechanisms enabling the user to manipulate the search strings. Of course, we already have an editor for working on strings (in fact we have two!), and this could be used. We could add such a search string editor as a special feature, or better it could be included in an environment that manages objects and object–object operations like find/replace.

II.4 Discussion

We've seen how a simple editor can be built up out of simple specification components using pointer spaces. Translation of mouse coordinates inwards, and cursor positions on display are achieved using the standard *back* and *fwd* maps. We've also seen how easy it is to add facilities such as find/replace. Not only was this easy, but the mechanism is identical for any operation defined on the underlying data type that can supply a *pull* function. This is very important, as it means that find/replace can be thought of as a *tool* used by the editor rather than an integral part of the editor. Other filters could be included dynamically in the editor's repertoire, for instance an intelligent spelling checker or a program pretty printer. This extends significantly the power of environments such as Unix which make heavy use of simple filters as tools.

References

Abowd, G., Bowen, J., Dix, A., Harrison, M. and Took, R. (1989). *User Interface Languages: A Survey Of Existing Methods*, Programming Research Group, Oxford University, Oxford, UK.

Abowd, G. (1990). "Agents: recognition and interaction models", pp. 143–146 in *Human–Computer Interaction – INTERACT'90*, ed. D. Diaper, D. Gilmore, G. Cockton and B. Shakel, North-Holland.

Alexander, H. (1987a). "Executable specifications as an aid to dialogue design", pp. 739–744 in *Human–Computer Interaction – INTERACT'87*, ed. H.-J. Bullinger and B. Shackel, North-Holland.

Alexander, H. (1987b). *Formally-Based Tools And Techniques For Human-Computer Dialogues*, Ellis Horwood.

Alexander, H. (1987c). "*Formally-Based Tools And Techniques For Human-Computer Dialogues*", TR.35, University of Stirling.

Alvey, (1984). *Alvey MMI Strategy*, published IEE.

Anderson, S.O. (1985). *Specification And Implementation Of User Interfaces: Example: A File Browser*, Heriot-Watt University.

Anderson, S.O. (1986). "Proving properties of interactive systems", pp. 402–416 in *People And Computers: Designing For Usability*, ed. M. D. Harrison and A. F. Monk, Cambridge University Press.

Ansimov, A.V. (1975). "Languages over free groups", *Springer Verlag Lecture Notes In Computer Science 32*, pp. 167–171.

Arbib, M.A. (1969). *Theories Of Abstract Automata.*

Bahike, R. and Hunkel, M. (1987). "The user interface of the PSG programming environments", pp. 311–316 in *Human–Computer Interaction – INTERACT'87*, ed. H.-J. Bullinger and B. Shackel, North-Holland.

Bancilhon, F. and Spyratos, N. (1981). "Update semantics of relational views", *ACM Transactions On Database Systems*, **6**(4).

Bornat, R. and Thimbleby, H. (1986). *The Life And Times Of Ded, Display Editor.*

Bundy, A. (1983). "The Computer Modelling Of Mathematical Reasoning", pp. 133–149 in , Academic Press.

Burstall, R.M. and Goguen, J.A. (1980). "The semantics of Clear, a specification language", pp. 292–332 in *Proceedings Of The 1979 Copenhagen Winter School on Abstract Software Specification*, Springer-Verlag.

Card, S.K., Moran, T.P. and Newell, A. (1983). *The Psychology Of Human Computer Interaction*, Lawrence Erlbaum.

Card, S.K., Pavel, M. and Farrel, J.E. (1984). "Window-based computer dialogues", pp. 355–360 in *Human–Computer Interaction – INTERACT'84*, ed. B. Shackel, North-Holland.

Card, S.K. and Henderson, D.A. (1987). "A multiple, virtual-workspace interface to support user task switching", pp. 53–59 in *Proceedings Of The CHI+GI*, ACM, New York.

Chi, U.H. (1985). "Formal specification of user interfaces: a comparison and evaluation of four axiomatic approaches", *IEEE Transactions On Software Engineering*, **SE-11**(8), pp. 671–685.

Cockton, G. (1986). "Where do we draw the line? – Derivation and evaluation of user interface software separation rules", pp. 417–432 in *People And Computers: Designing For Usability*, ed. M. D. Harrison and A. F. Monk, Cambridge University Press.

Collis, J., Malone, J. and Martin, M. (1984). *Spy © Science And Engineering Research Council*, Rutherford Appleton Laboratory.

Cook, S. (1986). "Modelling generic user-interfaces with functional programs", pp. 369–385 in *People And Computers: Designing For Usability*, ed. Harrison, M.D. and Monk, A., Cambridge University Press.

Cosmadikis, S.S. and Papadimitriou, C.H. (1984). "Updates of relational views", *Journal Of The ACM*, **31**(4), Association for Computing Machinery .

Coutaz, J. (1987). "PAC, an object oriented model for dialogue design", pp. 431–436 in *Human–Computer Interaction – INTERACT'87*, ed. H.-J. Bullinger and B. Shackel, Elsevier (North-Holland).

Dijkstra, E.W. (1976). *A Discipline Of Programming*, Prentice-Hall.

Dix, A.J. and Runciman, C. (1985). "Abstract models of interactive systems", pp. 13–22 in *People And Computers: Designing The Interface*, ed. P. Johnson and S. Cook, Cambridge University Press.

Dix, A.J. and Harrison, M.D. (1986). "Principles and interaction models for window managers", pp. 352–366 in *People And Computers: Designing For Usability*, ed. M.D. Harrison and A.F. Monk, Cambridge University Press.

Dix, A.J. (1987a). "The myth of the infinitely fast machine", pp. 215–228 in *People And Computers III – Proceedings Of HCI'87*, ed. D. Diaper and R. Winder, Cambridge University Press.

Dix, A.J. (1987b). "*Formal Methods And Interactive Systems: Principles And Practice*", YCST 88/08, D.Phil. thesis, Department of Computer Science, University of York.

Dix, A.J. (1987c). "Giving control back to the user", pp. 377–382 in *Human–Computer Interaction – INTERACT'87*, ed. H.-J. Bullinger and B.Shackel, North-Holland.

Dix, A.J. and Harrison, M.D. (1989a). "Interactive systems design and formal development are incompatible?", pp. 12–26 in *The Theory And Practice Of Refinement*, ed. J. McDermid, Butterworth Scientific.

Dix, A.J. (1989b). "Software engineering implications for formal refinement", pp. 243–259 in *Proceedings ESEC'89*, ed. C. Ghezzi and J. McDermid, Springer-Verlag.

Donahue, J. (1985). "Cedar: An environment for experimental programming", pp. 1–9 in *Integrated Project Support Environments*, ed. J McDermid, IEE Software Engineering Series.

Donahue, J. and Widom, J. (1986). "Whiteboards: a graphical database tool", *ACM Trans. Off. Inf. Syst.*, **4**(1), pp. 24–41.

Earl, A.N., Whittington, R.P., Hitchcock, P. and Hall, A. (1986). "Specifying a semantic model for use in an integrated project support environment", pp. 220–235 in *Software Engineering Environments*, ed. Ian Sommerville, Peter Peregrinus.

Ehrig, H. and Mahr, B. (1985). *Fundamentals Of Algebraic Specification 1*, Springer-Verlag.

Foley, J.D. and van Dam, A. (1982). "Fundamentals Of Interactive Computer Graphics", in , Addison Wesley.

Garlan, D. (1986). "Views for tools in integrated environments", pp. 314–343 in *Advanced Programming Environments*, ed. Reidar Conradi, Tor M. Didriksen and dag H. Wanvik, Springer-Verlag, Lecture Notes in Computer Science 244.

Goldberg, A. (1984). *Smalltalk-80, The Interactive Programming Environment*, Addison-Wesley.

Gosling, J. and Rosenthal, D. (1985). "A window manager for bitmapped displays and Unix", pp. 115–128 in *Methodology Of Window Management*, ed. F.R.A Hopgood *et al.*, Springer-Verlag.

Halasz, F., Moran, T. and Trigg, R. (1987). "NoteCards in a nutshell", pp. 45–52 in *Proceedings Of The CHI+GI*, ACM, New York.

Hansen, W.J. (1984). "User engineering principles for interactive systems", pp. 217–231 in *Interactive Programming Environments*, ed. D.R. Barstow, H.E. Shrobe and E. Sandewall.

Hardy, G.H. and Wright, E.M. (1954). *An Introduction To The Theory Of Numbers (third Edition)*, Oxford.

Harrison, M.D., Roast, C.R. and Wright, P.C. (1989). "Complementary methods for the iterative design of interactive systems", pp. 651–658 in *Designing And Using Human-Computer Interfaces And*, ed. G. Salvendy and M.J. Smith, Elsevier Scientific.

Harrison, M.D. and Dix, A.J. (1990). "Modelling the relationship between state and display in interactive systems", pp. 241–249 in *Visualisation In Human–Computer Interaction*, ed. P.Gornay and M.J.Tauber, Springer-Verlag.

Hekmatpour, S. and Ince, D.C. (1987). "Evolutionary prototyping and the human-computer interface", pp. 479–484 in *Human–Computer Interaction – INTERACT'87*, ed. H.-J. Bullinger and B. Shackel, Elsevier (North-Holland).

Hoare, C.A.R. (1969). "An axiomatic basis for computer programming", *Communications Of The ACM*, **12**(10), p. 576.

Hutchins, E.L., Hollan, J.D. and Norman, D.A. (1986). "Direct manipulation interfaces", pp. 87–124 in *User Centered System Design*, ed. D.A. Norman and S.W. Draper, Lawrence Erlbaum.

Hutchinson, T. (1926). *The Poems Of William Wordsworth*, Oxford University Press.

Ichbiah, J.D. *et al* (eds.) (1983). "*Reference Manual For The Ada Programming Language*", ANSI/MIL-STD-1815A-1983.

Jackson, M.A. (1983). *System Development*, Prentice-Hall.

Jones, C.B. (1980). *Software Development: A Rigorous Approach*, Prentice-Hall.

Jonge, W.d. (1983). "Compromising statistical databases responding to queries about means", *ACM Trans. On Database Systems*, **8**(1), pp. 60–80, ACM.

Joy, W. (1980). *Ex Reference Manual*, University of California, Berkeley.

Kernighan, B.W. and Ritchie, D.M. (1978). *The C Programming Language*, Prentice Hall.

Khosia, S., Maibaum, T.S.E. and Sadler, M. (1986). *Database Specification*, Deptartment of Computing, Imperial College (internal report).

Kiss, G. and Pinder, R. (1986). "The use of complexity theory in evaluating interfaces", pp. 447–463 in *People And Computers: Designing For Usability*, ed. M.D. Harrison and A.F. Monk, Cambridge University Press.

Kurtz, T.E. (1978). "The programming language BASIC", *ACM SIGPLAN Notices*, **13**(8), pp. 103–118.

Luckham, D.C. and Suzuki, N. (1979). "Verification of array, record and pointer operations in Pascal", *Transactions On Programming Languages And Systems*, **1**(2), pp. 226–244, ACM.

Macfarlane, J. and Thimbleby, H. (1986). *The TIN: An Approach To Powerful And Cheap User Interfaces*, University of York Internal Report.

Marshall, L. (1986). *Ph.D Thesis*, University of Manchester.

Millo, R.A., Lipton, R.J. and Perlis, A.J. (1979). "Social processes and proofs of theorems and programs", *Communications Of The ACM*, **22**(5), pp. 271–280, Association for Computing Machinery.

Monk, A. (1986). "Mode errors: a user centred analysis and some preventative measures using keying-contingent sound", *International Journal Of Man–Machine Studies*, **24**, pp. 313–327.

Monk, A.F. and Dix, A.J. (1987). "Refining Early Design Decisions with a Black-box Model", pp. 147–158 in *People And Computers III – Proceedings Of HCI'87*, ed. D. Diaper and R. Winder, Cambridge University Press.

Monk, A.F., Walsh, P. and Dix, A.J. (1988). "A comparison of hypertext, scrolling and folding as mechanisms for program browsing", pp. 421–436 in *People And Computers: From Research To Implementation – Proceedings HCI'88*, ed. D.M.Jones and R.Winder, Cambridge University Press.

Morgan, C.C. (1985). *The Schema Language*, Programming Research Group, Oxford.

Nelson, T.H. (1981). *Literary Machines: The Report On, and of, Project Xanadu, Concerning Word Processing, Electronic Publishing, Hypertext, Thinkertoys, Tomorrow's Intellectual Revolution, and Certain Other Topics Including,.*

Ossanna, J.F. (1976). *Nroff/Troff User's Manual*, Bell Laboratories.

Parker, J. and Hendley, B. (1987). "The Universe program development environment", pp. 305–310 in *Human–Computer Interaction – INTERACT'87*, ed. H.-J. Bullinger and B. Shackel, North-Holland.

Payne, S.J. (1984). "Task-action grammars", pp. 527–532 in *Human–Computer Interaction – INTERACT'84*, ed. B. Shackel, North-Holland.

Pfaff, G.E. (1985). *User Interface Management Systems*, Springer Verlag.

Pike, R. (1984). *The BLIT: A Multiplexed Graphics Terminal.*

Pressman, R.S. (1982). *Software Engineering: A Practitioners Approach*, McGraw-Hill.

QuickC, (1988). *C For Yourself*, Microsoft Corporation.

Rasmussen, J. (1987). "Cognitive Engineering", pp. xxv–xxx in *Human–Computer Interaction – INTERACT'87*, ed. H.-J. Bullinger and B.Shackel, North-Holland.

Runciman, C. and Hammond, N. (1986a). "User programs: a way to match computer systems and human cognition", pp. 464–481 in *People And Computers: Designing For Usability*, ed. M.D. Harrison and A.F. Monk, Cambridge University Press.

Runciman, C. and Thimbleby, H.W. (1986b). "*Equal Opportunity Interactive Systems*", YCS 80, Dept. of Computer Science, University of York.

Runciman, C. and Toyn, I. (1987). *Transformational Development Of Purely Functional Prototypes From* `PiE` *Interaction Models*, University of York.

Runciman, C. (1989). "From abstract models to functional prototypes", pp. 201–232 in *Formal Methods In Human–Computer Interaction*, ed. H. Thimbleby and M. Harrison, Cambridge University Press.

Shneiderman, B. (1982). "The future of interactive systems and the emergence of direct manipulation", *Behaviour And Informations Technology*, **1**(3), pp. 237–256.

Shneiderman, B. (1984). "Response time and display rate in human performance with computers", *ACM Computing Surveys*, **16**(3), pp. 265–286, Association for Computing Machinery .

Smith, D.C., Irby, C., Kimball, R., Verplank, B. and Harslem, E. (1983). "Designing the Star user interface", pp. 297–313 in *Integrated Interactive Computer Systems*, ed. Degano, P. and Sandewall, E., North Holland.

Stallman, R.M. (1981). "EMACS: The Extensible, Customizable Self-Documenting Display Editor", *ACM SIGPLAN Notices*, **16**(6), pp. 147–156.

Stefik, M., Foster, G., Bobrow, D., Kahn, K., lanning, S. and Suchman, L. (1987). "Beyond the chalkboard: computer support for collaboration and problem solving in meetings", *Communications Of The ACM*, **30**(1), pp. 32–47.

Stephens, M. and Whitehead, K. (1986). "The analyst – a workstation for analysis and design", pp. 104–117 in *Software Engineering Environments*, ed. Ian Sommerville, Peter Peregrinus.

Stoy, J.E. (1977). *Denotational Semantics The Scott-Strachey Approach To Programming Language Theory*, MIT Press.

Sufrin, B. (1982). "Formal specification of a display-oriented text editor", *Science Of Computer Programming*, (1), pp. 157–202, North Holland Publishing Co..

Sufrin, B. and He, J. (1989). "Specification, refinement and analysis of interactive processes", pp. 153–200 in *Formal Methods In Human–Computer Interaction*, ed. H. Thimbleby and M. Harrison, Cambridge University Press.

Sufrin, B.A., Morgan, C.C. and Sørensen, I. (1985). *Notes For Z Handbook: Part 1, The Mathematical Language*, Oxford, Programming Research Group.

Sun, (1986). *SunView Programmer's Guide.*

Suzuki, N. (1982). "Analysis of pointer rotation", *Communications Of The ACM*, **25**(5), pp. 330–335.

Thimbleby, H. (1980). "Dialogue determination", *International Journal Of Man–Machine Studies*, **13**, pp. 295–304.

Thimbleby, H. (1986a). "Ease of use – the ultimate deception", pp. 78–94 in *People And Computers: Designing For Usability*, ed. Harrison, M.D. and Monk, A., Cambridge University Press.

Thimbleby, H. (1986b). "The design of two innovative user interfaces", pp. 336–351 in *People And Computers: Designing For Usability*, ed. Harrison, M.D. and Monk, A., Cambridge University Press.

Thimbleby, H. (1987). "The design of a terminal independent package", *Software Practice And Experience*, **17**(5), pp. 351–367.

Thimbleby, H.W. (1983). "What you see is what you have got – a user engineering principle for manipulative display?", *ACM Proceedings Software Ergonomie*, pp. 70–84.

Thimbleby, H.W. (1984). "Generative user-engineering principles for user interface design", pp. 102–107 in *Human–Computer Interaction – INTERACT'84*, ed. B. Shackel, North-Holland.

Thimbleby, H.W. (1990). *User Interface Design*, ACM Press, Addison-Wesley.

Thompson, K. and Ritchie, D.M. (1978). *UNIX Programmer's Manual, Seventh Edition*, Bell Laboratories.

Took, R. (1986a). "The presenter – a formal design for an autonomous display manager", pp. 151–169 in *Software Engineering Environments*, ed. I. Sommerville, Peter Peregrinus.

Took, R. (1986b). "Text representation and manipulation in a mouse-driven interface", pp. 386–401 in *People And Computers: Designing For Usability*, ed. M.D. Harrison and A.F. Monk, Cambridge University Press.

Took, R. (1990). *Immediate Interaction: An Integrated Formal Model For The Presentation Level Of Applications And Documents*, PhD Thesis, Computer Science Department, University of York.

Vitter, J.S. (1984). ''US&R: a new framework for redoing'', *IEE Transactions.*

Wilson, M., Barnard, P. and MacLean, A. (1986). ''Using an expert system to convey HCI information'', in *People And Computers: Designing For Usability*, ed. M.D. Harrison and A.F. Monk, Cambridge University Press.

Wordstar, (1981). © *MicroPro International.*

Yankelovich, N., Meyrowitz, N. and van Dam, A. (1985). ''Reading and writing the electronic book'', *Computer*, **10**, pp. 15–30.

Yoder, E., Akscyn, R. and McCracken, D. (1989). ''Collaboration in KMS, a shared hypermedia system'', pp. 37–42 in *Proceedings CHI'89 Conference on Computer-Human Interaction*, ACM, New York.

Young, R.M., Green, T.R.G. and Simon, T. (1989). ''Programmable user models for predictive evaluation of interface designs'', in *Proceedings CHI'89 Conference on Computer-Human Interaction*, ACM, New York.

Yourdon, E. and Constantine, L.L. (1978). *Structured Design*, Yourdon Press.

Index

a

b

c

q

r

s

t

u

v

W

X

Y

Z